寻找银色光影

图书在版编目（CIP）数据

寻找银色光彩：2015中高龄时尚服饰研究/上海视觉艺术学院，上海中高龄时尚服饰研究中心主编.—上海：东华大学出版社，2015.10
ISBN 978-7-5669-0922-0

Ⅰ.①寻… Ⅱ.①上…②上… Ⅲ.①中年人—服饰美学—研究②老年人—服饰美学—研究 Ⅳ.①TS976.4

中国版本图书馆CIP数据核字（2015）第238107号

责任编辑：马文娟　李伟伟
艺术总监：彭　波
封面设计：薛小博

寻找银色光彩
——2015中高龄时尚服饰研究
XUNZHAO YINSE GUANGCAI: 2015 ZHONGGAOLING SHISHANG FUSHI YANJIU

上海视觉艺术学院　上海中高龄时尚服饰研究中心　主编
出　　版：东华大学出版社（上海市延安西路1882号）
邮政编码：200051　电话：（021）62193056
出版社网址：http://www.dhupress.net
天猫旗舰店：http://dhdx.tmall.com
发　　行：新华书店上海发行所发行
印　　刷：上海锦良印刷厂
开　　本：889mm×1194mm　1/16　印张：22.25
字　　数：784千字
版　　次：2015年10月第1版
印　　次：2015年10月第1次印刷
书　　号：ISBN 978-7-5669-0922-0/TS·651
定　　价：98.00元

谨以此书献给志在千里的风尚长者

编委会成员

名誉主任
龚学平

顾 问
胡延照　王荣华

主 任
李柯玲

副主任
袁 仄

委 员
包铭新　Kathryn.Hagen　李 俊　徐家华
陈 莹　刘 侃　孟祥令　邵家瑜

序

作为世界上唯一老年人口过亿的国家，老龄化已经成为我国日趋突出的社会问题。鉴于此，政府提出要"积极应对人口老龄化，大力发展老龄服务事业和产业"。而上海是我国迄今为止人口老龄化程度最高的特大城市，老年人口比例始终高于全国平均水平。

上海作为"时尚之都"，时尚产业发展迅速。而《上海市文化创意发展"十二五"规划》则明确指出，时尚产业是城市现代化和社会文明程度的重要标志，它折射着一个城市的历史与文化底蕴，同时也蕴含着时代气息。上海的老龄化趋势向社会提出了新的课题，同时也向产业与高校提出了新的课题。关注"银发群体"，建设"银发工程"，大力推进"银发产业"是高校亟需解决的问题，也是切实关爱这一社会群体的行动。

上海视觉艺术学院是一所应用设计类的高校。自建校始，学校重视教育与社会需求接轨。基于此，上海视觉艺术学院提出了"中高龄时尚服饰设计"这一社会亟需的课题。同时"上海中高龄时尚服饰研究中心"于2014年9月在上海视觉艺术学院时尚设计学院正式挂牌，并于同年成功举办了第一届中高龄时尚服饰国际论坛。论坛的成果集结成《寻找银色光彩——2014中高龄时尚服饰研究》一书已经正式出版。同时，论坛得到来自政府、协会、企业、海内外高校的大力支持与积极参与。

2015年第二届中高龄时尚服饰国际论坛将于10月召开，而研究中心的第二部中高龄服饰研究专著《寻找银色光彩——2015中高龄时尚服饰研究》将付梓成册。这部专著关注电子信息时代下中高龄群体的生活形态、服饰的智能性与功能化、时尚服饰的设计、消费行为与市场现状等领域。

希望"上海中高龄时尚服饰中心"的系列活动能够抛砖引玉：越来越多的海内外高校参与中高龄这一课题的研究，越来越多的企业能够关注并启动"银发工程"，从而使我国的"银发产业"向着积极健康的方向前进！

上海视觉艺术学院校长

2015年9月20日

001　**中高龄服饰智能化研究**

002　银色的"朝阳"产业　Silver Industry, Sunrise Industry　袁 仄

005　Baby Boomers' Acceptance of Solar-Powered Clothing 婴儿潮一代对太阳能服装的接受度研究　Chanmi Hwang　Eulanda A. Sanders

021　论可穿戴设备在中老年服饰中的应用与推广　The Application and Promotion of the Wearable Devices in the Elder People Clothing　陈 莹　唐 倩

025　基于"无障碍"理念的老年服装功能设计　The functional design of clothing for aged group based on the concept of Barrier-free　何清远　王云仪

032　基于可穿戴设备的中高龄功能性服饰研究　Middle-aged and Old People's Functional Clothing Research Based on the Wearable Devices　田亚楠　邢晓宇　洪正琳　尚笑梅

039　老年智能服装设计要点与现状分析　Design key points and present situation analysis of smart clothing for the aged group　王 莹　王云仪

046　老年智能化服装探讨　The discussion of intelligent clothing for the old　王诗潭　王云仪

053 康复治疗服装的现状及发展趋势　The present situation and development trend of rehabilitative garments　邢晓宇　洪正琳　田亚楠　尚笑梅

059 **中高龄服饰艺术设计研究**

060 OUTLET: The trend of the new millennium　新千禧年的流行：中高龄服饰研究　Leo Giovacchini

074 传统与时尚在中高龄女装图案中的表现　The expression of the traditional and fashion in the pattern of the mature womenswear　汪　芳

079 Interpreting Male Ageing and Fashion: Arts-Informed Interpretative Phenomenological Analysis　中高龄男性与时尚——基于艺术资料的解释现象学分析　Anna Maria Sadkowska

092 基于城市中高龄人群休闲方式的服装设计研究　Study on the Clothing Design for leisure style of urban elderly population　张　洁

100 中国传统图案在中老年服装中的应用　Application of Chinese traditional patterns in the middle aged and old people's garment　杨　韧　付少海　梁惠娥

107 浅析明代补子图案在中老年家居服设计中的应用　Preliminary analysis about the application of Buzi pattern of Ming Dynasty in the design of middle-aged and senior people's leisurewear　董稚雅　梁惠娥

115 女书文字的审美意蕴及其在中高龄群体服装中的应用　The aesthetic implication of Nvshu and its application in the design of elderly population the clothing　贾蕾蕾　梁惠娥

122	二维码个性化设计在中高龄服饰中的应用　Personalized design of QR code in application of Fashing Design of mature people　夏　俐
130	服"适"生活　智时代下的中高龄服装创新设计与服务的思索　Discussion on Fashion Design and Service for Mature People in the Era of Intelligence　刘　坤
136	浅谈流行元素在中高龄服饰定制设计中的运用　The Application of popular elements in the designation of mature people　李昌慧
142	清末民初江南传统女装面料纹样的研究　The research of traditional women's dress fabric and pattern of southern Yangtze River in the late Qing dynasty and early Republic of China　张　岚

153　中高龄服饰功能性研究

154	老年防摔功能服装设计分析研究　The Design and Analysis of the Aging Clothing with Fall Prevention Function　朱达辉　宇　锋
163	老年人服装功能性实现方法分析　Analysis on Implementation Methods of Functional Design of the Clothing for Old-aged Group　张梦莹　李　俊
170	Kinetic Garment Construction for variable, aging bodies　中高龄人体的动力学服装结构设计研究　Rickard Lindqvist
182	基于3D试衣系统的中高龄服饰设计　Middle and old-aged people's fashion design based garment 3D fitting system　贾镇瑜　吴　琰　潘琦明

189	社交型老年女性服装结构设计探讨　Study on the Structure Design of the Social Old Women's Garments　陈　萍　李　俊
196	电加热技术在老年人冬季保暖服中的应用　Application of electrically-heated technology in warm clothing for the old　许静娴　李　俊
204	老年人着装舒适性的影响因素分析　Discussion on the influential factors in the old people's clothing comfort　王雅芝
211	老年群体功能服装产品的设计现状及进展　The Research and Analysis of Function Garments for Elders　刘慧娟　王云仪
217	中老年功能性服装概述　Overview of the Quinquagenarian Functional Clothing　洪正琳　田亚楠　邢晓宇　尚笑梅

223　中高龄服饰消费行为研究

224	上海地区针织塑身内衣市场现状及其面向中高龄消费者的思考　Market Status of Knitted Slimming Underwear in Shanghai and Suggestion to Middle Aged Consumers Oriented　李　敏　王　鑫
231	Considering the Needs of Aging Baby Boomers　老龄化的婴儿潮一代需求研究　Marilyn J. Bruin Sauman Chu Lin Nelson-Mayson Juanjuan Wu Becky Yust
242	A Study on Fashion Brand by New Mature Women's Characteristics 基于中高龄女性生活特征的时尚品牌研究　Hyesoo Yeom

248	关于老年人的服饰消费行为——针织商品的多样性　The Diversity of knitting Fashion of Mature People　岩佐正树
255	Travel Jams: IntersectingStrategies for the Mature Chinese Female Traveler　旅游拥堵:中国中高龄女性的旅行策略研究　Kathryn.Hagen　Eulanda A. Sanders
262	新时代下银发一族时尚服装产品发展理念　Study on Fashion Cloting Products of Silver Gens under the New Era　刘侃
268	当代城市中老年女性化妆品消费观察　Study on Cosmetic Consumption of Urban Mature People in Contemporary China　赵宏伟
275	The influence of types of SNS advertisement by fashion brand, social distance and involvement on commitment　时尚品牌社交媒体广告的类型、社会差异与品牌介入对消费者承诺的关系研究　Dongeun　Choi　Sunjin　Hwang
281	Emotional Fit Project: Mapping the Ageing Female Form　情感与着装:中高龄女性服饰搭配　Anna Maria Sadkowska　Katherine Townsend　Juliana Sissons
299	浅析我国中老年服装市场发展　Analysis of the development of China's elderly clothing market　陈颖
303	Advanced Fashion　中高龄时尚分析　Zeshu Takamura
310	The Possibility of Community Type Clothes Making　社区型服装设计工作室的可行性研究　Kazuro Koumoto

319	The Universal Fashion Marketing To support customers enjoy fashionable and healthy life　人性化时尚营销　Mori Hideo
330	Understanding Korea Ajumma Way　韩国中高龄女性生活方式研究　Haesook Kwon
341	编后记

中高龄服饰智能化研究

银色的"朝阳"产业

◆ 袁 仄[①]　（上海视觉艺术学院，上海，201620）

Silver Industry, Sunrise Industry

◆ Yuan Ze　*(Shanghai Institute of Visual Art, Shanghai, 201620)*

1　夕阳里的"朝阳"

　　唐人李商隐诗词《乐游原》中的一句"夕阳无限好,只是近黄昏",流传甚广,脍炙人口。诗中无非表达了思古幽情的情绪,其诗寓意诗人热爱生活,执着人间的情怀,但现代人往往解读为老当益壮之慨,让老年人奋发,但亦不免有壮士暮年之无奈。

　　同样,在改革开放之初,中国业界流行一个说法:纺织服装产业是西方发达国家的"夕阳"产业。而当时恰逢国家开放,尤以服装业门槛低,于是国人把洋人的"夕阳"产业悉数拾来,开始了中国服装产业的"朝阳",终于将中国发展为世界服装大国。其实,所谓"夕阳"产业只是西方将产业末端的加工丢到了中国,而该产业的设计、研发、潮流的引领、话语权皆在他人的手中。何为"夕阳"? 只要人类还要穿衣,还要"臭美"（时尚）,那么这个产业不可能是"夕阳",而永远是"朝阳"。关键在于是否把握产品的研发、科创、流行话语权……

　　再用同仁用过的数据:第六次人口普查结果显示,2012 年中国 60 岁及以上人口占13.26%,相比第五次普查上升了 2.93 个百分点。预计,到 2050 年我国 60 岁及以上人口将占全国人口的 25%,达到 4 亿。社会老龄化已经是世界性的态势,中国尤甚。因为中国人口基数庞大,加之前期的生育政策导致人口老龄比例加剧。且不说将来约四分之一的中老龄人口的社会负担,就其消费能量就不可以小觑! 更不可能小觑!

　　而在老年消费需求中,34% 为服饰穿着,33% 为医疗保健,33% 为其他消费。即三分之一为服饰类消费,国家民政部社会福利和慈善事业促进司的数据表明,2010 年我国老年人口消费规模超过 1.4 万亿,到 2030 年将达到 13 万亿元。如此庞大的消费体量,作为产业来讲,绝不是"夕阳",恰恰正是"朝阳"产业。

　　在纺织服装以及未来智能领域,该领域可以孕育诸多创新项目……也许可以遐想:若中国能在中高龄时尚产业领域里积极领跑,若中国在部分服装或某些纺材或些许智能设计上掌握

[①]　袁仄,哲学硕士、文学硕士；上海视觉艺术学院教授,北京服装学院教授。E-mail:yuanzehuyue@126.con

核心技术，我相信，在未来的时尚产业里，中国由此获得更多的话语权也不是不可能的。

2　朝"前"还是朝"后"

近年来，中外影视热衷于"穿越"，所谓"穿越"即幻想未来或回到过去。耐人寻味的是，好莱坞式的"穿越"喜欢向前幻想，想象未来地球、机器人、智能生物，等等；而我们影视的创作者们则喜好往"后"看，往过往祖先生活的年代"穿越"一把，跑到清朝当一回公子、格格就乐不可支。

这样的思维模式表现在我们在中老年服装产业的发展思路上。长期以来，我们基本停留在依照老人的经济收入、保守审美的定式思维里面，加上对老年体型要求和价格的期许，这就使得"中老年服装"这个概念一直叫好不叫座，市场上少有企业愿意开发销售这类服装。包括产业领导、企业家，几乎都没有从宏观的经济眼光来审视该产业的"超前"性、"稀缺"性和"朝阳"性。

正是因为国人以往的思维局限，喜好往后思考，导致所有的人都视其为一项带有"公益性"的产业。如此思维，何有"朝阳"可言？

毋庸讳言，现在被思考为老年对象的通常都是20世纪三四十年代出生的人，他们经历过战争、经济不景、计划供给、极左思维等洗礼，使得他们那辈积淀了"节俭"的家风，固化了"左倾"的审美意识。尽管他们也曾青春年少过、也曾热情洋溢过，但岁月摧毁了他们时尚的信念和追求，他们是不幸的一代，但他们并不代表中国中高龄者的永远。

今天我们思考的中高龄时尚，应该向"前"看，50年代、60年代……逐步将是未来中国中老年消费的主体。我们的商家往往很早就预见到未来市场是"80后""90后"，却从来不愿意思考一下未来全国四分之一的消费人群将是"50后""60后"，岂不怪哉。

而"50后""60后"作为中老年消费者既有老一辈节俭、保守的某些惯性，但他们毕竟是步入改革开放的人群，他们毕竟享受到丰富的物质生活、出国游历、购房驾车……他们毕竟也赶上了改变生活方式的互联网时代，智能手机、网上购物、视频通话……这一切已经预示着未来老年公寓硬件的改变：强劲的WiFi、快捷的快递服务、维修……由此可以联想到他们的服装、服饰、鞋帽，等等，他们肯定不会需要老土的所谓"中老年装"，而Fashion、时髦、靓丽、酷帅依旧是他们的习惯用语。

有了向"前"看的思维，不难发现该产业绝对是"朝阳"产业，大有前途，大有"钱"途。

3　老龄化撬动科创

前不久，哈佛大学经济系教授、劳动经济学家理查德·弗里曼来沪参加第二届中国发展研究双年会，他对上海拟建科创中心、培养科创人才等议题发表了有益的见解，尤其提到了人口老龄化的话题。他明确指出人口老龄化不是科创建设的减速器。他说到，世界上许多国际大都市都面临人口老龄化问题，但中国有其特殊的方面，其人口红利并未消失……

其实，老龄人口的增加，虽然是劳动力人口的衰减，但反之就是这部分人口新需求的增长。具体而言，他们对养老、医疗、保健、社交、旅游，以及各类时尚类消费如服装、化妆品、饰品、功能化服装用品、电子化智能化产品……将会掀起新的消费人群的消费"人口红利"。

"银发"产业的崛起应该列入国家和上海市科创的战略部署之中，应该分专题的探讨未来这部分"人口红利"的利用。在研究这个"朝阳"产业中，值得格外重视的是智能服装的研究开发，这项研究在欧洲发达国家已经开始介入，中国应该同样予以重视，否则有可能再重蹈20世纪80年代的覆辙，即再次重复接盘所谓"夕阳"产业，替西方发达国家加工新一代的智能服装。在纺织服装业相对遭遇瓶颈，产业急需转型之际，必须马上开始对未来纺织材料和服装服饰的设计开发，如开发具有检测生理特征、健康管理、生活辅助、运动增强、轨迹跟踪、娱乐时尚等元素的新型面料或服装。德国海恩斯坦研究预测，智能服装是一个有数十亿人需要数百万种产品的市场；美国著名市场研究公司 VDC 的报告指出，到 2018 年，仅美国市场智能型纺织品价值将达 30 亿美元。

目前的手机行业已经进军智能、保健等领域，而纺织服装的这方面开发相对滞后。相信人们对服装服饰智能的需求将会与手机智能升级一样的迫切。譬如，人们（尤其老年人）其实十分期盼可以自动调节温度的毛衣秋裤，设想这种空调衣将会成为未来衣柜里的大部分。能够监测有病老人生理和行动的服装，也无疑可以延伸到儿童产品、旅行者的服装、运动员的服装、等等。其他，如袜子用上传感纤维；帽子配上导航芯片；内衣拥有减肥功效……凡此种种，大胆想象，勇敢创新，商机无限。

上海拥有纺织、服装、科技和设计的丰厚基础，该项银色的"朝阳"产业会成为产业经济转型的新引擎。传统的纺织服装更多需要跨界，在物联网、感知技术、集成电路、城市网络等新兴信息技术领域合作展开新的产业升级。愿银发的"朝阳"产业撬动中国服装业的转型升级，让未来的世界重新看待"MADE IN CHINA"。

Baby Boomers' Acceptance of Solar-Powered Clothing

◆ Chanmi Hwang[①]　Eulanda A. Sanders　*(Iowa State University, Ames, IA. USA, E-mail: chanmih@iastate.edu, sanderse@iastate.edu)*

Abstract: *This study explored Baby Boomers' acceptance of wearing wearable technology, specifically: (a) how Baby Boomers perceive the benefits and concerns of wearing solar-powered clothing, and (b) how Gen Y and Baby Boomers differ on their perceptions of and attitudes towards wearing solar-powered clothing. A convenience sample of 720 responses were obtained from one Midwestern institution in the United States and used for data analyses. The results of independent samples' t-test showed that Baby Boomers perceived the clothing to be more comfortable, aesthetically pleasing, and eco-friendly, and showed more favorable attitudes towards purchasing the clothing compared to the Gen Y group. The respondents were asked to provide their views about the benefits or concerns of solar-powered clothing. Benefits included usefulness, energy conservation, saving money and innovativeness while concerns included high price, risks, style/appearance, comfort, environmental impact and usability. Baby Boomers may be a key segment of the population for solar-powered clothing and the findings are useful to current researchers and apparel industry members who seek to devise strategies for sales and marketing of products which inherently require both technology and clothing attributes.*

Key words: *Baby Boomers, Wearable Technology, Solar-Powered Clothing*

婴儿潮一代对太阳能服装的接受度研究

◆ Chanmi Hwang　Eulanda A. Sanders　（爱荷华州立大学，美国爱荷华州，E-mail: chanmih@iastate.edu, sanderse@iastate.edu）

【摘　要】　这项研究探讨了婴儿潮一代对可穿戴技术中太阳能服装的接受度。特别是：（a）婴

[①] Chanmi Hwang, Iowa State University, Ames, IA. USA, Graduate Student, E-mail: chanmih@iastate.edu, sanderse@iastate.edu

儿潮一代是如何察觉到穿太阳能服装的好处,以及对穿太阳能服装的负面看法,(b)y世代和婴儿潮一代对太阳能服装有着怎样不同的认识与态度。该研究从美国的一个中西部机构中获得720份调查问卷,然后进行数据分析。独立样本t检验的结果显示婴儿潮一代认为太阳能服装舒适、美观、环保。与y世代的消费者相比,对购买太阳能服装有着更良好更积极的态度。这些调查对象提供了他们对太阳能服装的正反两方面看法。调查结果显示,太阳能服装的优点为实用、节能、省钱、创新;相比之下,负面的看法为高昂的价格、风险、外观风格、舒适性、环境的影响和可用性。对于太阳能服装来说婴儿潮一代可能是关键的消费群体。这项研究可为其他研究者、服装业的营售战略、市场营销所需的具体技术以及服装属性的定位有所借鉴。

【关键词】 婴儿潮一代;可穿戴技术;太阳能服装

（中文翻译：谈金艳）

1 Introduction

Traditionally, mature people were considered vulnerable consumers and they were treated as dependents that should be taken care of by others (Institute for the Future, 2002). However "a substantial portion of aging Baby Boomers who will become the 'New Elderly' do not fit these traditional perceptions" (Kim, Jolly, & Kim, 2007, p.316), and will change conventional notions about what it means to age in America. Baby Boomers are those born between the years 1946 and 1964, with the youngest being 51 years old and the oldest being 69 years old as of 2015 (Howden & Meyer, 2011). There are more than 78 million Baby Boomers in the United States of America who have the highest disposable income on average (Paul, 2003), and they control half of the wealth of the nation (Matorin, 2003). Thus, the improved health status with this economic prosperity led older adults to participate in diverse leisure and social activities (Carrigan, Szmigin, & Wright, 2004).

Specifically, with the rapidly increasing numbers in the mature population, the Baby Boomer consumer market has attracted attention from apparel retailers. Kumar and Lim (2008) state that Baby Boomers seem to be experienced users of modern technologies and they are more open to new technologies than previous generations. Thus, with rapid textile and technological innovations, electronics are becoming embedded into textiles and garments, and the Baby Boomer market is opening a new opportunity for technical fabrics and apparel companies. Despite of Baby Boomers' high purchasing power and responsiveness to changing trends (Yang & Jolly, 2008), academia and industry have neglected a this group in the market (McCann, 2009) and little has been done to address mature wearer's perceptions and attitudes towards wearing technology-integrated garments. Thus, studies on responses, attitudes, and consumer acceptance levels for this age group are necessary to aid the commercialization strategies for wearable technology, specifically solar-powered clothing. Solar-powered clothing has attracted researchers and industry because of its functionality and pro-environmental attributes and it is expected to be a major item for the future fashion industry (Cho, 2010).

In order to understand Baby Boomer consumers, this study first quantitatively compared Baby Boomers to Generation Y, the children of the Baby Boomers born between the years 1980 and 1994, with the youngest being 21 years old and the oldest being 35 years old as of 2015 (Howden & Meyer, 2011). Generation Y was chosen since most of the apparel industry especially wearable technology targets young people and it would be effective to compare how young and older people differ on their attitudes towards wearing such clothing. To further understand Baby Boomers' perceive the benefits and concerns towards wearing solar-powered clothing, content analysis was undertaken on the open-ended questions in the survey.

Therefore, the overall purpose of this study was to explore Baby Boomers' acceptance of wearing wearable technology, particularly solar-powered clothing. More specific study objectives were to: (a) examine how Baby Boomers and Generation Y differ on their perceptions and attitudes towards wearing solar-powered clothing, (b) explore how Baby Boomers perceive the benefits of wearing solar-powered clothing, and (c) explore any concerns that Baby Boomers have towards wearing solar-powered clothing. The results of this study can provide important information to apparel retailers about the nature of Baby Boomer and Generation Y consumers, which would help retailers in devising strategies for sales. This study may further provide guidelines to marketers in the solar-powered clothing business for how to increase positive attitudes and purchase intentions and how to approach customers across the two different age cohorts. The results may also show if there are any similarities or differences between these two generations and their attitudes towards wearable technology.

2 Theoretical Background

2.1 Generational Cohort: Baby Boomers

Age is used widely as a demographic variable that characterizes the adoption of technologies between two or more consumer groups (Morris & Venkatesh, 2000). Generational Cohort Theory states that individuals born in the same period would have similar preferences and attitudes in their adult years, since they had similar experiences (Petroulas, Brown, & Sundin, 2010). A study conducted by Yang and Jolly (2008) showed that the effect of perceived usefulness on attitudes towards use of mobile data services appeared greater for the Baby Boomer cohort than the younger generations. Moreover, Baby Boomers have become more affluent than other age cohort groups and tend to respond to changing trends (Yang & Jolly, 2008). Furthermore, Baby Boomers are a generation with high moral priorities and one of their priorities includes environmental issues (Smith & Clurman, 2007). In particular, they are becoming more aware about environmental issues and interested in health and wellness as the older cohort of this generation is approaching retirement (Worsley, Wang, & Hunter, 2011). Thus, for retailers pursuing solar-powered clothing, it is critical to understand the growing influence of this new aging group.

2.2 User Acceptance of Technology

Perceived usefulness and perceived ease of use from the Technology Acceptance Model (TAM) are the most frequently validated factors that influence consumers' attitudes towards purchasing technology and innovative products. Perceived usefulness is defined as "the degree of which a person believes that using a particular system would enhance his or her job performance" (Davis, 1989, p.320). This variable has shown as the most powerful predictor for intentions to use and the technology adoption (Venkatesh, Morris, Davis, & Davis, 2003). Perceived ease of use is defined as "the degree to which a person believes that using a particular system would be free of effort" (Davis, 1989, p. 320). Along with these two original TAM variables, perceived risk, "the uncertainty consumers face when they cannot foresee the consequences of their purchase decisions" (Schiffman & Kanuk, 2000, p.153), also plays an important role that influence consumers' attitudes towards purchasing innovative products. Apparel has been regarded as a product category having multi-dimensional risks and in general, and the uncertainty of the performance of the product can influence consumers' attitudes towards purchasing new apparel products (Park & Stoel, 2005; Sjoberg, 2000).

2.3 Functional, Expressive, Aesthetic (FEA) Consumer Needs

Along with usability and functionality of the technology, smart clothing should satisfy user needs with regard to fashion and self-representation (Sonderegger, 2013). Lamb and Kallal (1992) suggest three dimensions of clothing useful in assessing consumer needs and wants: functional, expressive, and aesthetic (FEA). The functional considerations for clothing include satisfaction with comfort and fit where physical comfort is "a mental state of physical well-being expressive of satisfaction with physical attributes of a garment such as air, moistures, heat transfer properties, and mechanical properties such as elasticity, flexibility, bulk, weight, texture, and construction" (Sontag, 1985, p.10). Previous studies show that physical comfort was an important factor influencing overall evaluation of the product (Frith & Glesson, 2004; Sontag, 1985).

Along with functional considerations, consumers are also concerned about expressive considerations (Stokes & Black, 2012) where it relates to the "communicative, symbolic aspects" of clothes and are based on the socio-cultural and psychological aspects of the dress (Lamb & Kallal, 1992, p. 43). Therefore, compatibility, "the degree to which the innovation is perceived as consistent with the existing values, needs, and past experiences of the potential adopter" (Ko et al., 2009, p. 240) is an important factor. Previous studies show that perceived compatibility affects attitudes towards purchasing the product (Shim & Kotsiopulos, 1994; Ko et al., 2009).

Lastly, aesthetic criteria are also an important factor that influence consumers' evaluations of apparel (Chattaraman, 2006; Eckman, Damhorst, & Kadolph, 1990). Previous studies show that intrinsic attributes of the product such as color, design, and appearance were important criteria in women's decisions in apparel selection during the interest phase of their purchase, because apparel is an important means of visual communication (Eckman et al., 1990; Fiore & Damhorst, 1992).

3 Method

3.1 Sample and Procedure

An online survey, including both open and close-ended questions was used for gaining a thorough understanding of the characteristics of Baby Boomers. A convenience sample of college students and faculty at a United States mid-western university was recruited for the web-based survey. The sample for this study consisted of 19–34 year olds and of 51–69 year olds, both male and female, who were in the bracket of the targeted age for Gen Y and Baby Boomers. After receiving Institutional Review Board (IRB) approval, 31,190 college students and 6,400 faculty received an invitation email to participate in the study. The email contained a letter of research introduction along with the survey URL and informed consent forms that assured confidentiality. To provide participants a clear understanding of solar-powered clothing, a detailed information page describing solar-powered clothing was supplied at the beginning of the survey. This page included various images of solar-powered clothing (e.g., Colon Jacket, Zegna Sports, Scottevest) with a description of solar cells and instructions on how to use the product; the description and images can be retrieved when participants completing the online survey.

3.2 Instrument Development

All scales employed seven-point Likert type items, ranging from *Strongly Disagree* (1) to *Strongly Agree* (7), except perceived comfort, perceived performance risk, and attitude.

Three items adopted from Davis' (1989) were used to measure perceived usefulness and another three items from Davis' (1989) were used to assess consumers' perceived ease of use. Three items of perceived performance risk were adopted from Grewal, Gotlieb, and

Marmorsteina (1994). Perceived compatibility was adopted from Ko et al. (2009) and attitudes were measured with seven-point semantic differential scales (MacKenzie, Lutz, & Belch, 1986). Perceived comfort and perceived aesthetic attributes, items were developed from previous studies (i.e. Huck, Maganga, & Kim, 1997; Eckman, Damhorst, & Kadolph, 1990). For perceived comfort, four semantic differential scales were developed based on Huck, Maganga, and Kim (1997). For perceived aesthetic attributes, items were adopted from Eckman et al. (1990). The survey also included two open-ended questions addressing perceived benefits and perceived concerns.

3.3 Data analysis

Quantitative and qualitative data were used to achieve the research objective 1 and research objectives 2 and 3, respectively. Quantitative analysis was performed for the data from close-ended questions, while content analysis was performed for the qualitative data from open-ended questions. This mixed methods analysis was deemed appropriate to meet the research objectives.

The Statistical Package for Social Sciences 19.0 was used to examine descriptive statistics and generational differences between Bab Boomers and Generation Y. Independent samples't-test was used for analyzing quantitative data for the research objective 1. For the research objective 2 and 3, two independent researchers conducted a content analysis by coding the qualitative data from the two open-ended responses of Baby Boomers' perceived benefits and concerns towards wearing solar-powered clothing. Main themes and sub-themes emerged from the coding process (Creswell, 2009). Differences in coding were negotiated by the researchers and an inter-coder reliability of 99% was achieved.

4 Results

4.1 Sample Profile

A total of 206 Baby Boomers participants completed the online survey. The results showed that 78.2% of the respondents were female and 21.8% were male. Age ranged from 51 to 69. The majority of the respondents were White/European American (95.6%), followed by African American (2.0%) and Asian (2.0%). Participants' education level ranged from high school graduate to doctoral degree, with the majority being master's degree (32.8%) followed by bachelor's degree (27.4%) and doctoral degrees (19.9%). About 38% of the Baby Boomers earned between $50,000 to $74.999 annually and about 25% of them earned between $25,000 to $49,999. Nearly 60% of respondents earned less than $25,000 annually and 3.2% earned over $100,000 annually. In terms of participants' awareness of solar-powered clothing, 95 (46.0%) of the 206 participants indicated they had heard about solar-powered clothing previously.

Table 4.1. Demographics of the Baby Boomers

Demographics	Frequency	Percentage of Sample
Age (n=206)		
69-65	15	7.3
64-60	73	35.4
59-55	73	35.4
54-51	45	21.8
Gender (n=206)		
Female	161	78.2
Male	45	21.8
Ethnicity (n=205)		
White/European American	196	95.6
African American/Black	4	2.0
Asian	4	2.0
Hispanic American or Latino	.5	2.9
Other	.5	3.4
Education (n=201)		
High school graduate, diploma or equivalent	20	10.0
Associate degree	20	10.0
Bachelor's degree	55	27.4
Master's degree	66	32.8
Doctoral degree	40	19.9
Annual Income (n=201)		
$25,000-$49,999	50	24.9
$50,000-$74,999	76	37.8
$75,000-$99,999	26	12.9
Over $100,000	23	11.4
Choose not to answer	26	12.9
Heard of solar-powered clothing (n=206)		
Yes	95	46.0
No	111	54.0

Note. The N varies because of missing data

4.2 Research objective 1: Perceptions and attitudes

An independent samples' *t*–test was conducted to examine any significant difference between Baby Boomers and Generation Y in relation to their perceived usefulness, perceived ease of use, perceived performance risk, perceived comfort, perceived compatibility, perceived aesthetic attributes, environmental concerns, and attitude.

As shown in Table 4.2, a statistically significant difference was found between the means of the two groups in perceived usefulness ($t(710)= -3.58$, $p<0.01$), perceived comfort ($t(714)= 3.37$, $p<0.01$), perceived compatibility ($t(716)=-3.46$, $p<0.01$), perceived aesthetic attributes ($t(428.65)= 4.15$, $p<0.01$), environmental concerns ($t(466.75)= 3.99$, $p<0.01$), and attitudes ($t(716)= 2.10$, $p<0.05$). The mean of the Baby Boomers was significantly higher for perceived comfort, perceived aesthetic attributes, environmental concerns, and attitudes ($m=4.21$, $sd=1.09$; $m=3.95$, $sd=1.39$; $m=6.39$, $sd=0.87$; $m=4.96$, $sd=1.24$) than the mean of the Gen Y group ($m=3.89$, $sd=1.14$; $m=3.45$, $sd=1.59$; $m=6.08$, $sd=1.08$; $m=4.74$, $sd=1.25$). The mean of the Baby Boomers was significantly lower for perceived usefulness and perceived compatibility ($m=3.76$, $sd=1.50$; $m=3.38$, $sd=1.37$) than the mean of the Gen Y group ($m=4.19$, $sd=1.41$; $m=3.78$, $sd=1.41$). For perceived ease of use, the results showed no significant difference between the means of the two groups ($t(715)= 0.001$, $p= 0.99$) where the means of the Gen Y group ($m=5.53$, $sd=1.20$) and Baby Boomers ($m=5.53$, $sd=1.16$) were similar for both groups. The results for perceived performance risks also showed no significant difference between the means for Gen Y ($m=3.97$, $sd=1.31$) and the Baby Boomers ($m=4.16$, $sd=1.45$), ($t(341.78)=1.61$, $p=0.11$).

Table 4.2. Independent Samples T–test for Baby Boomers and Gen Y

Variable	Generation	N	Mean	Levene's Test for Eaulity of Variances		t–test for Equality of Means		
				F	Sig	t	df	Sig
Perceived Usefulness	Baby Boomers	203	3.76	2.17	.14	−3.58	710	.00
	Gen Y	509	4.19					
Perceived Ease of Use	Baby Boomers	205	5.53	.69	.41	.00	715	.99
	Gen Y	512	5.53					
Perceived Performance Risk	Baby Boomers	204	4.16	4.71	.03	−1.61	341.78	.11
	Gen Y	511	3.97					
Perceived	Baby Boomers	204	4.21	.34	.56	3.37	714	.00

（续表）

Variable	Generation	N	Mean	Levene's Test for Eaulity of Variances		t-test for Equality of Means		
				F	Sig	t	df	Sig
Comfort	Gen Y	512	3.89					
Perceived	Baby Boomers	205	3.38	.43	.51	−3.46	716	.00
Compatibility	Gen Y	513	3.78					
Perceived	Baby Boomers	206	3.95	8.27	.00	4.15	428.65	.00
Aesthetic Attributes	Gen Y	513	3.45					
Environmental	Baby Boomers	206	6.39	11.05	.00	3.99	466.75	.00
Concerns	Gen Y	513	6.08					
Attitude	Baby Boomers	205	4.96	.03	.86	2.10	716	.03
	Gen Y	513	4.74					

4.3 Research objective 2: Perceived benefits

When asked about the *benefits* of using solar–powered clothing in an open–ended question, the majority of participants confirmed that their decision was grounded in perception of the product as *useful*. Participants expressed a various range of perceived usefulness of the solar–powered clothing including *convenience, outdoor activity, traveling,* and *emergency situations.* Specifically, the participants stated that it is useful since charging electronic devices is readily available for *convenience*: "Have readily available energy source"; "The obvious benefit would be keeping my phone charged and not worrying about finding someplace to plug it in."; or "Having ability to power up devices on the go or without the electricity source". Further, they stated that the clothing would be beneficial for *outdoor activities* (e.g., "The ability to maintain a charge during prolonged outdoor activities when a power source isn't available (making receiving phone calls while participating in events/activities; reading or listening to music while lying on the beach)"; "Using this to recharge phone on extended bike rides or camping trips. Also for hiking in the mountains."; "Ability to keep personal electronics charged while off the grid. I pursue many outdoor activities where such clothing would be beneficial."), Participants also stated that it would be beneficial when *traveling* (e.g., "Easy to recharge small items when traveling, especially in other countries with different electrical current and/or outlets."; "it would work without worry about a power source convertor; especially potentially helpful when traveling to and in foreign countries."; "Traveling somewhere without charging stations, or power available."), and some further stated that it would beneficial when they are in an emergency situation (e.g.,"Less likely to be unable to use my phone when I want to, and particularly if there is an *emergency*. I already carry a charger in my car for emergencies; to carry one with me would be nice."; or "emergency use".

Followed by the *perceived usefulness,* participants stated that the clothing would be beneficial for *energy conservation*: "using solar power to reduce carbon footprint"; "I'm very concerned about protecting our environment and if this solar-powered clothing can contribute to save our environment I think that's the major benefit for me."; "Feeling that I was helping the environment."; or "Advantageous to the environment by adopting a green technology, and increases efficiency of several tasks". The Baby Boomers also perceived the clothing to be beneficial for other functions such as *warmth*: "keeping devices charged and possibly warming clothing". *Saving energy* was another benefits that Baby Boomers perceived (e.g., "free charging of devices without paying for electricity"; "Less electric bill"), and they further stated that the clothing's innovativeness is beneficial (e.g., "new and innovative gimmack (sic)"; or "uniqueness of solar powered charging my device"). Table 4.3 summarizes the emerged themes of Baby Boomers' perceived benefits and concerns towards wearing solar-powered clothing.

Table 4.3. Views of Baby Boomers about Solar-Powered Clothing

Views of Baby Boomers about Solar-Powered Clothing	
Perceived Benefits	Perceived Concerns
Usefulness	High Price
• Convenience	
• Outdoor Activity	
• Traveling	
• Emergency Situations	
Energy Conservation	
Warmth	Risks
	• functional risks
	—maintenance
	—durability
	• health risks
Saving Money	Style/Appearance
Innovativeness	Comfort
	Environmental Impact
	Usability

4.4 Research objective 3: Perceived concerns

When asked about any concerns of using solar-powered clothing, the Baby Boomers expressed that they are concerned about the *high price* of the clothing, "It is would cost more than I would be comfortable paying"; "Too expensive for my budget."; or "The biggest negative factor for me is the price range. I can't see spending that kind of money for something that is such a specialized concept.". Followed by the high price, majority of the Baby Boomers were concerned about perceived *risks* of the clothing. The participants expressed a various range of risk that include *functional risk* with *maintenance* and *durability* concerns and *health risks*.

Specifically they were concerned about the *functional risks* (e.g., "Does it really work. How long would it work."; "may not work as advertised"; or "Damage to the cells, not enough sunlight hitting them to be productive.") as well as they expressed concerns for *maintenance* and *durability* (e.g., "Reliability and durability of the electronics"; "No serious concerns, other than how well it would withstand cleaning and wear."; or "Cleaning and how it might hurt the system?"). Some expressed that they were concerned about *health risks* (e.g., "Not quite sure about any health issues that might arise from having solar panels so near the body. Could it get heated and cause skin problems, etc."; "I am concerned about the overall safety of the product."; "Concerned there might be some adverse health issues.").

Followed by the perceived risks, *style and appearance* were concerns for Baby Boomers, "The designs look bulky. I don't believe they are easily integrated into my wardrobe"; "Not really concerned – but think the design and purpose is more for younger folks."; or "Incorporation of the solar panels into the design of the clothing items – aesthetics do count.". Further, the participants expressed that they were concerned about the *comfort* of the clothing (e.g., "Comfort. Flexibility."; "The flexibility of the cells and whether it would feel "'stiff'"; "Weight and comfort."), and were doubtful about the *environmental impact* of using the clothing (e.g., "It is doubtful that environmental impact of producing such clothing would be offset by the clean energy it produces."; "Using solar powered clothing will probably have no effect on the environment."

5 Conclusions

This study explored Baby Boomers' acceptance of wearing wearable technology, specifically solar-powered clothing due to the increasing focus and development of products by researchers (Cho, 2010; Schubert & Merz, 2010) and the pro-environmental attributes of the product. Specifically, this study first explored how Baby Boomers perceive the benefits and concerns of wearing solar-powered clothing and further examined the differences between Gen Y and Baby Boomer on their perceptions of and attitudes towards wearing solar-powered clothing; these two groups were selected, because both groups have been of significant interest to social psychologists as well as marketers in the past (Morris & Venkatesha, 2000). The findings are useful to current researchers and apparel industry members who seek to devise strategies for sales and marketing of products which inherently require both technology and clothing attributes grounded on tenants of Generational Cohort Theory.

Based on the TAM and FEA frameworks, seven consumer-oriented variables related to attitude towards purchasing wearable technology were selected to understand Baby Boomers' acceptance of wearable technology. The results showed that Baby Boomers expressed the solar-powered clothing to be more comfortable and aesthetically pleasing as well as having favorable attitude towards wearing the clothing. The results of the content analysis align with the previous studies (e.g., Venkatesh et al., 2003) where perceived usefulness was a major factor that influenced Baby Boomers' beliefs of forming favorable attitudes towards wearing the technology-integrated clothing. Indeed, for the older generation, perceived usefulness was also the most powerful factor for accepting the technology adoption as they expressed a various range of usefulness especially for traveling abroad, outdoor activities, and emergency purposes.

Further, as previous studies states the importance of environmental concerns for Baby Boomers (e.g., Sith & Clurman, 2007; Worsley et al., 2011), the participants viewed that solar-powered clothing would support saving energy and pro-environmental. Baby Boomers also expressed a various range of concerns about the clothing's technological aspects such as malfunctioning and durability and this shows the importance of the perceived risks in technology acceptance as discussed by Sjoberg (2000). Lastly, aesthetic attributes such as style and appearance are important factor that should not be ignored in the development of technology-integrated clothing since apparel is an important means of visual communication (Eckman et al., 1990). In this regard, the classification of markets relying on consumer features and corresponding commercialization strategies are necessary considering both technological and apparel attributes of the clothing. Also, identifying potential early adopters of wearable technology is needed, since the product is at the early commercialization stage. For marketers, highlighting the usefulness through sufficient publicity about functions for a convenient lifestyle is important. Previous studies found that price influenced consumers' attitudes towards and intention of buying (Kim & Chung, 2011), and the high cost of the solar-powered clothing may be a factor that influenced consumers' attitudes towards purchase. In this study, on average, Baby Boomers had a higher yearly income ($50,000 to $75,000) than Gen Y ($3,000 to $10,000), possibly causing Baby Boomers to have more favorable attitudes towards purchasing the solar-powered clothing compared to the younger generation.

By exploring Baby Boomers' beliefs and perceptions towards wearing solar-powered clothing, this study adds new insights about Baby Boomers as a key segment of the population for wearable technology market. This study suggests that the ageing population can be powerful consumers in the wearable-technology market in the future who consider functional aspects of the clothing to be both beneficial and cautious. Thus, the findings of this study are useful for both fashion retailers and researchers working on technology-integrated clothing where they can develop products considering Baby Boomers' concerns towards both technological and apparel attributes such as perceived functional risks and comfort to increase Baby Boomers' positive attitudes towards purchasing the clothing. Further, current society is concerned with the environmental crisis and thus promotes socially responsible (SR) practices by companies (Ottman, 2011). In this regard, apparel companies with SR activities may want to adopt solar-powered clothing to promote clothing with pro-environmental attributes and raise awareness of consumers. In addition, retailers can target the consumers by explicitly informing consumers about environmental attributes of the solar-powered clothing.

There are several limitations to this study. Generalization of the research findings is limited because of the use of a convenience sample within a limited geographical location. Future research may use a sample that is more heterogeneous in terms of income levels, education, marital status, ethnic background and geographic location to confirm the findings. To further examine wearable technology, other types of smart clothing should be examined, with both technology acceptance variables and FEA variables, and compared with the results of solar-powered clothing. Considering the nascent field of the topic, in-depth interviews to identify the most important perceived attributes considered by Baby Boomers in adopting wearable technology would be alternative approaches to explore in future research. In-depth interviews with member checking during analysis process and possible wear testing of prototypes should be conducted in future studies.

References

[1] Ajzen, I., Attitude structure and behavior. In A.R. Pratkanis, S.J. Breckler, & A.G. Greenwald (Eds.), *Attitude structure and function*. Hillsdale, NJ: Erlbaum, 1989:241-274.

[2] Antil, J. H. Socially responsible consumers: Profile and implications for public policy. *Journal of Macromarketing, 1984, 4*(2): 18-39.

[3] Bush, A., Martin, C., & Bush, V. Sports celebrity influence on the behavior intentions of generation Y. *Journal of Advertising Research, 2004, 44*(1): 108-18.

[4] Carrigan, M., Szmigin, I., & Wright, F. Shopping for a better world? An interpretive study

of the potential for ethical consumption within the older market. *The Journal of consumer Marketing, 2004, 21*(6): 401–417.

[5] Chattaraman, V., & Rudd, N. A. Preferences for aesthetic attributes in clothing as a function of body image, body cathexis and body size. *Clothing and Textiles Research Journal, 2006, 24*(1): 46–61.

[6] Cho, G. *Smart clothing: technology and applications.* Boca Raton, Florida: CRC Press, 2010.

[7] Community Banker *Bet on baby boomers.* Retrieved from, 2000. www.allbusiness.com/periodicals/article/974183–1.html

[8] Creswell, J.W. *Research design: Qualitative, quantitative, and mixed methods approaches.* Thousand Oaks, CA: SAGE Publications, 2009.

[9] Davis, F. D. Perceived usefulness, perceived ease of use, and user acceptance of information technology. *MIS quarterly,* 1989, 319–340.

[10] Eckman, M., Damhorst, M. L., & Kadolph, S. J. Toward a model of the in–store purchase decision process: Consumer use of criteria for evaluating women's apparel. *Clothing and Textiles Research Journal,* 1990, *8*(2): 13–22.

[11] eMarketer (2013, March 21). How digital behavior differs among millennials, genxers and boomers. *eMarketer Inc.* Retrieved from http://www.emarketer.com/Article/How–Digital–Behavior–Differs–Among–Millennials–Gen–Xers–Boomers/1009748

[12] Fiore, A. M., & Damhorst, M. L. Intrinsic cues as predictors of perceived quality of apparel. *Journal of Consumer Satisfaction, Dissatisfaction and Complaining Behavior, 1992, 5*: 168–78.

[13] Frith, H., & Gleeson, K. Clothing and embodiment: men managing body image and appearance. *Psychology of Men & Masculinity, 2004, 5*(1): 40.

[14] Frost, K. *Consumer's perception of fit and comfort of pants.* (Unpublished master's thesis), University of Minnesota, St. Paul, 1988.

[15] Grewal, D., Gotlieb, J., & Marmorstein, H. The moderating effects of message framing and source credibility on the price–perceived risk relationship. *Journal of Consumer Research, 1994, 21*(1): 145–153.

[16] Howden M. K., & Meyer A. J. (2011). Age and Sex composition: 2010. *US Census Bureau.* Retrieved October 23, 2013 from http://www.census.gov/prod/cen2010/briefs/c2010br–03.pdf

[17] Institute for the Future. (2002). *Demographics in the 21st century: Defining future markets.* Retrieved September 14, 2013, from http://www.iftf.org/docs/SR772_Demographics_in_the_21st_Century.pdf

[18] Kim, H. S., & Damhorst, M. L. Environmental concern and apparel consumption.

Clothing and Textiles Research Journal, 1998, 16(3): 126–133.

[19] Kim, H. Y., & Chung, J. E. Consumer purchase intention for organic personal care products. *Journal of Consumer Marketing, 2011, 28*(1): 40–47.

[20] Kim, H. Y., Jolly, L., & Kim, Y. K. Future forces transforming apparel retailing in the United States: An environmental scanning approach. *Clothing and Textiles Research Journal, 2007, 25*(4): 307–322.

[21] Ko, E., Sung, H., & Yun, H. Comparative analysis of purchase intentions toward smart clothing between Korean and US consumers. *Clothing and Textiles Research Journal, 2009, 27*(4): 259–273.

[22] Kumar, A., & Lim, H. Age differences in mobile service perceptions: comparison of Generation Y and baby boomers. *Journal of Services Marketing, 2008, 22*(7): 568–577.

[23] Lamb, J. M., & Kallal, M. J. A conceptual framework for apparel design. *Clothing and Textiles Research Journal, 1992, 10*(2): 42–47.

[24] Lee, H. H., Fiore, A. M., & Kim, J. The role of the technology acceptance model in explaining effects of image interactivity technology on consumer responses. *International Journal of Retail & Distribution Management, 2006, 34*(8): 621–644.

[25] MacKenzie, S. B., Lutz, R. J., & Belch, G. E. The role of attitude toward the ad as a mediator of advertising effectiveness: A test of competing explanations. *Journal of Marketing Research, 1986, 23*(2): 130–143.

[26] Matorin, J. Generation 'G': Baby boomer grandparents a growing market offering glittering opportunity. *Nation's Restaurant News, 2003, 37*(33): 26.

[27] McCann, J. (2009). End-user based design of innovative smart clothing. In J. McCann, & D. Bryson (Eds.), *Smart clothes and wearable technology* Boca Raton, FL: CRC Press, 2009: 4–24.

[28] Morris, M. G., & Venkatesh, V. Age differences in technology adoption decisions: Implications for a changing work force. *Personnel psychology, 2000, 53*(2): 375–403.

[29] O'Donnell, J. Gen Y sits on top of consumer food chain; they're savvy shoppers with money and influence. USA Today, 2006, 11: 3.

[30] Ottman, J. A. *The new rules of green marketing: Strategies, tools, and inspiration for sustainable branding.* Berrett-Koehler Publishers, 2011.

[31] Park, J., & Stoel, L. Effect of brand familiarity, experience and information on online apparel purchase. *International Journal of Retail & Distribution Management, 2005, 33*(2): 148–160.

[32] Parment, A. Generation Y vs. Baby Boomers: Shopping behavior, buyer involvement and implications for retailing. *Journal of Retailing and Consumer Services,* 2013.

［33］Petroulas, E., Brown, D., & Sundin, H. Generational characteristics and their impact on preference for management control systems. *Australian Accounting Review, 2010, 20*(3): 221–240.

［34］Renn, K.A., & Arnold, K.D. Reconceptualizing research on college peer group culture. *The Journal of Higher Education, 2003, 74*(2): 93–110.

［35］Rogers, E. M. *Diffusion of Innovations.* New York: The Free Press, 2003.

［36］Schiffman, L. G., & Kanuk, L.L. *Consumer behavior* (7th ed.) Englewood Cliffs, NJ: Prentice Hall, 2000.

［37］Schubert, M. B., & Merz, R. Flexible solar cells and modules. *Philosophical Magazine, 2010, 89*(28–30): 2623–2644.

［38］Shim, S. Environmentalism and consumers' clothing disposal patterns: an exploratory study. *Clothing and Textiles Research Journal, 1995, 13*(1), 38–48.

［39］Sjöberg, L. Factors in risk perception. *Risk analysis, 2000, 20*(1), 1–12.

［40］Smith Walker, J. & Clurman, A. *Generation Ageless: How baby boomers are changing the way we live today...and they're just getting started.* New York: Harper Collins Publishers, 2007.

［41］Solomon, M. R. *Consumer Behavior (7th ed.).* Upper Saddle River, NJ: Pearson Education, Inc, 2007.

［42］Sonderegger, A. (2013, September). Smart garments—the issue of usability and aesthetics. In *Proceedings of the 2013 ACM conference on Pervasive and ubiquitous computing adjunct publication,* 2013(1): 385–392.

［43］Sontag, M. S. Comfort dimensions of actual and ideal insulative clothing for older women. *Clothing and Textiles Research Journal, 1985, 4*(1): 9–17.

［44］Venkatesh, V., Morris, M. G., Davis, G. B., & Davis, F. D. User acceptance of information technology: Toward a unified view. *MIS quarterly,* 2003, 425–478.

［45］Worsley, T., Wang, W.C. & Hunter, W. Baby boomers' reasons for choosing specific food shops. *International Journal of Retail & Distribution Management, 2011, 39*(11): 867–882.

［46］Yan, R. N., Hyllegard, K. H., & Blaesi, L. F. Marketing eco-fashion: The influence of brand name and message explicitness. *Journal of Marketing Communications, 2012, 18*(2): 151–168.

［47］Yang, K., & Jolly, L. D. Age cohort analysis in adoption of mobile data services: gen Xers versus baby boomers. *Journal of Consumer Marketing, 2008, 25*(5): 272–280.

论可穿戴设备在中老年服饰中的应用与推广

◆ 陈 莹[①] 唐 倩 （上海工程技术大学，上海，201620）

【摘　要】 围绕可穿戴设备在中老年服饰中应用的主题，探讨了两者间功能性与需求间的对位关系；从接纳的内在动因基础、存在的障碍和积极因素三个方面分析了中老年群体在接纳可穿戴设备程度方面的问题；提出了这些观点和建议：可穿戴设备不仅是青年人享受的特权，也将是中老年群体的喜爱；以中老年群体为设计对象，开发更加适合该群体特征的智能化服饰是中老年服饰发展的新方向；可穿戴设备在中老年服饰中的发展除了经济上支撑和观念上的突破之外，还依赖于社会的推助与中老年自身努力学习。

【关键词】 可穿戴设备；中老年服饰；应用；推广

The Application and Promotion of the Wearable Devices in the Elder People Clothing

◆ Chen Ying   Tang Qian *(Shanghai University of Engineering Science, Shanghai, 201620)*

Abstract: *Based on the theme: the application and promotion of the wearable devices in the elder people clothing, the paper discusses the position relationship between the function and the need; from three aspects, it analysis the problem on the acceptance degree of wearable devices in the older age groups; put forward that the wearable devices is not only enjoyed by young people, will also be in the old people's favorite; It is a new direction of the development of elder people clothing that putting the elderly group as the design object, to develop the more suitable intelligent clothing for them; The development of the wearable devices depends on the help and push from society and the efforts in learning of elder people themselves.*

Keywords: *wearable devices; elder people clothing; application; promotion*

[①] 陈莹，女，教授，上海工程技术大学。E-mail：chenying39@126.com

引言

随着互联网技术的发展,使信息的迅速传递、集聚与获取成为可能;可移动技术支持下手提电脑、平板电脑的问世,使信息的随时获取与处理成为现实;而智能手机的流行则使信息的快速传播与随时获取成为普及。从某种角度看,普及的意义绝不亚于出现,因为它意味着大众的参与,意味着创新成果价值的极大化实现。随之而起的是可移动技术向可穿戴性转移。将智能化设备与服饰相结合,生成智能化服饰,这与随身携带电脑设备有了很大的不同:它是以服饰——人类"第二层皮肤"的形式,十分亲和地伴随着人们生活的全过程,时刻为穿着者提供各种关于健康、环境、生活、工作等所需要的数据与信息,极大地提高了人类的生活品质。以苹果智能手表的推出与流行为标志,可穿戴设备以优雅、时尚的形态走进了人们的生活。而这一切都在逐渐改变着人们的生活方式、行为方式、交流方式和思维方式,这种改变将会是深刻的、持久的,对于服饰领域来说,是一场新变革的到来。

1 可穿戴设备适合中老年群体

可穿戴设备以服饰的形式带着高科技的气息、时尚的气息步入了当代人的生活,似乎更受到年轻人、时尚达人的推崇和青睐。事实的确如此,但这并不意味着可穿戴设备只是年轻群体享用的专利。深入分析可穿戴设备的主要功能性特点,我们可以得到它十分适合中老年群体的结论。

1.1 充分满足中老年群体对健康把控和保身健体的需求

可穿戴设备能够随时提供各种身体健康情况的数据,提供运动健身的数据,为穿戴者的健康保驾护航,这个功能非常符合中老年群体维系身体健康的愿望与需求。

1.2 充分满足中老年群体对查询解惑方面的需求

中老年群体由于应对变化能力减弱,因此特别关心天气变化情况、道路交通情况等;由于有充分的时间从事喜欢的活动,故对诸如保健、购物、股市、旅游、时事新闻等信息给予特别关注。而可穿戴设备能够充分满足中老年群体对查询解惑方面的需求。

1.3 充分满足中老年群体对终生学习、展示风采、保持身心健康的需求

可穿戴设备正像智能手机一样,使用后会激发中老年人自觉学习的兴趣和热情,从而不断刺激大脑,达到预防老年痴呆、保持身心健康的效果。同时也通过穿戴智能服饰表明自身与时俱进的年轻心态和活力风采。

1.4 充分满足中老年群体对提示功能的需求

健忘是中老年群体的通病,是否吃药了?锁门了?带钥匙了?关煤气了?等等一系列问

题常常困扰着中老年人,令其紧张。可穿戴设备的提示功能可以很好地解决此类问题。

1.5 充分满足社会、家庭对高龄、障碍老人等实时爱心监控的需求

随着年龄的增长,人的各种功能都在减退,尤其是高龄老人、障碍老人发生意外、走失等状况的几率就比较高,可穿戴设备的定位查找功能,以及实时监控功能,能够帮助解决一些这方面的问题。

可见可穿戴设备或智能化服饰对中老年群体具有不可低估的意义。

2 中老年群体对可穿戴设备的接纳分析

2.1 总体上有接纳的动因基础

从总体来看,中老年群体对可穿戴服饰是有一定接纳程度的,最根本的前提是随时能够获取健康信息,很好地达到保护自己的生命和管理自己的生活的目的。另外,正如前面所提到的,可穿戴设备的多种功能性非常适合中老年群体的特殊需求。

2.2 接纳中存在的障碍

不可否认,虽然可穿戴设备能够满足中老年群体的特殊要求,但在接纳问题上也存在着各种障碍。

第一,会遇到老年人对新事物的接纳比较被动、迟缓,甚至于有着发自内心中的排斥心理,以至于会表现出不接纳的现象。

第二,使用可穿戴设备,会提高经济方面的支出,从而或多或少影响接纳。

第三,面对高科技智能服饰,不会用,而学习成本又将会很高,包括全面地去学习新的东西所花费的脑力、体力、精力和财力等,否则难以充分发挥其智能服饰的各种功能。

第四,即便是想学也很难有年轻人或合适的老师能耐下心来教。

相比较之下,后面两条可能是更为关键的问题,这依赖于社会的推助和中老年自身的努力。

2.3 接纳中的积极因素

尽管在接纳可穿戴设备上中老年群体表现出一些困惑与障碍,但也不乏积极因素去克服、去接纳。如前所述,可穿戴设备重要的核心价值是引导人们的健康生活,这是唤起中老年群体逐步接纳的根本所在。尤其是处于刚退休、低龄段的老年群体,他们是一支活跃,并对时尚和健康生活有着极大兴趣与追求的队伍,他们大多有使用电脑的经历,有漫游网络的经历。使用智能手机,玩微信已成为他们的生活交流常态,这是十分宝贵的。有着主客观的基础,再有年轻群体和社会力量的助推,加上这支队伍的带动,可穿戴设备在中老年服饰中有着广阔的发展前景。

积极因素还来自于年轻人对父母辈的关心与影响,这也在中老年群体接纳过程中发挥着相当重要的作用。例如,年轻人购买可穿戴设备作为送给父母的礼物,并教会他们使用,这直接导致老人们的欣然接纳!其中不仅有享受健康、时尚生活需求的因素,而且还有着情感上的幸福体验,连同将这份幸福体验分享给他人的心理。

3 设计开发适合中老年群体的智能化服饰

通过以上的分析与阐述,可以得出中老年群体对可穿戴设备总体上是持欢迎态度的,而且此种智能化服装特别适合中老年群体的需要。因此开发中老年智能化服饰将是一个重要的发展趋势。

当然设计开发要有针对性,应从研究中老年群体生活方式、行为方式、思维方式、形体特征入手,设计适合该群体特点的和个性化需要的智能化服饰。智能手机、微信平台如今在老年群体中得到不断的推广使用给我们提供了可借鉴的成功案例。这与专门设计研发适合中老年使用的智能手机的努力是分不开的。便于识别、方便操作、舒服穿戴、优化功能、美观大方、提高性价比等,将是设计开发所要关注的重点。

4 结论

总结以上论述内容,可以得出以下基本观点:

(1)可穿戴设备的兴起预示着服饰领域新变革的到来,它不仅是青年人享受的特权,也将受到中老年群体的喜爱。

(2)可穿戴设备是中老年服饰发展的新方向。

(3)可穿戴设备在中老年服饰中的发展除了经济上支撑和观念上的突破之外,还依赖于社会的推助与中老年自身的努力,要加强对智能设备使用的学习。

(4)努力开发适合中老年群体特性的可穿戴智能化服饰。

参考文献:

[1](美)雷·库兹韦尔.机器灵魂的时代[M].上海:上海译文出版社,2006.
[2]陈根.智能穿戴改变世界[M].北京:电子工业出版社,2014.
[3]陈露晓.老年人审美与休闲心理[M].北京:中国社会出版社,2009.
[4]秦鹏.可穿戴技术的优与忧[J].健康人生.2014(9).
[5]肖征荣,张丽云.智能穿戴设备技术及其发展趋势[J].移动通讯,2015(5).
[6]友文.让生活更精彩——走进可穿戴智能设备[J].计算机知识与技术,2014(1).
[7]Zang Rui. Study of the Elderly Clothing Based on Universal Design[J]. Progress in Textile Science & Technology, 2014(01).

基于"无障碍"理念的老年服装功能设计

◆ 何清远[1,2]　王云仪[1,2]①　（1.东华大学功能防护服装研究中心，上海，200051；2.东华大学现代服装设计与技术教育部重点实验室，上海，200051）

【摘　要】　我国已踏入了老龄化阶段，老年服装的多功能设计得到关注。本文基于"无障碍"理念分析探讨了老年服装的功能设计问题。从提高生活自理能力和防护能力两个方面，讨论了服装所能起到的积极作用，包括服装结构的优化设计、服装的智能化设计、防护材料的应用以及可穿戴设备的使用等等，并对其中存在问题和未来可能的发展进行了归纳。

【关键词】　老年服装；功能服装；无障碍；智能服装

The functional design of clothing for aged group based on the concept of Barrier-free

◆ He Qingyuan[1,2]　Wang Yunyi[1,2]　*(1. Protective Clothing Research Center, Donghua University, Shanghai, 200051; 2. Key Laboratory of Clothing Design & Technology, Ministry of Education, Shanghai, 200051)*

Abstract: *China has entered the stage of aging, the clothing for aged group has been universally concerned. Based on the concept of Barrier-Free, the functional design of clothing for aged group was analyzed. The paper discussed the positive role of clothing in two aspects, namely improving self-care ability and protective performance in their daily life, through optimizing pattern, intelligentizing clothing, applying protective materials and integrating wearable smart devices into clothing. Moreover, the present problems and the trend of functional clothing for aged group were summarized.*

Keywords: *Clothing for the elderly; Functional clothing; Barrier-Free; Smart clothing*

---

① 王云仪，女，教授，东华大学。E-mail: wangyunyi@dhu.edu.cn

引言

我国早在1999年就已进入了老龄化的阶段,据统计我国的老年人口的占比达到世界的20%[1]。目前,中国是世界上唯一一个老年人口超过1亿的国家,且正在以每年3%以上的速度快速增长[2]。老年人生活的问题引起了越来越多人的关注。老年人随着年龄增长,身体机能逐渐减弱,在生活中出现穿衣困难、视力下降、记忆力减弱、走失、摔倒等诸多问题,对其生活造成了许多不便。如今,服装的功能性不仅仅可以通过材料、结构处理等方法得到提高,可穿戴智能设备与服装的相结合为服装赋予了许多新功能。老年服装设计正朝着合体化、功能化、智能化的方向发展。

在20世纪70年代,美国建筑家罗纳·德梅斯(Ronald L Mace)提出了无障碍设计(Barrier-Free Design)。无障碍设计是基于对人类行为、意识与动作反应的细致研究,致力于优化其使用功能,为使用者提供最大的便利[3]。将"无障碍"概念引入老年服装设计中,"以老年人为本",基于老年人体工效学,分析他们日常生活中所遇到的问题,并通过服装材料、服装结构以及智能设备等多层面的方法,使服装能够为老年人生活提供更多的便利,提升老年人的生理自理能力,同时赋予老年服装健康安全功能和防护功能,使老年服装趋于多功能化[4,5]。本文就从提升自理能力和提高防护功能这两个目的入手,探讨老年服装功能设计的手段和技术。

1 提升自理能力的功能设计

随着年龄增大,老年人的视觉、听觉、触觉等感官功能以及肢体功能都有所衰弱[6],这些变化都对老年人的生活造成了诸多不便。为了使服装在老年人的日常生活中发挥积极作用,在设计中注入"提升自理能力"的思想,基于服装舒适性和工效性的相关理论,采用有效的设计手段,将是对老年服装进行创新性的功能设计的思路之一。

1.1 服装结构优化设计

老年服装的结构设计需要以老年人群个性化的人体工效学为基础进行。与年轻人相比,老年人的体型特征、肢体活动能力等均会有明显的差异,因此,针对这些特点进行老年服装的结构设计是重要而又必要的出发点。

(1)肢体活动性

老年人的生理机能退化造成了肢体活动能力下降,服装对人体活动的束缚性在老年服装设计中应加以关注,减小服装束缚性是结构设计的关键点之一。例如,在肩、肘、膝、背和立裆等重要活动部位,需要进行宽松的或伸缩性的放松量设计,降低袖山高、选用弹性面料拼接等都是可行的方法[5];人体在进行坐、蹲、弯腰等运动时,颈围、胸围、腰围、臀围、腹围都会发生明显的变化,因此这些部位应放出适当的松量,以满足着装人体肢体活动的需要[7];而就肩部而言,肩宽量会随着人体运动尺寸减小,所以不需要设置额外的松量[7]。

此外,老年人体型差异较大,采用褶裥、隐藏拉链等设计,既可以满足活动时所需的松量,也可以提高服装的合体性。

(2)使用便利性

由于肢体活动范围受限,穿脱服装的便利性对于老年人服装至关重要。例如,上衣采用前开襟的方式,更便于老年人穿脱服装,钮扣的颜色应与服装的颜色对比度较大,便于老年人区分[8]。对于久卧型的老人来说,如何便捷地穿脱裤子尤为重要,例如,可在裤子的内缝线处设计隐藏开口[6,9],设置两条隐形拉链,只需拉下两条隐形拉链至大腿位置,便可以在他人帮助下进行更换裤子[6]。

此外,通过一系列简易的设计,可以提高服装使用的便利性。比如为了使其能更方便地找到口袋,可以在袋口位置采用凹凸的触摸感较强的面料;口袋的位置和袋口的倾角符合手插入的动作;袖子的长短通过暗条等进行调节等。

1.2 服装智能化设计

通过将可穿戴智能设备与服装相结合,将为传统的服装增加额外的功能特性,为老年人的生命安全、身体健康、生活便利提供积极有益的作用。目前,在提升老年人生活自理能力方面运用的可穿戴设备主要具有体温调节、感官辅助和生理监测功能。

(1)体温调节

与青年人相比,老年人对温度的识别能力也有所减弱。研究显示,60岁以上的老年人几乎不能察觉到1℃的温度变化,当环境温度变化时他们并不能做出及时的反应[10],微胶囊相变材料等智能调温材料具有虽然调温的作用,但其调温范围有限,并不能很好地满足老年人的需求。智能调温设备的引入弥补了这点局限,它可以根据监测人体体温状态,对服装微气候进行调整,使穿着者始终处于舒适的状态。

(2)感官辅助

老年人的视力、听力等感觉都有所下降。通过智能设备的加入,减少感官功能减弱所造成的不便,起到一种辅助的作用。据统计,具有视力缺陷的人群中82%超过了50岁[11]。针对这个现状,Tonya Smith-Jackson等人提出了一种老年社交型智能背心的概念,具有定位寻路系统,帮助有视力缺陷的老年人进行关键人物、物体定位和指路的功能[12]。此外,牛津大学正在研发一种恢复盲人视力的智能眼镜,内置微小的照相机和袖珍型电脑,使佩带者能够清晰地感受到眼前的人和物[13]。这两种设计都为老年人提供了视觉辅助。

(3)生理监测

由于监护人并不可能长时间陪伴在老年人身边,不能及时了解他们的生理状况,导致错过救治的最佳时间,这种显现频频出现。具有生理监测功能的服装就可以解决这一问题,实现对老年人24小时的监护。这类服装可以监测血压、心跳、体温等一系列生理指标以及环境情况,并能将这些数据通过无线设备发送到医疗机构以及子女手中,能对老年人身体健康状态进行监测和评估[14-15]。美国佐治亚理工学院研发了将光导纤维传感器埋入衣料内,可监控环境温、湿度以及穿着者的心率、体温、呼吸、血糖、血压等指标,通过数据分析协助医务人员日常监护[10]。

2 提高防护能力的功能设计

功能服装可以作为保护人体免受伤害的一种工具,对于易受伤害的老年人来说,赋予服装防护功能,将对其生命安全提供更多的保障。多功能的材料和可穿戴设备的发展为提高服装的防护性提供了可能。

2.1 防护性材料的应用

首先,老年人皮肤的敏感性下降,且皮肤干燥易产生静电,会出现加重心脏病、心律失常以及血钙减少等现象[18],因此,服装不宜采用合成纤维类织物,而应采用天然纤维织物,同时,防静电功能的面料也是较好的选择。

其次,在常规的功能材料方面,也有不少适合老年人群生理特征的面料,例如,透湿性能优越、吸湿快干的面料;防寒且轻薄、负重小的面料;吸湿抗菌面料;具有警示和夜行安全保障功能的反光或发光面料等。

此外,一些特殊功能的材料也为老年服装设计提供了新的思路。例如,英国工程师理查德·帕尔默采用尖端纳米技术研制出一种抗冲击材料 D3O 凝胶。在静止或缓慢移动状态下,这种凝胶又轻又软,但受到高速冲击时会急剧变硬,起到保护的作用[16,17]。这种材料可应用于肘、胯、膝等部位的防摔功能性设计。

2.2 可穿戴智能设备的应用

(1)防摔倒功能

老年人摔倒发生频率高,且后果严重。随着老年人年龄增长,摔倒造成的致病、住院、死亡的概率升高[19]。防摔倒系统运用三轴加速度传感器、陀螺仪等传感器进行数据采集,对人体摔倒过程中身体受到的冲击和姿态变化进行分析,对人体摔倒的过程进行预测和检测,并驱动保护装置[20]。Li 开发了一种腰带式可穿戴设备[21]。当老人摔倒时,通过气囊快速充气,起到保护老年人臀部的作用。美国弗吉尼亚理工大学的工程团队设计出了一种电子裤,当监测到人体步态不稳定时,能发出提醒警告,使老年人注意地面,提高警惕性[22]。

(2)防走失功能

定位功能也可以应用在防止老年人走丢上。据权威专家保守估计,近年来我国大陆地区每年有超过 30 万名的老年人走失[23]。不能确认自己的位置,不能找到目的地或起始地点的位置是老年人走失的主要原因之一[24]。具备防走失功能的服装装置GPS定位系统,具有定位、呼叫应答通话等功能[25,26]。GPS用于确定走失老年人所处的位置,使其能够迅速被找到。此外,为了便于老年人操作,呼叫应答通话功能一般采用一键式拨号,限定唯一拨号对象。例如,夏泽军研发的防走失服装将信号发射接收装置、电子显示设备、定位导航模块、无线传输模块、电池模块、人机交互的输入口和输出口以及天线集成为一体,嵌入服装中[27]。当老年人走失时,监护人可通过手机获取老年人的经纬度数据,快速找到老年人位置。

3 小结

服装需要引进其他领域的研究成果。随着计算机、信息等技术的不断发展,服装功能化、智能化是一个必然的趋势。尤其是老年人活动本身就存在不便的情况,如何在不影响服装美观性、舒适性、活动性的前提下,将服装与智能设备相结合,是现在研究的一大难题。

3.1 服装结构优化

老年人体型特征存在着差异,根据各种体型定制的服装结构的研究需要持续性研究。由于许多功能服装需要与智能设备进行结合,服装的结构需要满足固定、隐藏智能设备的要求,且不能以降低服装的美观性、舒适性和活动性为代价。

3.2 智能设备微型化和柔性化

就智能设备而言,微型化、柔性化的设备能够减少服装的重量,更加贴合人体,能大大降低其对服装舒适性和活动性的影响。这也是可穿戴设备的发展趋势。开发智能设备的操作平台,用户可以根据自身的需要对功能进行量身定制。

3.3 连接方式革新

智能设备之间的连接方式以及其固定方式对服装的整体性能有着极大的影响。导电性的织物、发电织物相关研究并未成熟。连接方式的革新与市场化将促进智能设备与服装的融合。

基于"无障碍"设计理念的老年服装功能设计,旨在为老年人生活提供更多的便利。通过服装结构的优化、服装的智能化、防护材料的应用以及可穿戴设备的使用等方式,提高老年人生活中自身的自理能力与防护能力。服装的结构需要考虑到肢体的活动性以及使用便利性;防护材料的应用能使老年人更好地适应不同环境,起到保护的作用;可穿戴设备与服装的结合赋予了服装更多的功能。然而服装不仅仅是智能设备的载体,智能设备的加入也不能牺牲服装的服用性能,如何将两者更好地融合是未来所需深入研究的问题。

参考文献:

[1] 全国老龄工作委员会办公室.中国人口老龄化发展趋势预测研究报告[N].中国社会报,2006-2-27(6).

[2] 国务院办公厅.社会养老服务体系建设规划(2011-2015年)[Z].2011.

[3] 冬梅.无障碍设计原则中的人文主义精神[J].艺术百家,2007,05:88-94.

[4] 文娟.功能性服装设计在老年服装中的应用[J].艺术教育,2014,12:301-302.

[5] 王文娟.无障碍服装设计在老年服装中的应用[J].艺术教育,2015,01:246-247.

[6] 陈庚笙.老年人功能性服装的应用分析[D].华南理工大学,2014.

[7] 王宝环 中老年女性体型运动变化与服装放松量的设置[J].浙江纺织服装职业技术学院学报,2008,04:31-35.

[8] Çivitci Ş. An ergonomic garment design for elderly Turkish men[J]. Applied ergonomics, 2004, 35(3): 243-251.

[9] 李晖.老年服装的人性化设计研究[D].齐齐哈尔大学,2012.

[10] 王海毅,冯伟.智能纺织品在老年服装中的应用分析[J].江苏纺织,2008,12:55-57.

[11] World Health Organization. Visual impairment and blindness[J]. Fact sheet, 2011 (282): 2009-2010.

[12] Socially-smart computing to support older adults with severe visual impairments: Proof-of-concept.

[13] 英国将研发智能眼镜 可让视力障碍者"复明"[EB/OL].[2011-07-06] http://it.sohu.com/20110706/n312562924.shtml.

[14] 阎珺.老年人功能服装的研究现状综述[J].科学中国人,2015,12:179-181.

[15] 孔超.适用于老年群体的可穿戴设备的设计研究[J].大众文艺,2014,08:64.

[16] 英国科学家发明"D30"凝胶状物质[EB/OL].[2009-03-03].http://www.chinanews.com/gj/sjkj/news/2009/03-03/1585612.shtml.

[17] 英国研究凝胶军帽,能保护士兵免受子弹危害[EB/OL].[2009-03-02].http://it.sohu.com/20090302/n262542090.shtml.

[18] 王洁.老年人谨防静电伤身[J].解放军健康,2005,02:27.

[19] Corsinovi L, Bo M, Aimonino N R, et al. Predictors of falls and hospitalization outcomes in elderly patients admitted to an acute geriatric unit[J]. Archives of gerontology and geriatrics, 2009, 49(1): 142-145.

[20] 佟丽娜.基于力学量信息获取系统的人体摔倒过程识别方法研究[D].中国科学技术大学,2011.

[21] Li Q, Stankovic J, Hanson M, et al. Accurate, fast fall detection using gyroscopes and accelerometer-derived posture information[C]//Wearable and Implantable Body Sensor Networks, 2009. BSN 2009. Sixth International Workshop on. IEEE, 2009: 138-143.

[22] 美国发明预防老人跌倒的智能服装[EB/OL].[2009-10-28].http://news.apparelsos.com/2009/1028/77725.html.

[23] 王智慧.老人走失:一个不容避开的社会话题[J].今日科苑,2005,02:21-23.

[24] 陈妮,张彩华.老年痴呆患者走失行为的研究进展[J].护理学杂志,2013,01:88-91.

[25] 朱建新,高蕾娜,田杰,张世访.基于Zigbee技术的老年人防走失装置[J].计算机工程与科学,2009,05:144-146.

[26] 张陆,高文钑.老年人求助及防走失智能技术与方法[J].社会福利,2014,06:53-54.

[27] 夏泽军.一种防止老年人走失的服装[P].实用新型,CN103610246A,2014.

基于可穿戴设备的中高龄功能性服饰研究

◆ 田亚楠 邢晓宇 洪正琳 尚笑梅[①] （苏州大学纺织与服装工程学院，江苏苏州，215021）

【摘　要】　可穿戴设备是可以直接作为配件穿戴在身上的便携式电子设备，在软件支持下感知、记录、分析生命特征，极大地提高了我们的生活质量。随着物联网和移动互联网的发展，可穿戴设备与各类应用软件紧密结合，成为其新的发展趋势。本文通过搜集中高龄服饰功能性研究内容，与相关的可穿戴设备结合，提出将医疗式可穿戴设备应用于中高龄服饰功能性设计中，主要从可穿戴设备的定义、分类、应用领域以及医疗式可穿戴设备在中高龄服饰设计中的应用现状及前景等方面进行综合叙述与讨论。

【关键词】　中高龄服饰；功能性；可穿戴设备；医疗式可穿戴

Middle-aged and Old People's Functional Clothing Research Based on the Wearable Devices

◆ Tian Yanan Xing Xiaoyu Hong Zhenglin Shang Xiaomei (1. College of Textile and Clothing Engineering, Soochow University, No. 1, Shizi Street, Suzhou P.R.China, 215006) (Corresponding author's email: shangxiaomei@suda.edu.cn)

Abstract: Wearable device can be directly as accessories to wear in the body of the portable electronic devices, with software support perception, recording, analyzing the characteristic of the life, greatly improve the quality of our life. Along with the development of the Internet and mobile Internet, wearable devices combining with all kinds of application software as its new development trend. This paper propose to make medical type wearable devices used in functional clothing design, through collecting the middle-aged and old people's functional dress research content and combining with related wearable devices. This article mainly describe and discuss

---

[①]　尚笑梅，女，苏州大学副教授。E-mail: shangxiaomei@suda.edu.cn

the wearable device of the definition, classification, application field, and the medical type wearable devices – clothing design present situation and prospects for the application in the future.

Keywords: *Middle–aged and old people; Functional clothing; Wearable devices; Medical intelligence*

引言

随着我国经济的发展和计划生育的催化作用,我国逐渐进入老龄化社会,截至 2012 年底,60 岁及以上老年人口已达 1.94 亿,占总人口的 14.3%,占亚洲老年人口的 1/2,老年用品消费市场已达 1 万亿元,已全面跨入老年国行列[1]。而可穿戴设备基于可穿戴技术的发展,该技术主要探索和创造能直接穿在身上或是整合进用户的衣服或配件的设备的科学技术。事实上,早在 20 世纪 60 年代,美国麻省理工学院媒体实验室就已经提出利用该技术可以把多媒体、传感器和无线通信等技术嵌入人们的衣着中,可支持手势和眼动操作等多种交互方式。将可穿戴设备致力于在中老年服饰功能性设计方面会有广阔的发展前景与应用空间[2]。例如,基于针对医疗的可穿戴设备可以通过对中高龄人群人体内部各项指标的测量,为使用者及家人提供及时准确的健康信息,并在紧急情况发生时为使用者提供周围治疗场所的信息。本文主要就可穿戴设备对中高龄人群的服饰功能设计方面的推进及应用进行了相关的叙述与讨论。

1 可穿戴设备的发展现状

可穿戴设备是人类对智能设备革命性的创新,推动人与物联网的发展,它将现实世界和虚拟世界相融合,为用户提供便利的服务,带给用户实时的互动体验,满足用户的需求。在 2013 年随着苹果、谷歌、三星等科技巨头相继投入到可穿戴设备的开发中,可穿戴设备逐渐成为人们所关注的热点。智能眼镜、智能手环、智能手表的层出不穷,让人们看到了可穿戴设备带给生活的改变[3]。

1.1 可穿戴设备的定义

可穿戴设备目前还没有准确的定义,通俗地讲就是一款可以穿戴在用户身上、或可以与用户的衣物整合的、能够随身携带并具有实时感知、记录、分析等功能的电子设备[3,4]。在通讯如此发达的信息时代,可穿戴设备的应用领域包括健康保健、增强现实、医疗监测、社交娱乐、安全救助、商务媒体、能量转换等,可穿戴设备以其便利的交互方式,将智能设备应用到生活的各个方面,引起研究、开发和生产的巨大浪潮。然而,目前的可穿戴设备缺少人们愿意穿戴的舒适性,如何让用户穿戴设备时感到自然,是可穿戴设备设计研发的基础。

1.2 可穿戴设备的分类

可穿戴设备的主要类型：按照应用来划分可以分为日常应用型和目标应用型两大类，日常应用型的如谷歌眼镜、智能手环、运动鞋等和人体紧密接触的日常穿戴产品；以功能来划分则可以分为功能延伸型和功能创新型，常见的功能创新型如平板电脑和智能手机，功能延伸型如智能手表，它能够实现上网，打电话发短消息、拍照并与我们的手机平板相互通信；而依据数据处理的形式来划分又可以分为内部数据采集型和外部数据采集型[3]。

1.3 可穿戴设备的应用领域

穿戴式智能设备时代的来临意味着人的智能化延伸，通过这些设备，人可以更好地感知外部与自身的信息，能够在计算机、网络甚至其他人的辅助下更为高效率的处理信息，能够实现更为无缝的交流。应用领域可以分为两大类，即自我量化与体外进化。具体的应用领域划分如图1所示。

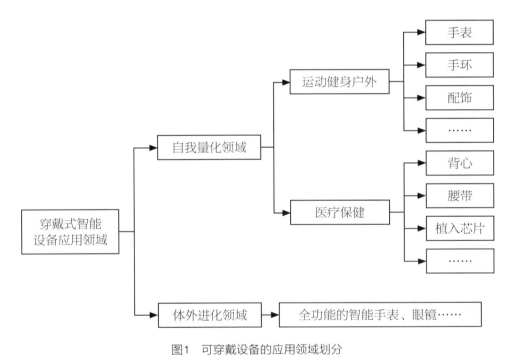

图1 可穿戴设备的应用领域划分
Fig.1 The application areas of wearable devices

1.3.1 自我量化领域

在自我量化领域，最为常见的即为两大应用细分领域，一是运动健身户外领域，另一个即是医疗保健领域。在前者，主要的参与厂商是专业运动户外厂商及一些新创公司，以轻量化的手表、手环、配饰为主要形式，实现运动或户外数据如心率、步频、气压、潜水深度、海拔等指标的监测、分析与服务[5]。而后者，主要的参与厂商是医疗便携设备厂商，以专业化方案提供血压、心率等医疗体征的检测与处理,形式较为多样,包括医疗背心、腰带、植入式芯片等。

例如,自我救助手环见图2a(图片摘自 http://www.smartcities.com.cn/),当用户落水或在游泳抽筋等紧急情况下,用户拉长手环,手环就

图2 自我救助手环及Hovding 隐形头盔
Fig.2 Self-rescue wristbands and Hovding stealth helmet

会迅速自动充气成救生圈,用户可通过救生圈自救或等待救援的到来。隐形头盔见图 2b(图片摘自 hovding),在衣领中内置安全气囊,当传感器检测到用户的动作超出常规幅度时,便会在极短时间内充满安全气囊,将使用者的头部包裹起来。与图 2a 需要手动救助的方式相比,这种自动监测危险并自动提供保护的方式,具有一定的优势,它不需要用户在危险时记忆任何操作来开启其防护功能。然而,该头盔需要电池给传感器供电,才能时刻监测用户的动作,电量不足或没电时,该防护功能可能会失效而引起防护失败的致命错误[4]。

1.3.2 体外进化领域

在体外进化领域,这类可穿戴式智能设备能够协助用户实现信息感知与处理能力的提升,其应用领域极为广阔,从休闲娱乐、信息交流到行业应用,用户均能通过拥有多样化的传感、处理、连接、显示功能的可穿戴式设备来实现自身技能的增强或创新。主要的参与者为高科技厂商中的创新者以及学术机构,产品形态以全功能的智能手表、眼镜等形态为主,不用依赖智能手机或其他外部设备即可实现与用户的交互。代表者如 Google、Apple 以及麻省理工学院等。

2 可穿戴设备于中高龄服饰设计的发展前景

目前,可穿戴设备的功能还仅仅停留于数据监测和基础信息的呈现层面,如果能和更高层面的健康保健和诊断医疗进行结合,将在人体健康领域发挥更大的作用。例如,当前智能手表等设备与移动互联网相连,除了可采集人体数据还可发送与接收网络信息。如果该信息与社会医疗系统或机构相对接,对人体数据进行分析后就能实现对用户身体健康的全方面管理,如分析、诊断、咨询甚至医院就医预约,这将使人体数据在医疗监测与预防领域的发展更加充分,也更人性化地贴合用户需求[1]。因此,将医疗式可穿戴设备应用于中高龄人群的服饰功能设计上去,将会拥有广阔的发展空间和社会价值。

2.1 中高龄服装市场的特点

据预测:到 2025 年,我国的老年人口将达 2.7 亿,占世界人口的 18.4%。由此可见,21 世纪

的中国老年市场将是世界最大的老年市场,或者说从市场规模的基础性要素——人口是市场的主体来看,中国的中老年服装市场具有不可限量的开发潜力[6]。随着社会的进步和经济收入的提高,中老年人正在逐步抛弃"重积蓄,轻消费""重子女,轻自己"的传统观念,花钱买健康、花钱买潇洒正成为现代中老年人的时尚追求[7]。目前,在国内中老年服装市场上只有较少数的商家在从事中老年服装的研发和销售,这是由于中老年体型多变和消费观念带来的影响。中老年服装市场数据统计表明,到2030年我国中老年人口的消费将达到13万亿元,尽管中老年人的数量在增多,然而我国从事中老年服装开发和制作的企业并不多,市场上能找到的高品质中老年服装也是少之又少,而真正能将可穿戴智能设备应用于中高龄服饰设计的实例更是少之又少。

2.2 可穿戴设备的应用现状

当前的可穿戴设备主要应用于人体数据的监测,可通过传感器采集人体的生理数据,如血糖、血压、心率、血氧含量、体温等。比如,三星即将推出的智能手环将专注于医疗领域,除了基础的运动和睡眠监测功能,还可以监测压力、湿度等环境数据。

图3　美信公司设计的Fit衫
Fig.3　The Fit designed by Maxim Integrated

2013年,美信公司(Maxim Integrated)试制出了嵌有多种传感器、能够测量生命体征数据的T恤,该公司称之为"Fit衫",如图3所示。"Fit衫"利用内嵌的传感器来测量心电图、体温及用户活动量等身体体征数据。2013年索尼申请了智能假发的相关专利,智能假发由动物和人体毛发制成,内藏电路板和传感器,可与手机或者眼镜等设备无线连接,同时内置三个触觉反馈制动器,用户在同手机等设备连接后,当有邮件或短信等通知时,假发会给佩戴者震动提醒。假发还将实现检测人体的健康状况,比如检测出汗情况、体温、血压等[1]。

目前上述产品的功能集中于人体数据监测上,人体数据的实时提供对人们的生命健康有着重大意义。例如,当前老年人只能通过定期医院检查才能对自身的健康情况进行了解,这可能导致慢性病发现不及时而错过最佳治疗时机。而可穿戴设备相比于当前的专业医疗设备更具便携性,有助于用户实时了解自身身体健康状况,发现身体异常时及时就医。

2.3 医疗式可穿戴设备的应用前景

近期频发的独居老人在居所离世许久才被邻里发现的新闻让社会的目光聚集到了独居老年人的健康生活,那么,如何让老年人的健康生活得到保证呢,医疗可穿戴设备是这个问题的有利解决方案[8]。医疗式可穿戴设备于中高龄服饰功能性设计的意义包括以下几点:

2.3.1 安全防护功能

随着我国经济的发展和计划生育的催化作用,人口老龄化的不断加剧,老年人的健康问题和安全保障问题日益突出。针对医疗功效的可穿戴设备可以通过对人体内部各项指标的测量,为使用者及家人提供及时准确的健康信息,并在紧急情况发生时为使用者提供周围治疗场所的信息(包括位置、距离、床位剩余量、值班医生信息及好评度等),以保证对紧急情况的有效救治。此外,日常随时观测身体指标还可以对一些慢性疾病起到预防作用,降低生病率及死亡率[9]。

2.3.2 医疗服务功能

通过医疗式可穿戴设备获得用户的生理数据,根据这些生理指标或信息,周期性的向用户或者用户的监护人推送健康报告,并依据这些报告向他们推荐附近的医院、健康机构等信息,包括针对用户的生理特征,擅长于这一方面诊断的专家或医院、当前时刻是否拥挤、医院的科室的位置等、出行的交通数据等[10]。

2.3.3 贴心保健功能

老龄化日益加剧的中国,很多子女并没有陪伴在父母的身旁,医疗可穿戴设备以及服务能够在子女繁忙的时候,为老人的健康安全保驾护航,同时又让子女省心。用户接收到的都是由专业人士经过分析得出的报告和建议。同时对于老人的健康管理,也提供了一站式的服务,省去子女大部分去询问医院,去查询保健礼品,去购买保健产品的时间[11]。

3 结语

可穿戴设备是人类对智能设备革命性的创新,推动人与物联网的发展,它将现实世界和虚拟世界相融合,为用户提供便利的服务,带给用户实时的互动体验,满足用户的需求。设计开发针对中高龄人群的医疗式可穿戴设备时,应该以中高龄人群的日常生活习惯为基础,运用可用性评估方法反复测试并改进设计方案中存在的可用性问题,提升可穿戴设备的可用性目标,并结合心理学、社会学、人机工程学、美学等理论学科,提升医疗式可穿戴设备的整体系统可接受程度,推动可穿戴设备的进一步发展,使其更好地为中高龄群体、为服装行业服务。

参考文献:

[1] 黄佳媛,张宏.基于可穿戴设备构建个人生态信息系统的探讨[J].现代传播,2015,37(2):139-142.

[2] 伍杰,王卓,曲刚.基于医疗可穿戴设备的商业模式设计[J].现代经济信息,2015,02:381.

[3] 戴婷.可穿戴设备的现状和发展趋势研究[J].企业导报,2015,08:69-61.

[4] 张元,骆雯,管倖生.可用性目标在可穿戴设备设计中的应用研究[J].包装工程,2015,08:72-75.

[5] 谢俊祥,张琳.智能可穿戴设备及其应用[J].中国医疗器械信息,2015,(3):18-23.

[6] 刘文永,徐晓慧.中老年服装消费行为分析及市场开发策略研究[J].针织业,2006,(10):24-26.

[7] 熊莹.中老年服装的市场前景分析及市场开发策略[J].黑龙江纺织,2014,(4):37-39.

[8] 王姝画.中老年服装市场的特点及策略研究[J].科技信息,2009,(14):236.

[9] 郭琪,杨勇强.可穿戴智能设备发展浅析[J].经济生活文摘,2013(3):539-540.

[10] 刘思言.可穿戴智能设备市场和技术发展研究[J].现代电信科技.20-24.

[11] 刘思言.可穿戴智能设备引领未来终端市场诸多关键技术仍待突破[J].世界电信.2013(12):38-42.

[12] 张洋.智能可穿戴设备的需求与机会[J].尚生活.2014(02):76.

[13] M. R. Daliri, "Automated diagnosis of Alzheimer disease using the scale-invariant feature transforms in magnetic resonance images," Journal of medical systems, 2012:995-1000.

[14] S. W. Tu, J. R. Campbell, J. Glasgow, M. A. Nyman, R. McClure, J. McClay, et al., "The SAGE Guideline Model:achievements and overview," ed: Elsevier, 2007:589-598.

[15] S. W. Tu, J. R. Campbell, J. Glasgow, M. A. Nyman, R. McClure, J. McClay, et al., "The SAGE Guideline Model:achievements and overview," Journal of the American Medical Informatics Association, 2007: 589-598.

[16] Y. Shahar, "The" Human Cli-Knowme "project: building a universal, formal, procedural and declarative clinical knowledge base, for the automation of therapy and research," in Knowledge Representation for Health-Care, ed: Springer, 2012:1-22.

[17] D. Isern and A. Moreno, "Computer-based execution of clinical guidelines: a review," International journal of medical informatics, 2008:787-808.

老年智能服装设计要点与现状分析

◆ 王 莹[1] 王云仪[1,2]① （1. 东华大学功能防护服装研究中心，上海，200051；2. 东华大学现代服装设计与技术教育部重点实验室，上海，200051）

【摘　要】　随着社会的老龄化，老年人成了社会中的重要且特殊的组成部分，智能服装在老年人中的应用势不可挡。为了能更好地为开发老年智能服装提供一定的理论依据，进而设计出满足老年人需求的服装产品，从研究分析老年群体在生理、心理、生活方式等方面的特殊性的基础上，本文提出老年智能服装的四项设计要点：紧扣生理特殊性，突出易用性；注重交互式设计，让服装更"活泼"；提升安全感，出行更无忧；加强功能性设计，实现多功能化。

【关键词】　老年人；智能服装；设计要点；现状分析

Design key points and present situation analysis of smart clothing for the aged group

◆ Wang Ying[1]　Wang Yunyi[1,2]　*(1. Protective Clothing Research Center, Donghua University, Shanghai, 200051; 2. Key Laboratory of Clothing Design & Technology, Ministry of Education, Shanghai, 200051)*（E-mail:wangying_ahaq@126.com, wangyungyi@dhu.edu.cn）

Abstract: *Along with the social aging,the aged have become an important and special part of the society,and the application of smart clothing for the aged is overwhelming. In order to provide theoretical basis for the development of smart clothing suitable for the aged,and design smart clothing that meets the demand of the aged,on the basis of analysis on the particularity of the aged in physiology,psychology, lifestyle and so on,this paper proposed four key points for the design of smart clothing for the aged:ease of use based on the physiological characteristics;interactive design to make clothing more "active"; security design to facilitate the outdoor activity; and multi-functional design.*

---

① 王云仪，女，东华大学教授。E-mail：wangyunyi@dhu.edu.cn

Keywords: *the aged group; smart clothing; design key points; present situation analysis*

引言

《中华人民共和国老年人权益保障法》定义60岁以上的人群为老年人,而最近,世界卫生组织对老年人的划分,则提出了新的标准,它将60岁到74岁的人群称为年轻的老年人,75岁以上的才称为老年人[1]。

一方面,相对于年轻人来说,老年人属于社会的弱势群体,他(她)们遭受意外伤害的概率要高于其他年龄群体,相比之下老年人更需要社会的关爱。设计开发满足老年人需求的产品,发展健全老年产业体制迫在眉睫;另一方面,随着人们对高品质、高创意、高智能化生活的追寻,智能产品逐步渗透到大众生活中的各个方面,服装的智能化也是其中重要的一个组成部分。因此,将智能设计的概念引入老年人群的服装设计,对于老年人的生命健康安全和医疗保健将具有重要的现实意义,并且能实现良好的商业价值,具有十分开阔的应用前景。

1 概述

1.1 智能服装

智能是一种技术,从最初的电子化技术和局部综合布线系统,到当今的将计算机技术、通信技术、信息技术的系统集成,以及将来的虚拟技术,都属于智能技术的范畴[2]。智能服装一般是综合采用了各种高科技技术、电子信息技术、生物医疗科学等相关知识,并且采用的材料、生产工艺、结构设计都相对特殊,智能服装通常集感知与反应于一体[3],因其特殊性,可以实现传统服装无法完成的功能。智能服装的出现毋庸置疑为服装产业的发展创造了新思路及新方向。

需要指出的是,智能服装的设计讲究以人为本,只有全方位了解消费者的需求,才能真正设计出以使用者为中心[4]的成功产品。本文将基于智能服装的基本设计方法和技术,结合老年群体的特殊性,提出并分析老年智能服装的设计要点。

1.2 老年智能服装现状

美国、日本和欧洲国家首先在智能服装方面地展开了探索性研究,近几年智能服装越来越多地受到各国研究者的重视,许多机构也开始致力于智能服装的开发。国外关于智能服装的比较有意义且成功的案例,主要集中在时尚类、运动休闲类、生理检测类、智能防护类等几大类。在进行智能化老年产品设计时会有各种高新技术被应用其中,如语音识别技术、健康检查、监测和管理技术、3G技术和远程医疗。目前,专属于老年人的智能服装研究并不多,而适用于老年人的智能服装研究主要有四大类:运动休闲、智能防护、健康监测和情绪监测,见表1。

表1 适用于老年人的智能服装
Tab.1 Smart clothing suitable for the aged group

	运动休闲	智能防护	健康监测	情绪监测
适用对象	喜爱运动及音乐类老年人	需要特殊防护的老年人	通常适用于患有慢性疾病的老年人	易发生情绪突变的老年人
代表产品	音乐外套外[5]	智能防摔倒服[6]	Vivo Metrics Life Shirt[7]	情绪监测服
优、缺点	优点：能够丰富老年人的生活；缺点：没有考虑老年群体的特殊性，易用性考虑不够	优点：与老年人的需求紧密结合，应用价值高；缺点：及时性、轻便化、有效性有待提高	优点：适时监测并发送老年人的生理信息至监护站；缺点：监测数据传输过程存在安全隐患	优点：及时了解老年人的情绪变化，采取缓解措施；缺点：情绪外露，可能会造成老年人的隐私压力

不难发现,老年人的智能服装功能表现为多样化,而不同类型的智能服装在表现出优点的情况下,也同时存在着一定的缺点,一部分原因是受到技术水平限制,另一部分原因则是目标消费者不明确,即它们不是专属于老年人的产品,导致在设计开始时就忽略了老年群体。

2 老年群体的特殊性

2.1 生理特殊性

老年人最典型的特殊性就是老,生理层面的老化也是最容易体现出来的。生理方面的老化基本归纳为以下五个方面：

（1）外貌形态上的老化

从外观上看表现为头发花白脱落、皮肤老化,皱纹出现、肌肉与骨骼功能弱化等。另外,值得注意的是皮肤的体温调节功能也会随着年龄的不断增长而逐渐减弱,这也使得老年人容易因为太冷而感冒,或是太热而中暑[8]。

（2）感官器官功能下降

由于老年人眼睛的晶状体弹性和睫状肌调节视觉焦点距离远近的能力下降,眼角膜随着年纪增长而变厚,退化成老花眼,即对于近距离的事物看不清楚。由于感官系统的下降还会出现老年性耳聋,味觉和嗅觉敏感度下降。另外,老年人感觉迟钝,皮肤的触痛、温觉减弱,表面的反应性也减弱,对不良刺激的防御等功能降低。这些都是因为随年龄增长,皮肤神经末梢的密度显著减少而造成的。

（3）脏器功能衰退

脏器功能的衰退,同造成外貌形态老化的原因一样,都是由于细胞功能的衰退造成。随着年龄的增长，老年人身体内部各种脏器功能和系统机制也开始下降。通常表现为肠胃功能紊乱、心脏病、高血压等常见老年人疾病。

（4）神经运动机能缓慢

医学上分析，是因为随年龄增加，细胞减少逐渐加剧所致。75岁老人组织细胞减少约30%之多。在老年人生理上主要有两个方面的表现：其一，细胞萎缩最明显的是肌肉，致使肌肉弹性降低、力量减弱、容易疲劳，肌腱、韧带出现萎缩僵硬，造成动作缓慢。其二，神经传导速度减慢，导致老年人对外界事物反应迟钝。

（5）记忆力衰退

随着年龄的增长，老年人的大脑皮层开始萎缩，脑细胞开始减少。60岁时大脑皮质神经细胞和细胞数减少20%~50%，小脑皮质神经数减少25%；70岁以上老年人神经细胞总数减少可达45%。老年人的记忆特点为：老年人的理解性记忆能力，逻辑性保持较好，但机械性记忆能力出现较明显下滑；再认能力衰退不明显，但老年人的再现记忆能力较差；生活早期记忆非常深刻，但对于近期发生的事情记忆比较模糊不清，通常表现为老年人总喜欢回忆过去。

2.2 心理特殊性

随着年龄的增长和身体状况不佳，老年人的心理也产生一些不同于其他年龄层次的表现，主要有以下四方面：

（1）易孤独

随着老年期的到来，不可避免地会出现离退休、丧偶、独居等情况，大多数老年人与社会的接触逐渐减少，与人打交道的频率降低，难免造成与社会一定程度的脱离，加上子女大多经常在外工作，因此，老年人经常无人倾诉和交谈，所以会常常感到孤独，再加上身体的年迈，各个部位开始出现各种毛病，就会产生焦虑的心情，严重的会产生自闭症和抑郁症。研究发现，老人的孤独感越强，生存质量越差[9]。美国医学家詹姆斯对7 000名美国居民做了长达9年的调查研究发现，在排除其他原因的情况下，那些孤独老人的死亡率和癌症发病率比正常人高出两倍。

（2）自信心匮乏

因为老年人的外貌发生了改变，以及各种感官的衰退和智力、体力的下降，使得他（她）们逐渐开始对于自己的形象没有了自信。老年人也常常在今昔的对比中感到已失去原有的自信和追求。再加上当今社会科技和文化的快速更新，多数老年人在退休之后开始与社会脱节。因为接受新事物难度和障碍，甚至一部分老年人会出现抵触新事物的心理，最终导致他们找不到眼前的生活和人生的意义。

（3）依赖性增强

依赖，通常是由不够自信的心理所致。依赖性增强会导致老年人做事愈发被动，凡事保持顺从的态度，而且情感十分脆弱，做事经常犹豫不决，遇到任何事情都想依赖别人去做，也喜欢让别人来做决定，觉得自己没有能力很好地处理。长期下去，就会产生情绪不稳的问题，感觉也会慢慢地退化。在老年人的老化过程中,有三种典型的依赖:经济上的依赖、生理上的依赖和社交上的依赖。

(4) 返老还童

许多老年人脾气和性格随着年龄的老去反而越发幼稚起来,时常表现出孩童的行为。如对事物表现出前所未有的兴趣和好奇心;主动要求别人过多的照顾和关怀等;此外,他(她)们也会通过言行举止和内心来表现出"我还很年轻",不希望人们把他(她)们当作没用的老年人,当作社会、家庭的负担。

2.3 生活方式特殊性

当人们步入老年之后,财富基本达到顶峰状态,同时生活节奏随着从工作上面解放而变得轻松、随意,生活方式也随之产生巨大的改变。鉴于人到老年后身体各项机能的衰退以及各类疾病的冲击,加重了老年人对健康的重视,从而使得运动[10]、旅游[11]成了老年人生活中的重要组成部分。另外,值得注意的是目前大多数老年人因退休以及与儿女聚少离多的缘故变得寂寞孤独,为了丰富日常生活,使得他们比工作时更加注重日常交际[10]。

3 老年智能服装设计要点

根据上文的总结,不难发现老年人从生理到心理以及生活方式上都和其他年龄段的人群存在着巨大的特殊性,然而这些特殊性又直接关系到设计者设计出的智能服装是否适用于老年人。因此,针对老年人群的智能服装的设计要点也应与此相对应。

3.1 紧扣生理特殊性,突出易用性

由于老年人的感官功能退化、各项身体机能衰退及其特殊的身型特点,老年人与年轻人对智能服装的需求大有出入。年轻人多追求时尚、精致及小巧,多数适用于年轻人的智能服装会对操作区域的功能按键进行统一、密集化处理,密密麻麻的小按键和功能标识混杂在一起,对于老年人的视力、触觉是极大的考验。这样的智能服装完全与老年人因生理特殊性而追求的易用性相违背,甚至可能会导致老年人因使用困难对智能服装望而却步。因此,对老年人来讲,只有智能服装设计紧抓他(她)们的生理特殊性,突出易用性,才能吸引更多的老年人成为智能服装的使用者和受用者。

3.2 注重交互式设计,让服装更"活泼"

多数老年人在心理上表现孤独、缺乏自信,而交互设计[12]可以调动使用者的参与,增加使用者对产品的喜好,进而可以缓解老年人在心理上存在的一些问题。另外,某些时候也可以通过交互式设计来提高易用性。如果为他(她)们设计的产品不仅不能够给他(她)们带来快乐,反而使他(她)们的自信心、自尊心受到了打击和伤害,使他(她)们本来就烦恼的生活变得更

加糟糕的话，就会更进一步导致老年人的负面情绪。因此，为老年人设计的产品必须要充分考虑他们的心理特征，激发他们使用产品的兴趣和乐趣，让使用过程变得简单、愉悦。

3.3 提升安全感，出行更无忧

大部分老年人都已经结束了职业生活，没有工作，老年人明显增加了大量的闲暇时间，走出家门参加各种社交活动以及旅游成了老年人晚年生活的重要组成部分。但是，基于生理和心理的特殊性，一些老人无法参与进去，如怕因腿脚不伶俐导致摔倒[13]、易走失[14]等。故，设计出更多能有效增强老年人安全感的智能服装意义重大。比如设计一件智能化导航服，老年人只要通过声控或手动按钮输入自己想要到达的目的地，然后智能服装通过卫星GPS定位导航系统的信息传输和解析语音导航到目的地，这样不仅能很好地解决老年人因记忆力减退而经常找不到回家的路的问题，也提升了老年人出行的自由感，并且在心理上使他们有一种独立感甚至是自豪感。

3.4 加强功能性设计，实现多功能化

智能服装本质属于一种特殊的功能服装[3]，人们使用它的主要原因基本在于它的功能。老年人使用智能服装的原因同样离不开功能，并且老年人对功能需求更大，例如防摔、定位、生理健康监护等。现在适用于老年人的智能服装功能设计还比较匮乏，而且缺少针对一些患病的老年人的智能服装。实现功能全面化是完善设计适用于老年人智能服装的设计的重要考虑因素。

4 结论

本文从分析老年人生理、心理及生活方式的特殊性出发，归纳总结出老年人在这些方面的典型特征，据此为设计适合老年人使用的智能服装提出了四个设计要点：紧扣生理特殊性，突出易用性；注重交互式设计，让服装更"活泼"；提升安全感，出行更无忧；加强功能性设计，实现多功能化，这些原则可以为后续设计者们提供设计依据。

未来面向老年人群的智能服装设计可以对设计细节加以更多的关注，例如，新兴的智能服装需要立足于老年人的生理特殊性，服装的功能按键的大小和按键力道要符合老年人手指操作的特点，并且能够提供明确的反馈信息；显示屏的角度设计成可调节的，以方便信息的查看和读取[15]；老年人感官系统退化及动作敏捷性差，开发具有火灾警报及求救功能的智能服装对独居老人生命安全意义重大；将智能材料与电子器件智能化结合起来，使得防护更加全面。

参考文献：

[1] 世界卫生组织网站 http://www.who.int/en/
[2] 王家跃. 老年产品设计中人性化、情感化、智能化的交互研究[D]. 山东轻工业学院，2008.

［3］陈庚笙.老年人功能性服装的应用分析［D］.华南理工大学,2014.

［4］田苗,李俊.智能服装的设计模式与发展趋势［J］.纺织学报,2014,02:109-115.

［5］http://informationtimes.dayoo.com/html/2008-10/20/content_349338.htm

［6］Amit Purwar, Do Un Jeong, and Wan Young Chung. Activity Monitoring from Real-Time Triaxial Accelerometer data using Sensor Network［C］. International Conference on Control, Automation and Systems, 2007: 2402-2407.

［7］VIVOMETRIC: Life-Shirt System［Online］. Available:http://www.vivometric.com/.

［8］姜晗.基于老龄化社会的老年人产品设计易用［D］.天津理工大学,2012.

［9］陈琪尔,黄俭强.社区老年人孤独状况与生存质量的相关性研究［J］.中国康复医学杂志,2005,05:363-364.

［10］朱宁,谢春萍.现代老年人生活方式初探［J］.苏州大学学报:工科版.2004.6.

［11］王凌云.南京市老年人旅游围观影响因素分析［J］.南京农业大学学报,2007.

［12］Jennifer Preece Yvonne Rogers, el en Sharp.交互设计-超越人机交互［M］.北京:电子工业出版社.2003.

［13］朱文娟,吴善玉.社区老年人跌倒恐惧的现状及其影响因素［J］.中国老年学杂志,2011,07:1225-1226.

［14］王智慧.老人走失:一个不容避开的社会话题［J］.今日科苑,2005,02:21-23.

［15］王猛.老年人应急性家用医疗产品的设计研究［D］.江南大学,2011.

老年智能化服装探讨

◆ 王诗潭[1]　王云仪[1,2]①　（1. 东华大学功能防护服装研究中心，上海，200051；2. 东华大学现代服装设计与技术教育部重点实验室，上海，200051）

【摘　要】　随着中国老龄化程度的加剧，老年群体的人身安全和身体健康得到了越来越多的重视，智能服装因具备感知、反应与反馈一体化的功能，在老年服装上的应用具有重要的开发意义与市场前景。本文分析了老年群体的生理特点，概括了老年服装智能化的机制。最后总结了智能老年服装的面料设计和技术设计两大要素，并对智能老年服装的未来发展进行了展望。

【关键词】　老年人；智能纤维；智能服装；可穿戴设备

The discussion of intelligent clothing for the old

◆ Wang Shitan[1]　Wang Yunyi[1,2]　(1. Protective Clothing Research Center, Donghua University, Shanghai, 200051; 2. Key Laboratory of Clothing Design & Technology, Ministry of Education, Shanghai, 200051) (E-mail:710796029@qq.com, wangyunyi@dhu.edu.cn)

Abstract: As the degree of aging is getting increasingly intense, more and more emphasis on their health and safety were achieved. Because intelligent clothing integrates perception, reaction and feedback function, application in the old people's clothing has important development significance and market prospects. Firstly, the physiological and psychological characteristics of the old were analyzed. Then, intelligent mechanism was summarized. Finally, two elements of the old people's intelligent clothing were summarized, recommendations for future research were also presented.

Keywords: The old; Intelligent fiber; Intelligent clothing; Wearable devices

　　第六次人口普查结果[1]显示，2012年中国60岁及以上人口占13.26%，相比第五次普查上升了2.93个百分点。预计，到2050年我国60岁及以上人口将占全国人口的25%，达到4亿。随着老年人口的迅速膨胀，老年消费市场的兴起将成为必然，而在老年消费需求中，34%为服

---

① 王云仪，女，东华大学教授。E-mail: wangyunyi@dhu.edu.cn

饰穿着，33%为医疗保健，33%为其他消费，可见服装和医疗保健的需求占了很大比重。如果能通过纺织技术、信息技术与微电子技术将服装与医疗保健等设备有效结合，实现服装的智能化，将占据很大的老年消费市场，更好地满足老年消费者对服装的多功能需求。本文基于老年人的生理特征变化，通过分析智能材料、可穿戴式设备与服装结合的可能路径，探讨老年服装的智能化发展空间。

1 老年人群生理特点

进入到老年时期后，人的身体形态和生理机能都会发生不同程度的变化与衰退，这些变化可以归纳为以下几个方面：

（1）体型改变

伴随着年龄增长，骨密度会随之下降，骨质疏松会导致躯干弯曲、驼背，身高迅速下降。肌肉与皮肤的松弛又会导致腰部、腹部和臀部的脂肪堆积，使得人体的外轮廓及不同人体部位之间的尺寸比例均会异于年轻人，整体呈现出不挺拔的老化感。

（2）体温调节能力减弱

人到了老年后，体温调节能力下降，加之人体活动量减少，受外界环境气候条件的影响程度加大。冬天由于代谢慢，体内产热减少，耐寒性差，会导致体寒感冒；夏天皮肤汗腺萎缩，出汗能力减弱，耐热性差，易产生中暑症状。

（3）免疫力差，易患病

随着年龄的增加，老年人身体的各器官、各组织会出现退化，抵抗外界病毒、细菌等微生物的干扰能力下降，尤其肾、心、肺等重要器官储备能力的下降，导致了一些老年慢性病，如冠心病、高血压、动脉硬化等。根据调查统计，北京市高龄老人慢性病总患病率为63.4%[2]，上海市高龄老人慢性病总患病率为72.4%[3]，对老年人的生命健康造成极大威胁。

（4）其他

体型的变化以及生理运动机能的衰退会导致他们四肢不灵活、行动缓慢，反应能力下降，易发生摔倒事件；皮肤缺乏水分，干燥松弛，在与服装接触时易产生静电；生理机能的减弱会使他们出现视力听力下降、记忆能力减弱等问题，这些均对老年人的日常生活造成了不便。

2 老年服装的智能化机制

服装的智能化，即服装不仅能感知外部环境或内部状态的变化，而且能通过反馈机制，及时地对这种变化作出反应。同时具有感知和反应双重功能是智能化服装的基本特征，感知、反馈以及反应是智能化服装的三大要素[4]。面向老年群体的智能服装是一个复杂的系统，其实现机制可以从以下两个角度加以考虑。

(1)纺织服装材料本身的智能化开发

智能纤维是指能够感知环境的变化和刺激并作出反应的材料,如调温纤维、形状记忆纤维、变色纤维等[5]。与传统纺织材料相比,智能纤维具有自我感知、自我修复、自我诊断等功能,将这些智能纤维与普通纤维交织或编入普通织物中制成服装,可以赋予服装一定的智能化特征。

由于身体内部各器官功能的老化,老年人对外界环境的适应能力下降,调温纤维制成的服装可以从一定程度上弥补他们身体抵抗力衰退的问题。而对于体弱多病的普遍性问题,形状记忆纤维能够发挥一定的保健作用。对于长期卧床的老年人来说,衣物更需要时刻保持卫生和干净,此时具有抗菌功能的纤维毫无疑问是首选。针对老年人容易跌倒的情况,则可在服装的肩部、肘部、膝部拼接"D30"耐撞击材料,减少摔倒对老人身体的伤害。

(2)微电子元件与无线通讯技术的介入

以较为"隐蔽"的方式,将一些微电子元件置入到传统的服装中,通过无线通讯技术及时传递信号,可制成智能化的"电子"服装。此类服装的智能化机制主要通过信息检测层、信息处理层和信息反馈层三层体系结构实现[4],如图1所示。信息检测层是人体与服装相互交互的第一个环节,包含微传感器等硬件装置,主要实现对人体生理信号的感知、获取并储存;信息处理层是整个系统的中枢,内置的微处理器将信息检测层所得到的模拟信号转化成数字信号,初步判断出穿着者的体征状况;信息反馈层是用户接口,这一层可以将微处理器分析得到的结果通过无线通讯技术传递给服装上的终端设备,并及时反馈给用户,以使服装的穿戴者根据这些结果适时地作出反应。

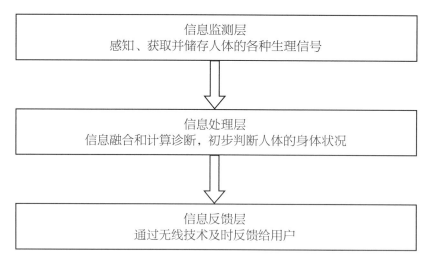

图1 电子服装体系结构
Fig.1 Smart clothing architecture

老年群体对健康的渴望是最普遍的心态,因此老年服装的智能化设计应以健康性和安全性为首要目的。对于一些患有慢性病或长期卧床的老年人,需要对其进行实时的监护以避免发生突发危险。然而不间断的人工监护会消耗大量的人力和物力,在服装上安装微传感器、GPS等电子器件,使之能够实时监控穿着者的身体情况、所在位置,并在出现异常情况时第一时间向监护者发出警告,确保救援的及时有效。

3 老年服装的智能化设计手段

3.1 智能面料设计

(1) 智能调温纤维

智能调温纤维是将相变材料技术与纤维制造技术相结合开发出的一类新型功能性产品,具有双向温度调节作用[6]。纤维中的相变材料会随着外界温度的变化产生固态——液态的可逆转化过程,在转化的过程中会吸收或释放热量,保持人体与服装微气候之间温度的恒定,达到穿着舒适的效果。早在1997年,调温纤维就已在美国实现了产业化,目前主要应用于军事、航天航空、医疗卫生及服装领域,如香港福田实业集团与美国杜邦公司合作[7],利用Outlast相变材料,生产出了具有温度调节功能的针织面料,并制成了"Fountian"牌温度响应型智能服装,已投放到高档服装专卖店中。

老年群体由于体温调节能力下降,受热容易中暑,受凉容易感冒。可在其服装中加入调温材料,通过材料相变的过程,促使人体达到热平衡。例如,人的足部离心脏最远,血液供应较差,俗话说"寒从脚起",对老年人来说,冬季足部的受凉会导致血管舒张功能失调,引发风湿病、关节炎等,使用智能调温纺织品制成的袜子或鞋衬能有效缓冲关节部位的温度差异,提高老年人对环境的热适应能力。

(2) 智能抗菌纤维

智能抗菌纤维是采用物理或化学方法将具有能够抑制细菌生长的物质引入纤维表面及内部,当细菌接近纤维时,具有破坏细胞膜作用的四级氨盐浸入到细胞内,与细胞核内的脱氧核糖核酸或RNA结合生成抗菌、杀菌的Ag^+、Zn^{2+}、Cu^{2+}等金属离子,生物体内的白血球,让髓过氧化物(MPO)的血红素蛋白质氧化产生过氧化氢、羟基、次亚氯酸离子、超氧化自由基等,即形成攻杀细菌的智能分子系统[8]。美国Nylstar公司[9]新近制造出了一种"智能聚酰胺纤维",将抗菌剂包藏在纤维内部,具有良好耐洗涤性,这种纤维可维持皮肤表面细菌的数量在正常水平,目前已投入到运动服装、内衣、医疗用品等领域。

老年群体由于生理机能老化,免疫力下降,对细菌、霉菌、微生物等比较敏感,尤其对于生活不能自理的老年人,得不到及时清理的人体排泄物会对他们造成极大困扰。智能抗菌面料可用于制作他们的贴身服装,如睡衣、内衣或短裤,可以起到抗菌杀菌除臭的效果,又具有卫生保健功效,针对一些过敏性皮肤病还能起到辅助治疗的功能。

(3) 导电纤维

导电纤维是铜、镍、银等金属纤维或将上述金属、碳以及最近出现的碳纳米管等的电子传导性粉末混合后的纤维,它可根据导电粒子间的距离和形成纤维粘接层间的界面距离来控制其导电性,主要用作防静电材料[10]。利用导电纤维制成的防静电面料,可以减少静电对皮肤的刺激,提高服装的安全防护性。

由于老年人的皮肤比较干燥,再加上他们心血管系统调节能力差,在衣物与皮肤摩擦时容易产生静电,造成血压升高。因此,可在其服装面料中加入导电纤维,减少静电积聚,消除静电危害。

3.2 智能服装技术设计

智能服装的技术设计主要针对"电子服装"而言,即将微电子技术、无线移动通讯技术、纺织技术、嵌入式技术等与服装结合,可实时地与人体进行交互,在服装中形成一套完整的信息采集、处理和反馈的系统。

(1)传感技术

传感技术主要以传感器为载体,通过传感器实现对人体生理信号及其他外部信息的检测,传感技术在服装上的应用主要经历三个阶段。最初阶段的电子服装主要是电子传感器与服装的简单结合,传感器体积大且功能单一,既影响舒适性和美观性又不耐洗涤。第二阶段的传感器主要向微型化发展,并与嵌入式技术、无线通讯技术有机融合,采用微细加工技术制成的微型传感器织入到服装面料中。最新的传感技术是基于纤维的传感技术,传感器直接由纤维和织物组成,如传导纤维、镀银纤维、织物传感器等,柔韧性和耐洗性都得到了很大改善,此类技术更好实现了人机交互的目的,是未来传感技术的发展方向。

目前,市场上可应用在老年服装上的传感技术主要实现两大功能:一是体外数据的采集,主要通过三维运动传感器或者GPS获取运动状况和周围环境状况;二是生理信息的监测,主要通过一些医用传感器监测心率、脉搏、体温、呼吸频率、血压、血糖、血氧、体脂含量等数据以实时判断人体的生理状况,提前采取预防措施。如加拿大的 **OMsignal** 公司[11]于2014年设计了一款可检测生命体征的T恤,T恤内置的多个微型传感器可同时检测心率、呼吸频率、呼吸量、心率变异性和消耗热量等多项生理指标,还可以测量行走的步数、步速、动作强度等数据,当这些掌控数据出现反常或者超标时,系统会通过第一时间向监护者发出提醒。

(2)无线通讯技术

传感器感知到的信息要传递给处理器进行数据处理与判断并最终将测试结果传输给使用者或监护者才能实现完整的智能化过程。传统的传输方式是有线传输,不具备移动性,各个电子器械之间、电子器械与用户终端之间都需要电源线连接,这会造成携带的不方便,影响正常生活。无线通讯技术主要运用电磁波信号来进行传播,很好地解决了远程信息的获取与移动中通信的问题,可以不受时间和地点的限制,快捷高效地直接传输信息,是服装智能化的关键技术。目前应用在服装上的无线通讯技术的主要体现在两个方面,一是通过具有无线收发功能的传感器节点,实时采集穿着者的生理指标信号,并将信号传到处理器,进行身体状况的判断。二是将处理后的数据全部输入到一个信息网络,如手机、手表、电脑等终端设备,当身体状况发生异常时会自动开启报警系统。

针对老年群体进行生理信息监控的智能服装,目前主要是通过卫星通信、全球移动通信、蓝牙、超宽带等进行射频电磁波传播实现无线传送生理参数信息。如东芝公司研发的

Lifeminder系统,该系统主要用来监测用户的身体健康情况、运动情况以及行为活动,通过蓝牙模块与PDA(掌上电脑)进行通信,并可通过PDA将监测到的数据传输到监护中心或远端医院。此外,针对老年群体设计的具有定位功能的智能服装就是在服装内配有个人局域网、GPS定位系统、电子显示器。GPS可获得具体的地理位置,并将数据通过个人局域网进行传输,个人局域网可以连接几个终端,如可传输到监护者的手机、服装上的小型显示器等。

(3)柔性电子技术

一般电子元件的制作材料都是刚性的硅,而与服装连接部位的支撑材料也是刚性的,弯曲折叠后会破坏脆弱的元件,导致性能失效。柔性电子技术是把脆硬的电子器件制作在软质弹性的基板上,其具有重量轻、可折叠、延伸性好、柔软的特点,很好地解决了刚性器件与柔性服装之间矛盾的问题,如柔性传感器、柔性电路板、柔性显示器等。

柔性电子技术在老年智能服装中的应用主要是电子器件的柔性化,如将各种微型电子元件都制作在一张柔性衬底上,在柔性衬底上设计缝纫孔,通过缝纫孔将这块柔性衬底与织物缝合。用于服装上的柔性衬底必须具备一定化学稳定性、力学性能和较高的黏附性,以更好地保护电子元件,常用于柔性衬底的材料主要有硅有机树脂和聚酰亚胺。Katragadda[12]设计了一款能与常规智能织物相结合的硅柔性外皮。该外皮可利用微显微技术制得,它由一系列"硅岛"组成,每个"硅岛"内有应变计,金属垫,传感器和电路等。将硅外皮缝入织物便能使服装具备感知反应的智能化。另外,还可以开发一些柔性触摸界面、柔性开关、柔性拉链等用于服装配件中。

4 总结

随着科技的发展,更多的智能技术及产品在老年服装的智能化设计上将发挥不可估量的作用。本文在分析老年生理特点的基础上,对老年服装的智能化设计进行了探讨,从面料智能化和智能化技术的应用两个方面进行了阐述。

智能服装是多学科、多种技术交叉作用的产品,研发具有实际医疗功能的可穿戴智能设备将是未来老年智能服装研究的中心和重点。目前的可穿戴智能设备主要应用在生理信号的检测和监护方面,其主要作用是记录人体相关数据与变化曲线,提前感知和反馈。而在实际的治疗方面,智能可穿戴设备发挥的作用还比较少,未来可研发针对老年群体常见问题能直接能达到治疗目的设备。

智能服装设计综合了多个学科的知识和技术,因此工程师与专业人员的有效合作联合研发是保证终端产品完整性和适用性的必由之路。目前开发智能服装的大多都是科技类公司,产品大多由工程师独立开发,专业人员很少参与其中。以具有生理监测功能的智能服装为例,显然需要服装设计、电子工程、医务等多领域专业人员的协同工作。未来的智能老年服装研发需工程师与专业人员的共同努力,将专业知识与先进技术结合才能设计出最实用安全的产品。

参考文献：

[1] 陈蕾.关于我国人口老龄化状况的概述与分析[J].呼伦贝尔学院学报,2014,22（2）:17-21.

[2] 李宁燕,姜鸥,平光宇,等.北京市广安门外社区高龄老人健康 状况的调查分析[J].中华全科医师杂志,2004,3（1）:37-38.

[3] 蔡静芳,孙慧娟,秦卫,等.上海市闸北区高龄老人慢性疾病现况调查[J].社区卫生保健,2005,4（3）:167-168.

[4] 蒙茂洲.基于柔性硅基薄膜技术的智能服装的研究[D].湖北:华中科技大学,2007.

[5] Tao, X, M. Smart Fibers, Fabrics and Clothing, Fundamentaland Application.[M].Cambridge England:Woodhead Publishing Linited, 2001. 1-6.

[6] 兰红艳,方磊.智能调温纤维的产生与发展应用[J].上海毛麻科技,2012,（1）:29-32.

[7] 刘娜.智能材料在服装上的应用[J].上海纺织科技,2011,39（7）:5-8.

[8] 王海毅,冯伟.智能纺织品在老年服装中的应用分析[J].江苏纺织,2008,（12）:55-57.

[9] 邵强,齐鲁.智能纤维及其纺织品的开发现状与展望[J].棉纺织科技,2007,35（10）:61-64.

[10] 杨艳玲,李青山.智能纤维的发展现状及应用前景[J].纺织科技进展,2006,（3）:17-22.

[11] 张如全,李建强,李德骏.电子服装的应用研究[J].服饰导刊,2015,3（1）:20-25.

[12] Rakesh, B, Katragadda, Yong, Xu. A novel intelligent textile technology based on silicon flexible skins [J]. MEMS,2007,（2）: 21-25.

康复治疗服装的现状及发展趋势

◆ 邢晓宇[①]　洪正琳　田亚楠，尚笑梅　（苏州大学，江苏苏州，215021）

【摘　要】　康复治疗服装作为智能服装装备受到很多老年人和病人的喜爱。据研究报道，穿着具有康复治疗功能的服装对病人术后运动能力的恢复有一定的积极影响。本文通过阅读大量文献，介绍了康复性服装的国内外发展现状和一些现有的康复性服装产品。目前应用较广的有两种，一种是将计算机技术应用于服装，通过运动治疗进行康复训练；另一种是利用服装压力促进血液循环，提高运动性能。此外，智能纤维也可有效地用于康复治疗，现处于研发阶段。

【关键词】　康复治疗；智能服装；老年人；运动治疗；服装压力

The present situation and development trend of rehabilitative garments

◆ Xing Xiaoyu　Hong Zhengling　Tian Yanan　Shang Xiaomei　*(College of Clothing Engineering, Soochow University，Suzhou, Jiangsu. China, 215021) (E-mail: shangxiaomei@suda.edu.cn.)*

Abstract: *Rehabilitation clothes are liked by a lot of old people and patients as smart clothing equipment. According to the report, dressing clothes with functions of rehabilitation treatment for patients have a positive impact on postoperative exercise capacity recovery. By reading a large number of literature, this paper introduced the developed situation of the rehabilitative clothes at home and abroad and some existing rehabilitation clothing products. At present, there two ways which are widely used, one is to apply computer technology to clothing for rehabilitation training by the exercise therapy. Another is using clothing pressure to promote blood circulation and improve athletic performance. In addition, the intelligent fiber can be effectively used for rehabilitation which is currently in development.*

[①]　邢晓宇，女，硕士研究生，苏州大学。E-mail：shangxiaomei@suda.edu.cn

Keywords: *Rehabilitation; Smart clothing; The elderly; Exercise therapy; Clothing pressure*

当前，我国已步入老龄化社会，且人口老龄化呈现加速发展的态势[1]。第六次全国人口普查显示，中国60岁及以上老年人口已达1.78亿，占总人口的13.26%，预计到2020年全国50岁以上人口的比例，将从2010年的24%攀升至33%。

随着人口老龄化趋势的加剧，人们对意外伤害、疾病所致的残疾、手术后的恢复等在治疗疾病、延年益寿等多方面的需求必随之增加，中国正处于健康产业快速发展时期，加快发展保健及家庭康复用品产业迎来了难得的历史机遇，市场潜力巨大。

老年群体是目前全社会所共同关心的群体，由于身体方面的原因，老年人在医疗保健方面的消费在其总消费中所占比重巨大[2,3]。现代医学领域的康复，主要是指身心功能、职业能力和社会生活能力的恢复，而康复治疗是一个从2000年之后才逐渐兴起的新兴职业。随着当今科技的发展，服装与人类生活紧密相连的同时已不单单满足御寒、保暖、遮体和美观等实用性功能，智能服装作为科技与电子信息技术及纺织服装发展的产物[4-6]，使得除了通过辅助器械，还可以利用一些特殊的功能性服装进行康复治疗，帮助病人快速恢复其日常生活活动能力，因而康复性服装成为老年人和病人的福音。

1 康复治疗服装在国内外的发展现状

康复治疗服装的研究，目前主要是以智能材料的研究为核心内容，并运用计算机等数字化技术研发适用于康复治疗的服装。我国康复治疗服的标准化研究起步较晚，几十年来，由于技术水平落后，康复治疗服及其标准化研究一直进展缓慢。我国康复治疗服标准化体系还存在很多不足，仍需完善。

国外特别是西方发达国家工业化革命较早，对老年人康复治疗可能遇到的问题了解比较深刻。经过近几十年的研究和积累，西方各国生产的康复治疗服有很大的改进，款式用途也越来越多。此外，用于康复治疗的产品主要是器械类，而康复治疗服装的在市场上却很少见，很多人甚至不知道这些产品的存在。

2 康复性服装的应用种类

2.1 运动治疗型康复治疗服

人们为了维持独立生活及适应生存环境而每天必须反复进行的一系列最基本的活动被称为日常生活活动能力，包括衣、食、住、行、个人卫生等基本动作和技巧。由于卒中患者偏瘫后存在肢体运动功能障碍，使生活自理能力下降，甚至完全依赖他人照顾，给患者及其家属带来沉重的负担。多项研究表明，康复训练能显著改善卒中偏瘫肢体的运动功能[7]，其中运动治疗是康复训练中最常用的增强肌力，肌肉耐力和恢复平衡能力的训练[8]。从20世纪90年代以来，康复治疗服成功地用于脑瘫、中枢神经麻痹的患儿,以后扩大应用到其他下肢功能障碍的病人,它能促进下肢血液循环,改进腿部新陈代谢,训练下肢的肌肉力量和运动协调性。

日本研究人员新开发出一款高科技"康复夹克",它能帮助手臂部分瘫痪的病人对上肢进行恢复性训练。研究人员开发出的这一"康复夹克"重仅1.8千克,适用于因中风等原因导致一只手臂瘫痪的病人。这种"康复夹克"包括由压缩空气驱动的人造肌肉。当病人穿上它活动自己健康的那只手臂时,安装在肘部和腕部的传感器就会搜集有关数据,并控制人造肌肉让瘫痪手臂完成和健康手臂同样的动作。病人穿上"康复夹克"对瘫痪手臂进行伸展和弯曲等训练,有助于他们重新恢复瘫痪手臂的活动能力[9,10]。2012年,日本还发明了一种全身可移动康复治疗服装[11](图1),可分别用于上肢和下肢运动。

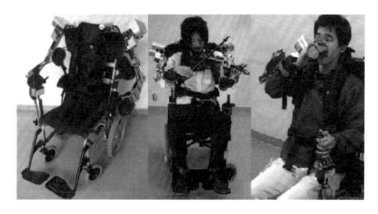

图1 康复治疗服
Fig.1 Rehabilitation clothing

俄罗斯研制出"REGENT"康复治疗服[12]由高强化纤制成,它是在宇航员腿部功能训练服的基础上经改进开发制成的,它将宇航员失重状态下,依靠弹力绳训练下肢肌肉的原理,用于下肢运动功能障碍的病人的康复治疗。全套康复治疗服分成几个基本部分:上装为超短背心,背心的长度在肚脐以上,胸前正中有拉锁便于穿脱,胸前设计了十字交叉强力条带,两边有垂直强力条带,下装为短裤,配有腰带,还有一副护膝,一双特制专用鞋。上下装之间由化纤强力条带、金属别扣,纵向地相互连接着,短裤以下用几根高强弹力绳和金属扣连接着护膝,护膝以高强弹力绳和金属扣连接脚上的专用鞋。其目的是将腿上和脚上的重力向上身分散,由身体带动腿部活动。病人在治疗时穿上它,可以借助弹力绳拉着患肢活动,调整姿势,转动身体等,以上装背心的一侧有襻,到同侧脚上的一只专用鞋,设计的最大垂直负荷可以承受40千克,而实际应用时一般不超过15~25千克。

北京交通大学研制了一种新型自动颈腰椎按摩服装[13],该技术采用先进的自适应模糊控制技术对特定穴位精确按摩,可对人体颈腰椎同时按摩。该产品结构简单,由按摩震动器、服装、控制板组成。通过按摩球的高频震动,对身体特定部位进行拍打。通过不同按摩球运动方式的组合,对身体部位进行揉搓。按摩球分布在颈椎及腰椎特定穴位,以保证按摩的有效性。该按摩服装可自动感知人体状况并精确按摩穴位,对颈腰椎疾病具有预防和治疗作用。

2.2 压力服装应用于康复治疗

服装压是指服装垂直作用于人体所产生的压力,主要包括重量压、束缚压和面压。压力服装对人体施加压力,进而提高运动性能并且有利于身体健康,其中包括增加血液循环,塑造形体,协助医疗后伤口的愈合等[14]。

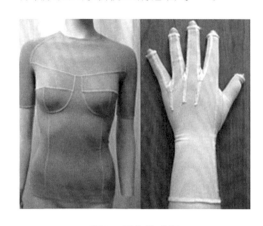

图2 压力治疗服
Fig.2 Clothing pressure for treatment

在医疗领域,压力服装(图2)可以辅助增生性症痕的愈合、预防静脉血栓、防止骨位异化、提高关节的灵活性等。1970年以来,压力服装就用来治疗增生性泡痕,原理是压力服装对伤口施加一定的压力,从而限制伤口处的血流、氧气和营养供给来控制伤口处胶原合成,进而可以加快伤口的回复,并减少痉挛发生的概率和进行外科手术的需要。理想的压力服装应该给受压部位提供适度的持续压力,例如,对于医疗用压力服装需要一天穿着23小时,并且一直持续到症痕愈合,一般情况下需要6个月。因此对于压力服装要定期测试其产生的服装压并及时调整。

2001年11月,在美国心脏协会举行的年会上,一位意大利研究者指出在乘飞机长途旅行时,长时间的坐姿会导致腿部血液凝固,危害健康。在研究过程中,他对833名试验者进行了测试,其中一半数量的试验者穿着长度到膝盖以下具有压力作用的长筒袜,另一半没有穿。结果显示:4.5%没有穿压力长筒袜的试验者出现腿部血液凝固现象,而穿着压力长筒袜试验者中只有0.24%出现这种现象。由此可见,长筒袜的压力作用可以减少由于长时间乘飞机旅行而产生的致命性血液凝固现象,袜子的压力作用有助于腿部的血流畅通。

2.3 智能纤维应用于康复治疗

目前,在医疗保健领域中具有药物释放功能、修复创伤功能以及屏蔽血液的智能纺织品正在被开发。智能纤维[15-17]作为药物释放体系的载体材料,集传感、处理及执行功能于一体,在药物释放体系中起着关键的作用。利用智能纤维作药物释放载体的研究已取得实质性进展。例如,将药物置于PNIP AA m接枝的PVA凝胶纤维中,当温度在20~30℃之间变化时,会自"开启、关闭",从而自动控制药物的释放。此外日本仿效人体胳膊的肌腱,研制了一种由外部温度变化而伸缩的智能材料可以使肌肉萎缩者的功能得到恢复。

浙江大学发明了一种移动智能康复治疗颈腰膝关节痛治疗仪[18,19],包括控制模块、微磁振装置、红外发热装置,如图3所示。其中,控制模块包括电源模块、调频模块和调温模块,红外发热装置包括多个发热元件和过热保护模块。控制模块与微磁振装置、红外发热装置相连,通过调频模块和调温模块分别控制微磁振装置的频率和红外发热装置的温度。这种智能康复治疗仪主要采用一种仿生纳米碳纤维复合材料(CFRP),可有效地康复或治疗颈腰关节痛等慢性炎症的症状,可普遍用于家居的服装、内衣、床被及携带式颈腰带等,不仅便于携带和洗涤,更具有智能变换、持续性生物磁振和远红外热敷控制的理疗特性。

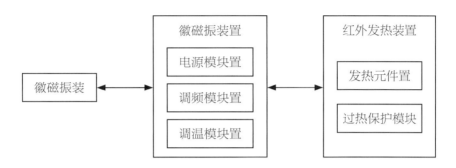

图3 智能康复治疗仪结构框图
Fig.3 Structure diagram of intelligent rehabilitation therapy apparatus

3 总结

康复治疗服装的研发正在逐步走向成熟,可以帮助行动不便的老年人进行运动治疗,或者通过压力服装促进血液循环、加快伤口愈合和提高运动性能。将智能纤维应用于康复治疗有很大的潜力,但目前在技术方面还有很多问题没有解决,很多智能纤维还没有产业化。由于缺乏有力的宣传和推广,很多老年人和瘫痪的病人未能使用康复治疗服,身体恢复比较缓慢。因此,我国必须多方面加强康复性服装的发展,一方面推广已有的康复治疗服装产品,另一方面研发新的康复治疗服,使需要康复治疗的老年人切身体会到具有康复治疗服装的功能,帮助他们快速恢复生活能力。

参考文献:

[1] 苏永刚,吕艾芹,陈晓阳. 中国人口老龄化问题和健康养老模式分析[J]. 山东社会科学, 2013,04: 42-47.

[2] 晁思达. 论人口老龄化对我国消费结构的影响[J]. 中国证券期货,2012,05: 199.

[3] 程逸群. 对开发老年服装市场的几点思考[J]. 中国纤检,2011,02: 78-79.

[4] 朱一帆. 服装业未来的智能化发展趋势[J]. 现代装饰(理论),2011,06: 26-27.

[5] 衣卫京,陶肖明,王杨勇,等. 用于应变测量的柔性导电织物开发[J]. 纺织导报,2013, 11: 75-78.

[6] 王海毅,冯伟. 智能纺织品在老年服装中的应用分析[J]. 江苏纺织,2008,(12).

[7] 王万利. 脑卒中偏瘫患者接受系统康复护理干预对肢体运动功能及日常生活能力的影响[J]. 中国临床康复,2005,21: 14-15.

[8] 朱晓军,王彤,陈旗,等. 规范化康复训练对卒中偏瘫患者日常生活活动转归的影响[J]. 中国脑血管病杂志,2007,06: 254-259.

[9] http://www.100md.com/html/DirDu/2006/11/01/26/28/74.htm

[10] 林莉.可当"医生"的智能服饰[J].科学之友,2007,(12).

[11] Tanaka E. Iwasaki Y, Saegusa S, et al. Gait and ADL rehabilitation using a whole body motion support type mobile suit evaluated by cerebral activity[A]. 2012 IEEE International Conference on Systems, Man and Cybernetics (SMC2012)[C]. 2012, 3286-91.

[12] http://www.ca800.com/news/d_1nrusj6oani9b.html

[13] 新型自动颈腰椎按摩服装[Z].北京交通大学,2008.

[14] 岳文侠.紧身衣针织面料服装压与延伸性关系的研究[D].东华大学,2012.

[15] 刘娜.智能材料在服装上的应用[J].上海纺织科技,2011,(7).

[16] 张华.智能纤维及纺织品的研制及其在医疗卫生领域的应用[J].化工新型材料,2007,(9).

[17] Giorgino T, Tormene P, Lorussi F, et al. Sensor evaluation for wearable strain gauges in neurological rehabilitation[J]. IEEE Transactions on Neural Systems and Rehabilitation Engineering, 2009, 17(4): 409-15.

[18] 浙江大学.一种移动智能康复治疗颈腰膝关节痛治疗仪[P].CN201410115526.4,2014.

[19] Zschenderlein D, Reichmann V, Mohring U, et al. Textile Sensors and Actuators for Prevention and Rehabilitation[J]. Tm-Technisches Messen. Plattform fur Methoden, Systeme und Anwendungen der Messtechnik, 2013, 80(5): 1-6.

中高龄服饰艺术设计研究

OUTLET: The trend of the new millennium
新千禧年的流行：中高龄服饰研究

◆ Leo Giovacchini[①] 　（*Instituo Europe di Design, Italy, Florence*，*10th July 2015*）

---

① 　Leo Giovacchini，Instituo Europe di Design, Italy, E-mail: leo.giovachini@sigmagi.it

Talking about OUTLETS!

The outlet is a very important theme today for businesses and consumers. Actually, many veterans of the fashion sector have no knowledge about the origins of outlets, why they have prospered and in what form. They have no knowledge about how they function or operate. Above all, they do not grasp how important they are for the manufacturers, the multibrands and, as a result, for consumption and consumers. Today, outlets are a very important business but many fashion sector people do not appreciate their significance and just consider them a second-class category.

This is particularly true for the Chinese market. I have many contacts on this subject for joint venture collaboration in China. Unfortunately, it is becoming more and more difficult because they don't know how to proceed, how to buy, how to sell or even how to implement the project.

I chose this theme because I live all of these considerations every day. My aim was to shed some light on this sector and broaden general knowledge about how it functions. I sincerely hope that this brief article of mine will contribute to creating a significant image of the world of OUTLETS, to appreciate its resources and to enable everyone to form their own opinion about whether it is a positive or negative phenomenon overall.

Leo Giovacchini

OUTLET: The Trend of the new millenium

For four years now, I have been the CEO of an enterprise operating in the sector of stock sales and OUTLETS, founded in Florence in 1972 and today considered one of the most important in Italy.

I was previously CEO for some important brands, so my position involved duties that were exactly the opposite of my duties today, even though they were in the same fashion sector. During these four years, I have seen how this secondary world of fashion has almost become the primary force, how different it is from the other and I have become familiar with the mechanisms, however incomprehensible they may be, that drive the world of OUTLETS.

All of this evidence made me very curious and I wanted to know more to understand the mechanism that feeds and develops it so rapidly. I also felt that it was right to share this information with others. Although I base my analysis on the Italian market, I am quite certain that it corresponds to the same situation on all other markets, aside from subtle differences of organisation. My company is active everywhere in the world. In particular, it sells directly to OUTLETS and establishes e-commerce agreements to supply seasonal quantities of products acquired from manufacturers that manage brand licences and high-end multibrand stores. These purchases all refer to collections of the previous season, never to the current season. Our revenues fluctuate in the range of 20–24 million Euro/year. This corresponds to a flow of 400–500 thousand garments per year through our warehouses. We have 3 outlet's shop (2 in Florence, 1 in Bologna), too. We are therefore a wholesale point of reference for this type of activity.

The word "OUTLET" was coined in the United States and later was used everywhere in the world to define this type of sales.

The foundation of this type of sale derives from the need to expand sales and therefore to increase production of the manufacturers.

Inside the company, a vendor sales space was reserved for registered employees and their families, with sales prices just slightly above the production costs. In Italy, this is called a "company store", in the United States it is known as an OUTLET.

Evidently, both the company and the consumers appreciated this form of sales. It was soon clear that this new type of sales had the potential to grow so the company opened its own point of sale to the public as well. Whether called a FACTORY OUTLET or simply a FACTORY STORE, these were probably the first form of OUTLETS. Most were managed directly by the manufacturer and they sold the goods produced by the company itself to the public. There being no intermediaries or representatives involved, FACTORY OUTLETS were able to offer their clients prices discounted by 30% to 50%.

Generally, they were set up in available space inside or near to the factory itself. Later, special structures with parking facilities were created near towns or localities that were easy to reach.

It generated a real buying fever. On the weekends, consumers went to visit cities where the companies they preferred and the FACTORY OUTLETS of their favourite brands were located to go shopping. Today there are still some sales structures managed directly by the producers.

In this context, even though the word OUTLET does not have a well-defined significance, it does provide a clear point of reference. OUTLET is a synonym for a point of sales where prices are much lower than in stores or boutiques. Initially, consumers thought the merchandise sold in OUTLETS had defects or was less than perfect, what is normally called second choice, but that is not true. That was simply a theory to justify the price difference between OUTLET and RETAIL.

In many cases, they serve to sell off cancelled orders, production over-runs, sample collections, etc. In general, they sell everything that cannot be sold through the traditional distribution. Some important brands understood the importance of this market and prepared special production runs for their OUTLETS with materials remaining in the warehouse. These were no longer current or past collections. It was production made expressly for the OUTLETS, where the only important factor of the sale was the brand name. In the late 1980's, an important brand near Florence understood this new business long before the others. It built an important structure where it placed its products for men and women. Inside, the products available were made especially for the OUTLET! In no time, this new point of sale became incredibly important and became the leader of OUTLET sales. A visit to this OUTLET was actually included in the tourist itineraries and became the most important destination for Italians and tourists from around the world to buy this brand. Every tourist who came to Florence visited this OUTLET. Considering the success, many other famous brands rushed in to do the same.

Evolution of the FACTORY OUTLETS created another new form of sales: the STOCK HOUSE.

The term STOCK HOUSE means OUTLETS that do not produce what they sell, but buy the merchandise directly from the manufacturers at discounted prices. The discount is passed on to the consumers, who can pay up to 60% less than traditional retail prices for a product.

How can the STOCK HOUSES buy at such low prices and how do they find discounted goods?

There are different methods, but all are based on the fact that some companies do not have the possibility or the interest to sell their merchandise in an alternative system such as a FACTORY OUTLET. That led to a new type of purchasing. The STOCK HOUSES bought lots of merchandise called stock from the manufacturers. They would open negotiations to buy a certain number of garments at a fixed price per unit. The colour or size is of no importance when the lot is stock. The STOCK HOUSES or the trading companies that invented this work buy lots called stock directly from the factories. They may be defective merchandise, sample collections, cancelled orders, end of season remainders from shops, merchandise from bankruptcies and so forth. Any system will do to get the products! This led to the creation of many small businesses of two or three partners at the most. This activity did not have a great reputation for reliability or respect. The people who did this kind of business were unjustly considered almost illegal or, at least, people to watch very carefully. Payment had to be made immediately and in cash. In many cases all this created movement of money that evaded tax control because they bought and sold without any invoices! Many companies did not sell their products in this way because they thought they might harm their brand image. It was certainly a period of confusion because this type of sales needed its own new identity.

Thus, it was possible to find everything and get good deals in the STOCK HOUSES. You just had to know how to make your selection and be lucky to find the right size. STOCK HOUSES imposed themselves on the market. Many consumers could buy their products, even from past seasons and from important brands they would never be able to afford through traditional channels. It was a race to find the garments and accessories of the big brands! It is only natural to think that, in most cases, remainders are products that were not sold so they were probably not successful but it did not matter. The discounted price made even these products important and beautiful. The important thing was to wear brand name garments.

The great development and success of the STOCK HOUSES attracted the attention of big investors and entrepreneurs, who with a far-sighted view of the market and their intelligence saw a future business in this trade: the OUTLET VILLAGE.

OUTLET VILLAGES are the most up to date expression of outlets. Also born in the United States, they landed in Europe in the 1990s, and about ten years later in Italy. Substantially, they are large shopping centres composed exclusively of small OUTLETS, like little shops. OUTLET VILLAGES are more open, with structures that in some way attempt to recall the architecture of a village or small city with some reference to the territory where they are built. OUTLET VILLAGES are usually conceived, built and managed by companies with big capital resources. The investments are very high and so it is necessary that products always be available in assortments of colours and sizes. The only way to achieve this objective is to allow the manufacturers to manage their OUTLETS directly and thus guarantee sufficient assortment. This is the fundamental point that distinguishes OUTLET VILLAGES from STOCK HOUSES, private OUTLETS or similar. The manufacturer's advantage in using the OUTLET VILLAGE format can be summarised as follows:

for the manufacturer

a) the possibility to sell remainders, reducing the costs of distribution and selling directly to the final customer

b) better protection of the image and brand that it produces

c) increased visibility of the brand and the hope of orienting customers toward the collection of the current season

d) direct management of remainders without having to give them to the stock house circuit

e) customer loyalty because they can buy in a direct point of sale

f) stylish use of end of series

for the consumer

a) consumers immediately perceive the OUTLET VILLAGE sales formula and the possibility to immediately verify the quality/price ratio of the merchandise they want to buy

b) the ample selection of brands offered in the same location and the possibility to compare the same or similar products by different brands

c) the possibility to evaluate the effective discount with respect to the real cost in standard shops

d) pass a few hours looking around even if they don't need to buy anything

The latest novelty offered by the OUTLET VILLAGE is the possibility to spend a day or half a day visiting the shops and have a meal or drink a cup of coffee. It is an opportunity to spend some free time with friends or family. Many OUTLET VILLAGES also organise events such as concerts, shows, theatrical events and so forth. Their aim is obviously to attract as many visitors as possible with the intent that many of them will make purchases.

The OUTLET VILLAGE sector has given highly specialised companies great success. They must be well prepared and the competence they must bring to their work is somewhat different from what is required for traditional shopping centres. The following are among the most important specialists in Europe in this sector:

a) MAC ARTHUR GLEN, the leading English company for development and management of designer outlet villages. It currently manages 5 OUTLET VILLAGES in Italy with 700 shops and revenues of over 800 million Euro (data 2013).

b) FREEPORT, an English company, has 4 OUTLET VILLAGES.

c) VALUE RETAIL, a British group, FIDENZA VILLAGE

d) NEINVER, a Spanish real estate group, CASTEL GUELFO, MILAN

e) SILVIO TARCHINI, a Swiss entrepreneur, IL FOX TOWN on the border between Switzerland and Italy

There are others that operate only on the Italian market, as follows: European Fashion

centre, Fashion District, AWG and Factory Outlet Development.

In the OUTLET VILLAGES, there are many brands, but not the top brands, the most sought after and the most expensive. Their presence in this reality would be a big risk for their image and positioning. That does not mean that they do not have the same requirements. They too have remainders or unsold goods that need to be capitalised. With scepticism and great prudence, the major brands have created a sort of consortium near Florence with official articles of association and regulations to create a very exclusive TOP OUTLET VILLAGE, where only certain names will be allowed to enter: The Mall. In this shopping centre there are only the top brands! The Mall is known around the world now and it is the favourite destination of all top brand consumers. They have even set up a non-stop bus service at no charge between Florence and The Mall and return.

Today this is an exclusive business worth millions of Euro!

FENDI **GIORGIO ARMANI** **G U C C I**

Loro Piana   Salvatore Ferragamo   VALENTINO   TOD'S

HOGAN   Dior   BOTTEGA VENETA   **BURBERRY**

MOSCHINO.   **BALENCIAGA**   roberto cavalli   **SAINT LAURENT** PARIS

TOM FORD **LANVIN** **SAINT LAURENT** PARIS   Pomellato

Gucci manages the entire consortium.

The growth of E-COMMERCE in the fashion market was an unexpected phenomenon for the speed of its growth. Really none of the operators in Europe expected it. Just a few years ago, they were sceptic about the possibility of developing this sales channel for fashion. The scepticism was based on the fact that a garment or a shoe cannot be tried on or even touched as is normal during traditional sales. So, they made some timid trials based on the mechanisms of their predecessors: mail order catalogues.

The mail order channel in Europe was widely diffused in Germany, the largest market after the United States, whereas it was never a great success in Italy.

The arrival of Internet led to development of a technology that could take a sale right into the home of the customer, regardless of the type of product, and particularly, garments and accessories. Unexpectedly, during the first few months of the year 2000, a revolution was introduced in the field of e-commerce. Use of the web to offer consumers the opportunity to buy the garments of the previous season at discounted prices, especially those of the biggest names. This was the case of the biggest Italian online store: YOOX. This idea was born in the company I am leading now. Thanks to the vision of a few persons and the availability of so many stock products with a variety of sizes and brands, the idea occurred to propose them on the web. Success has continued at such a great rate that it is now the most important Italian site. In 2014, Yoox invoiced revenues of 570 million Euro. In recent months, it has concluded a merger with another group, Pret–a–porter (Richemont), raising revenues to one billion two hundred thousand Euro, making it one of the most important e–commerce web sites.

Today there are colossal e–commerce companies with stratospheric revenue, all trading on the Internet. Customers make their selections, order, try on their purchases and return the items, all at reduced prices with respect to the boutique.

This brings us to the situation today.

It may be that the economic crisis made consumers poorer, it may be that buyers wanted to be more aware of the buying decisions they made, it may have been the desire to create a more personal style forgetting to wear only accessories and products of the current season. In any case, it is a fact that in the last few years the only distribution channel that has not suffered any crisis and has confirmed two–figure annual growth of revenues is the one including outlet sales, FACTORY OUTLET VILLAGES (FOC) and e–commerce.

The reason for such attractive prices is simple:

OFFICIALLY, the outlets sell products of the current collection or the previous season.

IN THEORY, those products are no longer available in the boutiques and are not included in advertising or fashion shows.

This is why even the most important and famous brands have adopted the outlet philosophy, creating a distribution channel that in THEORY will not overlap the traditional single or multibrand stores in the city, but will run parallel and enable the enterprise to sell off surplus and capitalise stocks. In THEORY, it should reach and attract new customers that would not be accessible otherwise for economic reasons.

In Italy, the outlet concept was developed much later than other European countries or the United States but during the last ten years the gap has been substantially eliminated, perhaps even surpassed. There are three important reasons, in my opinion:

a) the preference for products made in Italy

b) strong flow of tourists

c) the frenetic search for the best–priced brand products.

A similar situation has also been developing for some time in China. During the last four or five years, the growth trend of OUTLET/multibrand stores has been particularly strong. The construction of numerous shopping centres is continuing and all have assigned or reserved thousands of square metres for the OUTLET/multibrand stores acquired by vendor companies. These companies have thus launched a massive search for brand products known in China to fill the sales spaces.

My company has had meetings and contacts with many Chinese groups to launch collaboration, but we have succeeded in concluding solid, on-going agreements only with a few.

Unfortunately, however, almost all of them continue to make the same error. They search only for products of famous brands! That is not at all possible because the top brands saw the potential a long time ago and manage their own outlet businesses directly in China.

There are very limited quantities of top brand products available on the market. These are remainders that my company buys from qualified multibrand stores at the end of the season. Often there is no assortment of colours or sizes for these products precisely because they are remainders. In any case, the supply would not cover the demand of such a vast market. This is why it is not possible to collect top brand products on the market without targeting the current season (the parallel market), but in a completely different price range.

My company, which works in this sector specifically with the most famous brands, must deal with these issues every day. The Chinese groups should focus their attention not only on the famous brands, but also the second lines, on emerging brands that are young and new, which are already known in Italy and Europe in any case. The business of tomorrow will focus on these new brands, not only on the famous brands. The format of these projects are always the same: well-known brands, search for a partner that will supply the products, delivery on consignment, split sales 50/50 and so forth. These conditions are rarely acceptable to a company working in this sector. In the OUTLET world, buying and selling is always concluded right before delivery of the merchandise! The possibilities to implement these projects are therefore very limited unless agreements are made directly with the brand owners or licensees and, in any case, only if they have not already created their own direct sales organisation in China. Considering that the Chinese market is huge and has enormous potential, the base cannot be limited to top brands only. That would certainly be the simplest and quickest procedure to attain the objective but, unfortunately, it is not feasible.

Another point that does not coincide with the true European OUTLET concept concerns the location and type of shop. In Europe, the location must be very commercial and situated in points where average-standard consumers are numerous. The shop must be simple and practical. It must not have a strong image: it should be decorous but not luxurious. The customer must be completely at ease, feel free to touch the garments and check the price. Above all, the shop type must not intimidate the customer. To be frank, the shop must give the idea of being a place where products are inexpensive even if they have fancy labels or important brands. The display of brand names should be discrete and not excessively emphasized. Otherwise, the brand owners might create problems!

In China, on the other hand, they tend to use important locations that are new and modern, important furnishings, sometimes very luxurious and focus attention on the brand names. In fact, in China they don't use the word OUTLET, but almost exclusively the term multibrand store. This tends to confound the type of sales! All of this could easily create problems for the true distribution of the brand and create misunderstandings even if all is done according to the rules. It will be wise to rely on some specific precautions or basic common sense when undertaking this form of trade, to avoid such problems. Although my company is always available to provide guidance, it is often difficult to explain the specific dynamics of this type of activity.

In conclusion, I believe we must ask ourselves some questions. Our answers will provide even more incisive insight into the Outlet phenomenon and we hope this will be of use to the new operators. The answers however are valid only for the European market and not the Chinese, where the outlet concept is often substituted by the multibrand store. None the less, I am certain that Chinese operators can and will benefit greatly from the European concept and relevant experience.

1) Why do most outlets sell clothing?

The garment and accessories sector has some particular characteristics that make outlets convenient for consumers and manufacturers.

SEASONALITY. Garments can only be used depending on the season, and not vice versa.

TRENDS. Designers are very careful to change the cut, colours, fit, materials, etc. at every season, clearly to induce consumers to refurnish their closets at the same rhythm.

MARKUP. The passage from producers to resellers and all the related expenses generates a very high markup, especially for the best-known brands.

This means that the shops, the companies and the brands must get rid of unsold garments at the end of the season. The risk is that the warehouse will remain full of products that cannot be sold the following year because trends and fashion have changed. They hold sales, but sometimes they are not sufficient to dispose of the remaining merchandise. It was a godsend that people thought of an alternative form of parallel sales meaning the OUTLETS, STOCK HOUSES or OUTLET VILLAGES with the capacity to dispose of remainders by offering attractive discounts to recover a part of the expenses.

2) Why is the creation of new outlets or factory outlet villages always an occasion for polemics and diffidence?

They are usually situated on the major commercial routes or places that have a well-supplied basin of potential users. Outlets are often considered a serious threat to the commercial activities of existing shops. Factory outlet villages are thought to pull many people away from the city centre, thus damaging existing businesses. In many case, the construction of a factory outlet village might be considered damaging to the landscape itself.

On the other hand, a factory outlet village will create new jobs and will certainly become a tourist attraction for the territory where it is built.

3) Is it true that outlets propose mainly defective merchandise?

NO, this is absolutely not true! Most outlets propose only merchandise in perfect conditions and without defects. There may occasionally be merchandise with small defects, but it is always identified and proposed with extremely generous discounts. Outlets are generally quite careful to guarantee that customers are aware of what they are buying, and displays this type of merchandise in separate sales sectors.

4) Is it true that unknown brands are often found in outlets?

YES, it is true! The important brands are displayed to attract consumers, who may prefer other products for many reasons. The price, even with the discount, may still be excessive so customers may look around and settle on another brand, even if unknown, because it satisfies their taste and the price is right for their budget. The display of branded garments is still the main attraction. This is certainly the main force drawing the consumer into the outlet by following the brand and then purchasing other products.

5) What happens in outlets during the period for sales and remainders?

Just like traditional shops, outlets too need to dispose of remainders. In this case, they offer further discounts that can even reach 80%. In most cases, they recover only the cost. Sometimes they go below cost but capitalise the remainders. Lately, many outlets apply a single price to all products in the intent to get as much material out of the door as possible. After applying all the possibilities, what can be done with the products that are obsolete, have terminated, and completed their turn in the sales mechanism? Whether a traditional shop, a multibrand, factory outlet or outlet, after an 80% discount, what is left? The very last chance: they are sold for very low prices to market pedlars. They in turn sell in local and country markets, etc. A single price and take it as is!

6) Is it true that outlets are just a marketing ploy and actually, there are no real savings?

It is undeniable that everything that rotates around this world is a marketing ploy. It is also true that the term outlet has been scandalously used and abused, contributing to suspicion by consumers. It is also true that the manufacturers produce garments specifically for the outlet sector, perhaps branding them differently to differentiate the top line. Sometimes it is therefore difficult to know whether there were any savings because there is no point of comparison. For example, if a dress of brand A costs 100 Euro and the outlet price is 50 Euro, it seems like a good deal. But, if brand A exists only in the outlets, are we still certain about the savings? Reliable outlets, as in the case of my company that manages three outlets, leave the original labels attached to the garments so the consumer can see with his own eyes that the garment cost 100 Euro and he bought it at 50 Euro. In conclusion, in outlets and stock houses you can find just about anything! So let's not be fooled by those who claim that outlets are a scam, nor by those who claim they make great deals in the outlets. The truth, as so often happens, is in the middle and nothing can take the place of each buyer's final opinion.

7) What impact have the new outlets and factory outlet villages had on unemployment?

A recent study of the sector has established that one of the few sectors that has led our economy in these last few years has been the sector of outlets and factory outlet villages. In 2014, there was a boom of Asian tourists (+13%), spending an average of 1042 Euro each. All looking for the market of concentrated luxury but focused on outlets and factory outlet villages. Both have led the RETAIL sector with vitality and were responsible for a 30% increase in hiring. This all demonstrates the strong demand and development of this type of sales.

8) What interests and resources does an outlet or a multibrand store constitute for a brand?

They are very limited but also very important. They both make it possible to eliminate remainders, to capitalise a part of the unsold goods and thus to gather new energy for their own future. In the case of a multibrand store, the procedure terminates with the sale of remaining garments at a price equal to or slightly less than the price paid for it. In the case of branded merchandise, where and how to sell remainders is a crucial decision. A mistaken sale can produce a negative impact on the brand image. This is why first and second brand lines manage these sales directly through factory outlet villages.

9) What tricks do outlet sales use?

There are no tricks but many brands, even the most important, produce garments and accessories expressly for their outlets as if they were part of their normal collections of the past, whereas they are not. They have their labels and are produced by the brand, but they are made expressly for the outlet. Obviously, this has become a big source of earnings.

10) What will outlets represent in the future?

Based on their development in recent years and considering the economic crisis we are going through, I believe that this type of sales, together with e-commerce, will undoubtedly continue to expand at a great rate. By now all consumers, and I mean to include those with sufficient economic means, visit and buy in outlets. Today the trend is the search for the best price!

11) How have outlet sales modified the philosophy of buying?

Thanks to the outlets, the philosophy of buying has been turned upside down. In most cases, no one buys a product anymore because it is necessary. Now consumers buy a product because it has a pedigree and a great discount on the price. It must be said however, that today many consumers have even greater hunger for discounts. Outlets also offer discounts of 30%, 40%, 50% and 70% with respect to the initially discounted price, so many customers start asking when the discounts will apply right after the article goes on sale! The market is spoiled and used to all forms of discounts. Consumers today wait for the discounts to do their shopping.

12) How is buying conducted to purchase the merchandise for outlets?

Buying is done in a traditional manner. The companies like the one I direct are simply "COLLECTORS". They have specialised personnel with profound knowledge of the product. These employees visit multibrand stores and manufacturers at the end of the season to gather remaining merchandise. With regard to multibrand stores, the negotiation of the discount is based on the RETAIL price, and varies according to the remaining quantities, sizes and, above all, the brands. In the case of manufacturers, the negotiation is based on the WHOLESALE price. Obviously, the advantage of buying from the manufacturer is to have products available in different colours and sizes. Payments for merchandise, once the discount is agreed, are made immediately upon delivery. Obviously, each company has its own sources of supply that it will try to conserve with attention and great confidentiality. Lately, there have been and there still are problems to obtain important quantities. Sales are not going well, so orders are inferior and remainders are scarce. Manufacturers are facing the same sort of difficulty. Production runs are more limited so remainders are also reduced. This situation has created an anomaly. With less quantity available, buyers are offering less discount to succeed in getting the product. This is a negative situation for the outlets and a positive situation for the sellers.

After acquisition, the products are consigned to the company outlets and sold.

Undoubtedly, each person involved in the work cycle would give a different answer, but one thing is for sure and no question about it:

the development of OUTLET sales, by whatever name you want to call it, was truly a phenomenon of the first decade of the second millennium. It changed consumers' philosophy, and it produced labour resources and huge revenues around the world!

传统与时尚在中高龄女装图案中的表现

◆ 汪 芳[①] （东华大学，上海，200051）

【摘　要】　随着服装对消费者的定位细化，中高龄女装设计中的图案元素，也成为产品设计中不容忽视的设计要素，而作为中高龄女性，经历与人生积淀形成了其对着装特有的审美诉求。文章围绕着中高龄女装的图案表现进行了梳理，并对图案中的传统与时尚之间的关系进行了阐述，旨在对中高龄女装的设计具有借鉴作用。

【关键词】　中高龄女装；传统元素；时尚表现；图案；配色

The expression of the traditional and fashion in the pattern of the mature womenswear

◆ Wang Fang　*(Donghua University, Shanghai, 200051)*

Abstract: With the garments position on the customers more and more definite, the elements of the patterns appeared in the design of the mature womenswear become the key factor which cannot be ignored in the field of product design. However, the experience and accumulation of aged women's lives form their special aesthetic appeals for their dresses. The article focus on the expression of the patterns on the mature womenswear and describes the relationship between the tradition and fashion that appeared in the pattern so as to make reference to the design of mature womenswear.

Keywords: *mature womenswear; traditional elements; expression of fashion; pattern; color matching*

[①] 汪芳，东华大学服装·艺术设计学院，副教授，服装艺术设计系主任，染织艺术理论与设计方向硕士生导师。
E-mail: wfsnoo@163.com

引言

如今,中高龄女性的服装已逐渐摆脱了女装大码化的现象,而其款式与图案等设计的诸要素也日渐明确,产品的成功品牌与案例虽还很不足,但发展的趋势也成为行业的关注点,对其设计投入也是业界的共识,应运而生的中高龄女装品牌也逐渐被消费者认知。

承载着岁月与传统的中高龄女性,依然有着对时尚的心理需求,尤其在时尚大行其道的今天。能带着过去进入今天,无疑是中高龄者的重要生活态度。在传统中加入时尚元素,在时尚中注入传统的要素,传统与时尚的并存与交融,也成为这个群体的服装文化特点,在服装的图案设计中尤为呈现明确的表征。

1 传统与时尚之于中高龄女装图案的关系

中高龄人口的增长,以及该人群的经济地位改善,中高龄服装的产业已成为不可忽视的发展行业。唐人有"东隅已逝,桑榆非晚",而"爱美之心人皆有之",正是在这种情景下,中高龄女性对待着装虽不及年轻女性追求多变与标新立异,但对服装的品质与美感的追求依然有着自己特有的一份执着。

图案是服装设计的重要要素,也是最能直观地传递传统与时尚的设计要素之一。每个国家民族的文化,都能以图案的形与色传递,同时,流行文化也以其特有的图形、色彩被人认知。中高龄女性的人生经历与社会角色决定了其审美样式,她们骨子里有对传统文化的审美倾向,但又不希望彻底传统而显得落伍。时尚的光鲜与冒进,虽不能让中高龄女性全盘接受,但其强势也多少会影响到她们的审美而表现在对服装的选择上。传统与时尚的并存,也就成为中高龄女性服装的显著特征,其相互作用,演绎出中高龄女装的别样风采。

2 图案的图形特征

服装图案的特征是由图案的题材选择、风格定位、造型表现、材料与工艺样式等诸要素构成与呈现的。以下重点从中高龄女装图案的图形及色彩特征两方面来阐述传统与时尚的具体表现。

2.1 传统图形与时尚表现

漫长的世界服装史为今天的服装图案设计积累了丰富的图像资源,也是中高龄女装创作的源泉。期间,传统图形的时尚表现是图案显著的特点,这种基于传统图案的题材之上进行元素的变异或取舍等的变化,使图案在传统中有着时尚气息,在时尚的表象下又蕴含着传统的韵味。

2.1.1 大朵百花纹

花卉纹是女装图案中最具传统和经典的题材,是女性一直钟爱的服装题材。泛指题材多样的百花纹,在清代的女装中极为盛行,也是西方传统宫廷壁毯中常见的题材。汇集各种花卉的百花纹,以花枝簇拥,蔓草与枝叶穿插,形成一派丰富多彩的富贵繁荣的景致。中国传统中以其寓意对中高龄女性对晚年生活的祝愿,也是对家族的兴盛与吉祥美好的祝愿。有别于传统服装图案中对百花纹的表现,现今的中高龄女装的百花纹多以满底构图,循环式连接,花卉以多品种的大朵花,或写实或写意,结合对比色或协调色表现,用于中高龄女性的春夏休闲装。体现了中高龄女性性格开朗、生活积极的一面。中高龄女性大多体态丰盈,大朵百花纹图案设计的服装恰当地结合服装款式,可以弥补体型的不足,以迎合当今以瘦为美的审美心理,使穿着者更显年轻与活力。

2.1.2 独枝花卉纹

独枝花是折枝花的一种表现形式,描绘植物上截取花与枝叶部分的图案,强调完整花朵与折枝的造型关系。折枝花是服饰面料设计的一种重要造型图案,遍及世界各地各民族的服饰设计中。唐代诗句就出现对织绣图案的折枝花描写,中国的明清时期,折枝花图案已发展成一种极为普遍的服饰图案,同时也是欧洲18世纪"中国风"服饰的常用图案。有别于传统样式的折枝花,现今的中高龄女装常以独枝、定位的形式将图案表现在休闲风格的裙装、上衣等春秋或夏装中。独枝花卉纹可表现出鲜明的轮廓、自然的动感,加上定位的排列形式,更使图案显得别致而生动,结合胶印、数码印,刺绣等工艺手段,使穿着者更显情趣与朝气。

2.1.3 变异佩兹利纹

佩兹利纹源自克什米尔,而由苏格兰佩兹利小城发展起来的著名织造工艺表现的披肩图案,图案以藤本植物的涡旋造型构成基本图形,寓意吉祥美好,绵延不断,因其细腻、繁复、华美的造型而数百年来被广泛地运用于服饰产品中。而今出现在中高龄女性服装上的佩兹利图案,不再拘泥于原有的细密与紧致,而保留原型的涡形曲线,任意加入花草或几何元素,循环连续排列,夸张套色的对比使其在繁复中获得一份概括的简练,又不失原有的华美,使佩兹利纹在时尚中透着传统的艺术魅力,尤为彰显知性女性的内涵与华贵气息,为中高龄女性四季钟爱的休闲服装图案。

2.1.4 紧致细密暗纹

这里泛指排列紧凑、图案细小、色调弱对比的图案。内容可以传统中常见的吉祥纹、条格纹、千鸟纹、花卉纹等具象与抽象纹为题材,却打破原有的图案样式。以循环连续的排列方式,呈现出整体感强、调性统一却不失细节的图案特色。精致细密的暗纹图案最易结合光亮的丝绸锦缎,或是质地柔软细密的羊毛材料,并以提花工艺来表现材料的质感,凸显图案的精致与内敛的华美感。适宜于中高龄女性的礼服等正装图案的表现,以体现中高龄女性的典雅与端庄,还拥有女性干练的美感,是职业与知性中高龄女性的常见服装图案。

2.2 传统配色与时尚表现

女装的色调通常以黑色、白色、灰色、棕色、米色、藏青、酒红等色调构成最基本的色彩,并在每一季的流行色中循环出现,也是图案配色的基本色系。而中高龄女装的图案却有着自己特有的时尚与流行,期间涉及到这个年龄层特有的心理与生理因素,决定了她们对色彩的喜好与选择。

2.2.1 青花瓷色系

青花瓷色源于中国的传统青花瓷的花纹釉色,而泛指蓝白色系搭配的图案色彩。蓝与白色的图案可追溯到极为普及的中国传统型版印花的经典样式——蓝印花布。今天,继承了蓝与白的图案色彩搭配的中高龄女装却与传统的图案样式有着极大的不同。首先,保留了原来的套色格式,图形元素有极大突破,如以随意的几何形替代传统的花草纹,使传统乡村感的图案更具有都市的气息;也可以用图案的色彩面积的改变来凸显时尚感:加大白色或蓝色的比例,拉开形的色彩面积的对比,以突破传统的色彩均衡感;或结合印染工艺的特点,以吊染而形成渐变色来表现图案的样式,获得生动自然的色彩韵味。结合天然棉麻丝等材质的面料,以呈现质朴、端庄、宁静、雅致的图案气息,以适宜中高龄女性的精神气质,表现在夏装的衣裙设计中。

2.2.2 水墨色系

源于中国水墨画的水墨色,以墨色的黑和宣纸的白构成其的主要用色,而墨色的浓淡又形成了丰富的灰色。有别于传统的黑白色系图案,水墨色系具有一个重要的图形要素是色块的晕化形成的渐变关系。柔和、温润的过渡色使黑色与白色获得了自然的连接,也改变了传统黑色的厚重与肃穆感。水墨色系图案配色适宜在面料上的直接手绘,也可以用手绘图案稿进行接版,数码印制面料。面料以真丝或棉质等轻薄的为首选,如柞蚕丝绸、雪纺、玻璃纱等,以有效地体现色彩的透明感,达到水墨的韵味。以水墨色系表现的服装图案,具有严谨中不失温婉的特性,最适合有内涵与底蕴的中高龄女性穿着,且亦透出当今流行的"文艺范"气息。

2.2.3 金银色系

金银色因光泽感成为调和色中特殊的色彩,金色对应暖色调的调和,银色对应冷色调的调和。金色和银色在面料上的运用可以追溯到中国的西汉长沙马王堆的出土丝织品,而金色、银色以刺绣的样式更是传统中常见的图案表达形式。如今,在图案中金银色多采用金银粉染料印花、烫金箔印、仿烫金浆印等技术,其色泽光亮醒目,尤其与抽象形或文字复合搭配,极具时尚感。而中高龄女装更多地以金银纱线来进行提花表现图案,将金银色以小面积色块或线状对图案的花形进行勾勒与点缀,提花工艺下的纱线交织,使金银色或隐或现,在整体的图案配色中显得华丽而不张扬。此类设计多运用于秋冬季中高龄女装图案中。金银配色在中高龄女装印花图案中也常有表现,多以斑驳的色块配以其他中低明度的色彩,金银色块因斑驳而使其不显孤立地融合于图案色调中,使色调在华丽中多了几分粗犷与随意的艺术气息,为具有个性魅力的中老年女性喜好。

2.2.4 对比色系

对比色系是配色中重要的调式,虽然它是和谐的反向,但却是能调动视觉关注力的有效要素。在对比色系的概念下,因颜色的色相、饱和度、明度及形的布局、面积的要素,可呈现千变万化的色彩关系,印染织造业的发展也使极致的对比色获得实现。中高龄女装的对比色系多源自带有民族感的异域风格色调,与年纪轻的女性消费者的服装所不同的是,中高龄女装中的对比色调多定位在明度较低的色相对比,或者说是以低明度的要素使对比色获得了和谐感。期间也多加入黑色、灰色、白色、褐色等中性的色彩加以统一调和,或是以对比色的一方为主面积,缩小另一方对比色的面积,使图案呈现深沉稳重的整体调式,却又不乏有跳跃的对比色彩,获得不张扬的耐看色调,以符合中高龄女性心理与审美特性。

3 结语

时代的变革下,今天的中高龄女性已不再是传统社会"马大嫂"("买汰烧"的谐音)的单一生活角色,职业化、旅游、社交、娱乐等社会角色与活动给中高龄女性的生活注入了活力与色彩,"老来俏"的衣冠样式,正是其在着装上更多更高的需求体现。文化与艺术修养的提高,是这个群体的服饰文化提高的前提。在印染与织造业日新月异的今天,科技与工艺技术成为图案的艺术实现的助推剂,支撑着艺术创作。中高龄女装的图案设计为我们年轻的设计师提供的天地宽阔,令人期待。

参考文献:

[1] 葛彦. 中老年女性服装购买行为的研究[J]. 江苏丝绸, 2011.
[2] 张丽霞. 浅析老年人消费心理与消费行为[J]. 当代经济, 2013.
[3] 张渭源, 王传铭. 服饰辞海[M], 北京: 中国纺织出版社, 2011.
[4] 熊莹. 中老年服装的市场前景分析及市场开发策略[J]. 黑龙江纺织, 2014.
[5] 汪芳. 寻找银色光彩: 2014中高龄时尚服饰研究. 解读中国传统中高龄女装图案特征(c)// 上海: 东华大学出版社, 2014.

Interpreting Male Ageing and Fashion: Arts-Informed Interpretative Phenomenological Analysis

◆ Anna Maria Sadkowska[①] *(Nottingham Trent University, United Kingdom)*

Abstract: *Drawing on my doctoral project in Art and Design at the Nottingham Trent University, United Kingdom, in which I explore older men's experiences of fashion and clothing, I use this article to demonstrate how, by adapting a hybrid Arts–Informed Interpretative Phenomenological Analysis methodology, my work has the potential to further the existing fashion research and design methodologies. In this vein, this paper presents a critical and reflective review of the research process, summary and commentary of the body of work completed to date and selected initial findings.*

Key words: *fashion; men; ageing; interviews; personal inventories*

中高龄男性与时尚
——基于艺术资料的解释现象学分析

◆ Anna Maria Sadkowska （诺丁汉特伦特大学，英国）

【摘　要】　笔者借鉴本人在英国诺丁汉特伦特大学艺术与设计的博士生项目，探索老年男性对于时尚和服装的经验。本文通过调整所提供的艺术资料，运用解释现象学的分析方法进行研究。该研究有可能让现有的时尚研究和设计方法更进一步。最后，本文提出具有批判性与反思性的研究过程。论文主体是初步调查结果和总结。

【关键词】　时尚；男人；衰老；访谈；个人总结

（中文翻译：范振毅）

[①] Anna Maria Sadkowska, Nottingham Trent University, United Kingdom. E-mail: anna.sadkowska2012@my.ntu.ac.uk

1 Introduction

The relationship between fashion and ageing has recently become topical, especially within the context of the current socio-demographic changes such as the development of ageing populations and maturing of the so-called baby boomers generation. These global phenomena have resulted in a stronger link between fashion and aging than ever before. Twigg (2013:1) describes the relationship between fashion and ageing as *"sit [ting] uncomfortably together"*. Yet in recent years it has been studied from various perspectives including physiology, sociology and design studies. Indeed, no matter what discipline we operate within, this relationship seems to be analyzed mainly from the angle of bodily deterioration distorting and dimming the complexity of the embodied lived experience (Biggs, 2002). The postmodern approach to ageing redefines these constrained perceptions of growing old and by placing the emphasis on the individualities (Powell and Gilbert, 2011) encourages the alternative ways of exploring what it is to age.

Together with the shift in the way fashion and ageing is theorized and in parallel to its academic conceptualization, we witness the developments of various creative enquiries within the field of the cultural gerontology. It can be argued that these initiatives are mutually dependent and that the experience of ageing through the lens of fashion and clothing can now be identified as well-established in the on-going academic debate and changing social discourse. These transitions are especially visible within contemporary media, where initiatives such as Fabulous Fashionistas TV documentary, AdvancedStyle and ThatsNotMyAge blogspots all explore 'the beauty' hidden within old age. The fact that they are echoed by academic events such as (a)Dressing the Ageing Demographic symposium held at the Royal College of Art or Mirror Mirror: Representations and Reflections on Age and Ageing held at the London College of Fashion suggest that fashion and ageing is no longer a taboo subject. However, what is common in all these initiatives is that they mainly focus on women, neglecting the topic of older men and their experience of growing old.

Although some recent studies (e.g. Krekula, 2007; Henrard, 1996; Russel, 2007) have started to explore the relationship between gender and ageing, the topic of older men and their relationship with fashion and clothing is still a relatively new field of enquiry. Twigg (2013:19) observes that *"older men are [still] largely disengaged from fashion as a cultural field"*. Consequently, this paper seeks to contribute to this field of study by presenting selected findings from Arts-Informed Interpretative Phenomenological Analysis study with a small sample of British older men.

1.1 The ageing male body

Ageing is a complicated process of biological but also psychological and social becoming, unique and individual to each human being. Individuals in society are constituted by the intersection of one's past, present and future. It is precisely in the context of the mundane everyday practices that the phenomenological understanding of the lived body can emerge. Within this research the embodied experience of ageing has been explored through the lens of fashion.

In 1995 Featherstone and Wernick in the Introduction to the book *"Images of Ageing: Cultural Representation of Later Life"* argued that the lived body was a subject of the academic neglect when it comes to studies of ageing, which they related to the legacy of the Cartesian body and mind split. This dualism, in their opinion, influenced not only the lack of studies exploring the bodily experience of ageing but, more importantly, the hegemony of bio–medical approaches, which resulted in the production of studies which were *"data rich and theory poor"* (ibid:1) and argued for the need of new, alternative approaches in the field. Similar critique can be found in Faircloth (2003:1) who points out that *"we are lagging behind the people that really count in any academic discipline that enhances the 'social'– individuals and their lived place in society"*. This research seeks to explore the social character of the "lived" ageing male body. The various phenomenological methods document participants' individual relations with the everyday objects such as fashionable clothes; the analysis and interpretation of data gathered extend the understanding of how their ageing identities are negotiated on both personal and societal levels.

Likewise, Turner (1995) proposes the everyday "phenomenological body" as a method of exploring the body in illness, while Arber, Davidson and Ginn (2003) argue for the prior life course to be the only reference through which the current experiences of older people can be explored. The same authors point out that there is a need to rebalance the existing framework, which tends to focus more on the disadvantaged position of older women, overlooking the nuances of the ageing "new masculinity". This study argues for the possibility of utilizing the "phenomenological body" concept for creating the new interpretative approach to explore the lived experience of fashion by older men.

1.2 The fashioned male body

The relationship between body and fashion can be described as an intimate symbiosis, where both elements are interdependent and benefit from each other. The predominant intention of designing fashion clothing is for them to be worn and presented by and on human bodies. An underlying motive of clothing the body in fashion can be the creation of the intended gendered body image. Thus, the fashioned body can be considered as the "joint effort" between the clothing and the wearer. As indicated by Tseelon (1995, 2001) there is a significant lack of studies of the clothed body that are grounded in individuals' own accounts of their lived experience.

The role of fashion and clothes as the communicators and mediators between self and society, have been recognized by many scholars (Craik, 1993; Entwistle, 2002; Entwistle and Wilson, 2001; Crane, 2000; Kaiser, 2012). Weber and Mitchell (2004:4–5) describe the women's dress stories collected by them as *"a research method"* and *"a method of inquiry into identity process and embodiment"*, while Twigg (2009:93) argues that *"[clothes] offer a useful lens through which to explore the possibly changing ways in which older identities are constituted in modern culture"*. Thus, from a sociological point of view, clothes can become the key to analyse and understand a whole array of tensions between personal and social factors of which self–identity is composed. A phenomenological approach, with its emphasis on practice and experience, enables *"un–locking an understanding of what it means to be a human person situated within and across the life course"* (Powell and Gilbert, 2009:5). When it comes to fashion and clothing particularly, phenomenology provides the possibility to uncover the lived experience of ageing through the lens of fashion, and to establish the interrelation between the stories of individuals, objects and the times they inhabit.

One of the most profound gender stereotypes is that of men not being interested in their appearance, and consequently that fashion is predominantly a female domain. Craik (1993) relates this to the 20th century equation of masculinity with sober and modest clothing and puritanical appearance, which largely explains the contemporary marginal academic interest in men's fashion comparing to parallel women–oriented studies. Drawing on this, sociologist Tim Edwards (2011:42) discusses the lack of academic studies in men's fashion as a result of three historical factors: supremacy of studies in haute couture over the street fashion, development of women's movements which prioritized the role of women's dress, and in what he describes as *"the gendered development of fashion itself"*.

Phenomenological explorations of the lived ageing body in relation to clothing are rare in the contemporary research agenda, but they are not entirely absent (e.g. Twigg, 2009). However, interest in fashion and older men remains marginal. As Kaiser (2012) points out the expression "clothes maketh the man" should be understood as much in terms of what it says as in what it does not say. And it precisely does not say "fashion maketh the man"; while in western society men are "allowed" to do clothes, which often become the symbol of their masculinity and social power, they are not entitled "to do" fashion, which implies feminine characteristics of frivolity and superficiality (Kaiser, 2012; Edwards, 2011). Therefore it is justified to say that the scope of this research operates within two strong social stereotypes: that of the biased bio–medical perception of the ageing gendered body and that of men not being interested in fashion.

Understanding older men's experience of fashion is only possible through reference to their prior fashion experiences. Men entering midlife and later life in UK in the early years of the twenty-first century had a very different fashion life course compared with older or younger generations. Older men, born after World War II, baby boom teenagers whose adolescence coincided with the development of the youth-oriented subcultures such as Mods or Rockers; maturing with the evolution of menswear into men's fashion influenced by the concept of the New Man, represent very distinctive attitudes and consciousness towards their appearance including fashion and clothing. As Susan Kaiser wrote in "*Fashion and Cultural Studies*" (2012:172):

Intersecting, embodied subject positions are not just about who we are becoming; they are also about when and where we are becoming. Time and space are abstract and yet crucial-concepts that shape how we style-fashion-dress our bodies, and what we know (and how we know it) about ourselves in relation to others.

The initial findings of this project show that the contexts of the time, space and others play a significant role in the narrations of all of the participants, often being critical for the process of unfolding their fashion identities. Therefore Kaiser's argument is crucial for the purpose of my work, which aims to explore the lived experience of the homogenous sample (age, race, sexual orientation, social status, location, interest in fashion) of older men, who identify themselves as appearance-conscious.

2 Methodology and data: Arts-Informed Interpretative Phenomenological Analysis

My methodology is informed by two emerging approaches: in developing an in-depth understanding of the small and homogenous sample of older men's experience of aging in context of fashion and clothing, I am conducting an Interpretative Phenomenological Analysis study; by extending the interpretative strategies through various creative practices and producing artefacts as a valid form of knowledge dissemination I am undertaking Arts-Informed Research.

Interpretative Phenomenological Analysis (IPA) is a qualitative approach to research concerned with participants' personal lived experience, developed by Jonathan Smith as an alternative to descriptive psychology (Smith, 1996). Finlay (2011:140) identifies three "touchstones" of IPA: a reflective focus on subjective accounts of personal experience, an idiographic sensibility and the commitment to a hermeneutic approach.

Arts-informed research is an emerging approach located within the expanded qualitative paradigm. At the heart of this alternative qualitative approach lies the enhancement of the human condition through creative processes and representational forms of inquiry (Cole and Knowles, 2008). At the same time arts-informed researchers aim at reaching beyond academic audiences in order to make scholarship widely accessible. Knowles and Luciani (2007:xi) asserts:

We cannot stress more the importance of accessibility in research, in communicating complex understandings through multiple or alternative media for purposes far beyond mere artistic fancy and pleasure, and personal gratification.

It is this possibility of utilizing art making as a valid research practice in order to enhance the internally consistent research process that I pursue in this project.

The final elements of methodology are the methods selected as the means to conduct the research. In the following section I will discuss the chosen methods of collecting data: semi-structured, in-depth interviews, personal inventories and practical exploration.

2.1 Interviews

IPA researchers choose methods that allow them to "(...) *invite participants to offer a rich, detailed, first-person account of their experiences*" (Smith et al., 2009, p. 56). Therefore semi-structured, in-depth interviews are the most common research method used in IPA, which acknowledges also the importance of the interviewer's role in the process of listening and reacting to the interviewees' given accounts. The choice of this method is also supported by its growing recognition within the fields of art and design (Gray and Malins, 2004; Martin and Hanington, 2012; Crouch and Pearce, 2012; Ni Chonchuir and McCarthy, 2007) and fashion (Bugg, 2006).

2.2 Personal inventories

Martin and Hanington (2012:102) notice that "[i] *nterviews can be made more productive when based around artifacts* (...)". The use of certain types of objects as the stimuli for conversation is especially relevant to my research which aims at exploring the various and complex types of relationship the participants have developed with fashion, represented by different types of fashion-related artefacts such as garments, accessories or photographs. Although the personal inventories method is not typically used in IPA research, it focuses on encouraging participants' autonomy in selecting significant artefacts (Csikszentmihalyi, 1991; Odom and Pierce, 2009) and it is consistent with the locus of the IPA concern.

2.3 Practical explorations

In 1993 Christopher Fryling identified three modes of design research: research into design, research through design and research for design. Since then there has been attempts to further specify the relation between research and design and its potential role in knowledge production (Archer, 1995; Jonas, 2007; Fallmann, 2007; Cross, 2007). Niedderer and Imani (2008) notice that there is insufficient understanding of how to attach relation between tacit/explicit knowledge to the research methodology, regardless of intensified focus on qualifying practice as a research method. Martin and Hanington (2012:146) describe research through design as a research method which is "*constituted by the design process itself, including material research, development work (...) [a] nd the critical act of recording and communicating the steps, experiments and iterations of design.*" Practical explorations within the field of fashion in the research process have become a method allowing me to advance and articulate the unfolding interpretations of the data gathered through interviews and personal inventories.

2.4 Sample and procedure

The participants in this study were 5 fashion–conscious men recruited through the "word of mouth", contacts established through my supervisors and peers. Each potential participant was then contacted via e–mail or phone, which gave me a chance to introduce myself as a researcher, explain the interview scenario and generate the initial rapport. This involved the instruction to prepare/bring along 3–4 fashion–related artefacts that carry special meaning to participant such as garments, accessories, textiles or photographs.

All participants were white British males born between 1946 and 1964 (tab. 1). Participants were interviewed by the researcher either in their own homes (n=2) or rooms within the Nottingham Trent University (NTU) settings (n=3). The interview schedule included 6 questions/topics for discussion including participants' personal definition of the term "fashion", describing your current relationship with fashion; the time when they felt really good about the way you looked and their perfect fashion item.

Each interview lasted between 90 and 120 minutes, was recorded and transcribed verbatim by the researcher or the qualified typist. After each interview the set of research observation notes were completed in a research reflective diary.

Table 1. Sample characteristic and interview settings

Name*	Age	Occupation	Data of interview	Place of interview
Eric	60	Artist	20/05/13	His home
Grahame	61	Social care worker	24/05/13	His home
Henry	53.5	Academic	30/05/13	NTU
Ian	58	Company director	05/06/13	NTU
Kevin	63	Lecturer	05/07/13	NTU

* All names were changed to protect participants' anonymity.

2.5 Analysis

Many sources (Langdridge, 2007; Smith et al., 2009; Willig 2001; Finlay, 2011) offer guidelines on how to analyze data in IPA. Yet, since IPA's main concern is with personal lived experience, the meaning of it and how participants make sense of it, those outlines are often flexible and not prescriptive. In my analysis I followed the guideline provided in Smith et al. (2009): reading and re-reading of a transcript, initial noting, developing emergent themes, searching for connections across emergent themes, moving to the next case sand looking for patterns across cases.

No software was used at any stage of this analysis; I used a hard copy of each transcript in landscape with wide margins on either side, allowing me to complete the exploratory coding and to note the emergent themes. Each transcript was re-read at least 4 times before I embarked on exploratory coding, and another 3-5 times before I begin to distinguish the emergent themes, separate for each case. This stage was completed for all 5 transcripts. Furthermore, throughout each stage of the analysis I utilized various practical explorations/ artful practices in order to enhance my creative understanding of the phenomenon under study.

3 Results: The Dis-Comforting theme

The results to-date include a set of themes capturing the richness of the male ageing phenomenon in relation to fashion and clothing, their written interpretation and corresponding fashion artefacts and films. I will present this, using as an example the "Dis-Comforting" subordinate theme, which is a part of the "Learning Fashion" superordinate theme, jacket and film. This theme is touching upon the development of a unique system of values by participants in the past where the physical comfort was often compromised for the sake of a fashionable look.

The "Dis-Comforting" theme is significant not only in terms of the psychological but also physical effects on some of the respondents. Those participants shared sharp memories of the caused by clothes uneasiness or even physical pain. Their past willingness to sacrifice the bodily comfort was significant for the way those participants developed their present expectations towards fashion and it is often reflected in their current fashion behaviors, especially the negotiation between the physical and mental comfort. It is this negotiation that I found the most puzzling and inspiring, the same time, as a researcher and understanding of which I tried to enhance through practical engagement with the form of a second hand man's jacket.

Rather surprisingly, this theme was present in the accounts of all of the participants. One participant, Henry (53) discussed it especially openly and vividly both as part of his past, and current fashion practices, traces of which can be found in the following extract:

When I am wearing clothes... comfort... noooo I don't think so. As long as I can get my increasing size into it, I am happy. Comfort... no. Style first. I will breathe in, and fasten anything, and live with it for the day, if I have to. I would probably consider a corset, if I had to get into something. If I had to but... no. It is shape and form and colour. I am not sure about the order. Shape, style and colour is what is important for me.

Henry, 488–494

In this extract Henry explains the role of the dis–comforting practices in his life. By reflecting on his current practices he draws some interesting convergences between past and present, in which his current habits are rooted in those previous practices discussed by him elsewhere in his interview. While there is an underlying assumption that similar practices are typical behaviors for younger rather than older individuals, Henry's account can be read as the demonstrating assertion of his body–ability to perform youth activities.

It appears that for Henry it is important to highlight the fact that those practices are not yet beyond him that he still can shape his body in the manner that he wishes to. This can be interpreted in a twofold way. For once Henry might indeed still feel young and capable of dis–comforting his own body. Secondly it feels as Henry might try to persuade the interviewer and himself that this is his corporeal reality. And although elsewhere in his interview Henry points out his body changes, this account can be understood as a specific contradiction to the image that Henry is scared to present and accept – the image of an older, less capable body.

What is also compelling in this extract is Henry's use of an expressive parallel to a corset, which can symbolize, especially for women, bodily oppression and limitation. Henry says, "*I will breathe in, and fasten anything, and live with it for the day, if I have to. I would probably consider a corset, if I had to get into something*". In this sense Henry further subscribes to the dis–comforting practices. But in order to more deeply exploit the hermeneutic potential of this extract it is important to look on Henry's interview as a certain entity in which he several times expresses his belief in fashion being a problematic field for men to enter and to be associated with. This suggests that for Henry fashion became a way of his own resistance against gender stereotypes. As a result, Henry intentionally chooses to use as example a typically feminine garment.

Finally, it is worth noticing that there is also an interesting change of temporal referents from second to third conditional present in Henry's account. This subtle change from Henry saying, "*I will breathe in, and fasten anything, and live with it for the day, if I have to*" to " *I would probably consider a corset, if I had to get into something*", reveals a lot about his relation to the actual possibility of those events. While in the first sentence Henry signals that he talks about real and possible situations, in the second sentence those situations are impossible and 'unreal'. In this sense, Henry's fight with gender stereotypes discussed in the previous paragraph is somehow limited to the verbal account only, rather than actually wearing this garment. I sought to explore this further through a series of practical experimentations resulting in the creation of the Dis–Comforting jacket (fig. 1) and film (fig. 2).

Fig 1　The Dis–Comforting Jacket, corsetry spiral wire (Photo: Fraser West, 2014) a. front view; b. back view

Fig 2　a–b The Dis–Comforting film (stills), corsetry spiral wire wrapped around the male model's waist (Photo: Fraser West, 2014)

4　Conclusion and recommendations

Firstly, my research has demonstrated that fashion and clothing can constitute valid lens through which the experience of male ageing can be interpreted. This however is dependent on the participants' profiles and their unique relationships with fashion and clothing. The sample for this study was homogenous in their open interest and to some extent, fascination with fashionable clothes that was continued throughout their lifecourses. While their experiences add new perspectives to our understanding of the ageing phenomenon, they cannot be generalized and at any point extended to groups or populations. It is therefore recommended to replicate the study in different settings, with men representing different generations and different attitudes towards fashion as much as to extend it to a larger sample.

Secondly, as the initial findings suggest, the developed trans-disciplinary methodology allows for the creation of the interpretative written accounts that can potentially offer new insights of the under-researched phenomena of male ageing and older men's experience of fashion and clothing. This is possible by merging activates of interpretative writing and making within one innovative fashion research approach. In this the developed methodology allows me for the production of corresponding to written accounts artefacts that can potentially carry cultural significance equal to written texts. This approach can encourage the developments of alternative fashion research strategies and furthermore to broaden empirical, theoretical and reflective practices within this field.

References

［1］ ARBER, S. and DAVIDSON, K. and GINN, J. (eds.) Gender and Ageing. Changing Roles and Relationships, 2003.

［2］ ARCHER, B. The Nature of Research. Co-Design, January, pp. 5–13, 1995.

［3］ BIGGS, H. The Ageing Body. In: EVANS, M. and LEE, E. (eds.) Real Bodies: A Sociological Introduction. London: Palgrave, 2002.

［4］ BUGG, J. Interface: Concept and context as Strategies for Innovative Fashion Design and Communication. Unpublished thesis (PhD), University of the Arts London, 2006.

［5］ COLE, A. L. and KNOWLES, J. G. Arts- Informed Research. In. KNOWLES, J. G. and COLE, A. L. (eds.) Handbook of the Arts in Qualitative Research. London: Sage. pp. 55–70, 2008.

［6］ CRAIK, J. The Face of Fashion: Cultural Studies in Fashion. London: Routledge, 1993.

［7］ CRANE, D. Fashion and Its Social Agendas: Class, Gender, and Identity in Clothing. London: The University of Chicago Press, 2000.

［8］ CROUCH, C. and PEARCE, J. Doing Research in Design. London: Berg, 2012.

［9］ CSIKSZENTMIHALYI, M. Design and Order in Everyday Life. Design Issues, 8 (1), pp. 26–34, 1991.

［10］ EDWARDS, T. The clothes maketh the man: masculinity, the suit and men's fashion. In: EDWARDS, T. Fashion in Focus: Concepts, Practices and Politics. London: Routledge, pp. 41– 64, 2011.

［11］ ENTWISTLE. J. The Dressed Body. In: EVANS, M. and LEE, E. (eds.) Real Bodies. A Sociological Introduction, Basingstoke: Palgrave, pp. 133–150, 2002.

［12］ ENTWISTLE, J. and WILSON, E. Body Dressing. Dress, Body, Culture. Oxford: Berg,

2001.

[13] FAIRCLOTH, C. A. (ed.) Aging Bodies: Images and Everyday Experience, 2003.

[14] Oxford: Altamira Press.

[15] FALLMANN, D. Why Research-oriented Design Isn't Design-Oriented Research: On the tension Between Design and Research in an Implicit Design Discipline. Knowledge, Technology & Policy, 20, pp. 193–200, 2007.

[16] FEATHERSTONE, M. and WERNICK, A. (eds.) Images of Aging: Cultural Representations of Later Life. London: Routledge, 1995.

[17] FINLAY, L. Phenomenology for Therapists. Researching the Lived World. Oxford: Wiley-Blackwell, 2011.

[18] FRYLING, C. Research in Art and Design. London: RCA, 1993.

[19] GRAY, C. and MALINS, J. Visualizing Research. A Guide to do Research Process in Art and Design. Surrey: Ashgate, 2004.

[20] HARRE, R. Social Being. Oxford: Blackwell, 1979.

[21] HENRARD, J.C. Cultural problems of ageing especially regarding gender and intergenerational equity. Social Science & Medicine, 43 (5), pp. 667–680, 1996.

[22] JONAS, W. Design Research and its Meaning to the Methodological Development of the Discipline. In: MICHEL, R. (ed.), Design Research Now, Basel: Birkhauser, pp. 187–206, 2007.

[23] KAISER, S. B. Fashion and Cultural Studies. London: Berg, 2012.

[24] KNOWLES, J.G. and LUCIANI, T. C. The Art of Imaging the Place of the Visual Arts in Qualitative Inquiry: An Introduction. In: KNOWLES, J.G. and LUCIANI, T. C. and COLE, A, L. and NEILSEN, L (Eds.) The art of visual inquiry (Vol. 3, Arts-informed Inquiry Series). Halifax: Backalong, 2007.

[25] KREKULA, C. The Intersection of Age and Gender: Reworking Gender Theory and Social Gerontology. Current Sociology, 55 (2), pp. 155–171, 2007.

[26] LANGDRIDGE, D. Phenomenological Psychology: Theory, Research and Method. London: Pearson, 2007.

[27] MARTIN, B. and HANINGTON, B. Universal Methods of Design. Beverly: Rockport, 2012.

[28] NI CHONCHUIR, M. and MCCARTHY, J. The enchanting potential of technology: a dialogical case study of enchantment and the Internet. Personal and Ubiquitous Computing, 12(5), pp. 401–409, 2007

[29] NIEDDERER, K. and IMANI, Y. Developing a Framework for Management: Tacit Knowledge in Research using Knowledge Management Models. In: Undisciplined! Design

Research Society Conference 2008, Sheffield Hallam University, Sheffield, UK, 16–19 July 2008.

[30] ODOM, W. and PIERCE, J. Improving with Age: Designing Enduring Interactive Products. In: Proceedings of the 27th International Conference on Human Factors in Computing Systems, CHI 2009, Boston, MA, USA, April 4–9, 2009.

[31] POWELL, J. and GILBERT, T. Phenomenologies of Aging – Critical Reflections. In: POWELL, J. and GILBERT, T. (eds.) Aging Identity: A Dialogue with Postmodernism. New York: Nova Science, pp. 5–16, 2009.

[32] RUSSEL, C. What Do Older Women and Men Want: Gender Differences in the 'Lived Experience' of Ageing. Current Sociology, 55 (2), pp. 173–192, 2007.

[33] SMITH, J. Beyond the divide between cognition and discourse: Using interpretative phenomenological analysis in health psychology. Psychology & Health, 11 (2), pp. 261–271, 1996.

[34] SMITH, J. and FLOWERS, P. and LARKIN, M. Interpretative Phenomenological Analysis: Theory, Method and Research. London: Sage, 2009.

[35] TSEELON, E. The Masque of Femininity: The Presentation of Women and Everyday Life. London: Sage, 1995.

[36] TSEELON, E. Ontological, Epistemological and, Methodological Clarifications in Fashion Research: from Critique to Empirical Suggestions. In: GUY, A. and GREEN, E. and BANIM, M. (eds.) Through the Wardrobe: Women's Relationships with their Clothes. Oxford: Berg, 2001.

[37] TURNER, B. Medical Power and Social Knowledge. (2nd ed.) London: Sage, 1995.

[38] TWIGG, J. Clothing, Identity and the Embodiment of Age. In: POWELL, J. and GILBERT, T. (eds.) Aging Identity: A Dialogue with Postmodernism. New York: Nova Science, pp. 93–104, 2009.

[39] TWIGG, J. Fashion and Age: Dress, the Body and Later Life. London: Bloomsbury, 2013.

[40] WEBER, S. and MITCHELL, C. (eds.) Not Just Any Dress. Narratives of Memory, Body, and Identity. Oxford: Peter Lang, 2004.

[41] WILLIG, C. Introducing Qualitative Research in Psychology: Adventures in theory and method. Maidenhead: Open University Press, 2001.

基于城市中高龄人群休闲方式的服装设计研究

◆ 张 洁[①] （上海视觉艺术学院，上海，201620）

【摘 要】 本课题以城市中高龄人群为中心，通过研究该人群休闲方式特点，分别从中高龄服装的风格、色彩、款式、面料等角度，分析阐述中高龄服装的设计要素与方法，力求为中高龄服装设计实施提供理论依据。

【关键词】 城市中高龄人群；休闲方式；服装；设计

Study on the Clothing Design for leisure style of urban elderly population

◆ Zhangjie *(Shanghai Institute of Visual Art, Shanghai, 201620)*

Abstract: The topic is researching on the clothing design of urban elderly population. According to the characteristics and demand of leisure style and demand for the elderly population, the paper which is respectively from the view of old clothing color, styles, fabrics an so on, elaborates the design method and elements of the elderly clothing, is in order to provide theoretical basis for the elderly clothing design and implementation.

Keywords: urban elderly population; leisure style; clothing; design

1 休闲已成为当代社会重要特征之一

随着我国社会、文化、经济、科技水平的飞速发展，人民生活水平不断提高，休闲逐步成为当代社会的重要特征之一。人们通过具有人文性、文化性、社会性和创造性的休闲活动，不仅能增强自身体质，保持身心健康，提高生活质量。还能通过休闲增加社会交流与互动的机会，进一步推动和谐社会的构建与发展。

[①] 张洁，上海视觉艺术学院时尚设计学院，硕士、讲师。E-mail: 896089166@qq.com

1.1 休闲的定义

汉语中的休闲,休——休止劳作、休假、休息;闲——空适、悠闲的状态。休闲是指在非劳动及非工作时间内以各种"玩"的方式求得身心的调节与放松,达到生命保健、体能恢复、身心愉悦的目的的一种业余生活。

根据休闲的特点与作用,可将其分为三类:休闲时间、休闲活动、精神休闲。休闲时间,即1天24小时中,除去劳动、睡眠等其他必须活动所盈余的时间。休闲活动,指在完成劳动、家庭及其他社会义务之后,不仅是自己能够随心所欲地活动,也是提供休息、恢复、自我实现、精神性再生、知识提升、技术开发、社会活动参与的机会。精神休闲,则是将休闲视为一种主观性的心理与精神上的休闲状态。

1.2 休闲活动类型

休闲按照不同的角度,其分类不同。从空间的角度,可以分为室内休闲活动与室外休闲活动;从社交的角度,则可以分成独自进行的个人休闲与社会性休闲。还有考虑休闲体验深度的情况,根据休闲体验结构的特征,可以分成轻度休闲与真正的休闲。

2 我国城市中高龄人群的休闲需求

2.1 我国人口老龄化趋势

我国于1999年进入老龄社会。2010年第六次全国人口普查显示,我国60岁及以上老年人口达1.78亿,占总人口的13.26%。截至2012年5月,中国60岁以上老年人口达1.85亿,作为世界上唯一的老年人口过亿的国家,中国"银发"增长速度之快在世界人口大国发展史上前所未有。

2.2 中高龄人群的定义

国际老年学会(1951)认为处于在人类老龄化过程中出现的生理性、心理性、环境性变化与行为变化相互作用的复合形态过程中的人群为老年人。

国际上将老年人的定义的细分认为55岁以上进入老年期,并将其分为四个阶段:"初老年群体(55~64岁)""中老年期群体(65~74岁)""年长期群体(75~84岁)""老衰期群体(85岁以上)"。在中国,60周岁以上的公民为老年人。本课题的中高龄人群亦是中老年人,定义为50岁以上人群。

2.3 中高龄人群休闲的必要性

美国老年学家N.R.霍曼等认为,老年人参加休闲活动,取代以前的劳动角色,为未来的生活提供新的源泉,这样,老年人在晚年将会更加自信,精神更加愉快,生活也将更加满意、幸福。

中高龄人群的休闲是其自我实现的最后机会。从繁忙的社会工作事业中退出，有计划地规范生活，从事积极的休闲活动，可以增强身体机能，增加中高龄人群社交机会，有助于这一人群的身心健康，延缓衰老。

3 城市中高龄人群的休闲特征

随着我国社会经济的发展，医疗卫生条件的改善，我国人均寿命得到显著提升，老年人口数量也在迅速增加，老龄化进程越来越快，老年群体的传统思想观念也随之发生改变，作为拥有闲暇时间最多群体的老年人。越来越愿意投入到休闲运动的队伍中去，期望借此来维持或增进身体健康、丰富自己的晚年生活。

我国城市中高龄人群休闲活动主要分为体育休闲、文娱兴趣活动、学习活动、公益活动、旅游度假等。

3.1 体育休闲

以锻炼身体为目的，且体力消耗较小，花费较少的的体育休闲类项目。主要有散步、长跑、健美体操、跳舞、骑车、太极拳、钓鱼、乒乓球、羽毛球、篮球、足球、网球、排球、保龄球、游泳、滑冰、滑雪、登山等。

3.2 文娱兴趣活动

文娱兴趣类活动，根据中高龄人群的性别、文化背景、职业，有针织刺绣、菜肴烹饪、剪裁、电影鉴赏、戏剧鉴赏、美术鉴赏、音乐鉴赏、电子游戏、网上消遣、乐器演奏、象棋、读书、打扑克、书法、跳棋、麻将、体育观览、绘画雕刻制作、音乐会、唱歌等。

3.3 学习活动

学习活动是休闲生活的重要组成部分，是闲暇活动文明化、科学化的标志，具有深刻的进步意义和社会价值。自学或者在单位、俱乐部、专门学校以及用电脑、网络、广播学习。

3.4 旅游度假

旅游主要有当日旅游、回老家省亲、国内家庭旅行、国外观光旅行、独自旅行等。

3.5 公益活动

这主要指针对老年人、儿童、残疾人服务活动等。

4 城市中高龄人群的服饰需求特征

4.1 城市中高龄人群服饰基本需求

4.1.1 功能需求

中高龄人群因其年龄的增长,身体各部位机能都在逐步衰退。日常生活中对于服装实用性要求更高。一些能给穿着者带来舒适感受、改善中老年身体状况、方便其日常生活的服装款式或面料受到这一人群的青睐。例如天然面料,如棉、麻、丝、毛等,以及有温控、抗菌、磁疗等改善身体状况的高科技面料更受到中高龄人群的欢迎。

4.1.2 品质需求

城市中的中高龄人群,相对于其他地区人群,更懂得追求生活品质。对服装的要求,舒适性和实用性的传统需求以外,还关注服装的其他方面。例如服装的面料是否舒适、健康、环保;版型是否合身或修身、服装工艺是否精致、服装的售后服务是否有保障等。

4.1.3 时尚需求

现在的城市中老年人群,比其他地区中老年更易接受新事物。各种微信、QQ 等网络通讯社交方式的普及,使得城市中的老人们对在于时尚流行的关注远高于过去。城市里中高龄人群,越来越讲究个人形象。服装设计是否时尚美观;修饰和美化其身型特征;服装色彩搭配是否能展现气质、使面色更健康、更年轻等,对于服饰的审美与时尚需求不断增加。

4.1.4 品牌需求

追求高品质生活的城市中高龄人群,对于服装的品牌要求不断增加。有别于过去强调物美价廉、节俭实用的购物观念,现在的中高龄人群注重名牌,追求时尚,更渴望展示个人气质风采,满足个人物质和精神追求。服装名牌效应能帮助提升自我认同感,满足其补偿心理。

4.1.5 精神需求

人到老年,其社会角色和经济地位也由主导变为辅助,生活时间结构和空间结构的变化,一般老年人生活上由社会生活转为家庭生活,使余暇时间增多,单独居住的比例不断增长,心理需求也发生了很大的变化。由于生理的变化,心理方面老年人也容易出现消极孤僻、自卑压抑的情绪。通过服饰展示其个性风采、提升自我认同感,满足其消费欲望和补偿心理。

4.2 不同休闲项目的服饰要求

根据城市中高龄人群主要参与休闲项目的不同特点,将中高龄服饰用途分为以下几类:户外运动、居家休闲、社交聚会、展示表演等。

4.2.1 户外运动

注重健康养生的城市老人，对于康体类的户外活动较为热衷。多去室外晒太阳、呼吸新鲜空气，有助于身心的健康。在城市的公园绿地散步、慢跑、骑行、舞蹈、钓鱼等户外活动是多数中高龄人群常见的休闲项目。或个人或组团去郊游或者国内外旅行，也是越来越多城市老人的户外休闲的重要方式。

这些休闲活动，因在户外，且有一定的体力消耗，服装应适应与满足气候变化的需要的同时，需要具备运动装的基本特征。服装款式宽松舒适、利于运动；面料具备吸湿排汗、防风防雨、御寒保暖等功能。服装的色彩上可适度明亮，鲜艳、纯度高或者对比色搭配，有一定的有诱目性，可激发穿着者积极参与活动的热情和积极性，同时提升户外活动的安全性。

4.2.2 居家休闲

中高龄人群因身体机能的衰退，大多数休闲活动以静为主。其中居家看电视、上网、看书、读报是这类人群常见的休闲内容。还有城市老人多有一定文化修养或精神追求，如写字、画画、养花养鸟等怡情类的室内休闲活动，也是居家休闲的常见方式。

以静为主的室内休闲方式，多以个人、家庭或小群体为单位，追求平和、自然、放松的休闲感受。这类休闲方式在服装的需求上，以舒适性要求为主。服装色彩多给人平静感受的以柔和舒适的中性色，或者色彩偏深的浊色为主。这类色彩能够帮助着装者放松心情，同时也能具备一定的耐脏抗污的实用性。款式要求宽松舒适、没有束缚感。大领口的设计、减少拉链、扣襻、腰带等使用频率。宽松大裆的裤装，使久坐的老人下半身没有束缚。

4.2.3 社交聚会

中高龄人群的社交聚会活动，相对于年轻人，或在职人员频率较低。但这个年龄阶段，从社会工作中退出，独立的家庭生活较为孤独。大多数中高龄人渴望社交、希望展现自我价值，需要得到其他人的关注与肯定。因此，中老年人对于社交聚会类休闲活动格外重视，如参加朋友生日宴、子女的婚宴及老同事、朋友、亲戚聚会等。在这类社交中，常常通过自我着装与形象来展现自我价值。

中高龄人群社交聚会的着装，要求体面、展现个人形象气质。根据社交目的与场合，服装正式或隆重程度不同。正式场合服装，需要展现服装品质，体现中老年人的个人身型、气质与品位。男性以西装、中装等款式居多，女性合体修生的套装或裙装为主。强调面料质感与花色。着装在色彩上，男性以黑色、深蓝色等稳重色为主；女性除了较为稳重的深色系和无彩色系，中绿色、宝蓝色、玫红色、大红色等，纯度较高鲜艳的色彩，是敢于追求时尚品位，展现自我的城市的中老年女性偏好的色彩。一般聚会场合，强调中老年人的亲和力与群体性。在着装上，要求体现自然、亲和力的休闲装为主。通过服装造型展示个人形象气质时，但不能强调或突兀，否则会跟群体形成差距，影响社交沟通过程。

4.2.4 展示表演

中高龄阶段人群,退出社会生产事业,需要寻找新的方式展现自我价值,获得社会及他人的肯定。在中老年群体活动中,通过社区、社团、公益活动等平台,展示自身特长,满足自我时间的需要。中老年的团体舞蹈、音乐戏曲、话剧、书画等多种表演展示方式,向社会团体,展现自身价值。

中高龄人群展示表演活动,因具备表演性质,服装从色彩款式深可以适度夸张。但仍以舒适性为主,美观性、装饰性要求增加。根据活动项目与主题,服装风格特点不同。如太极拳展示穿着宽松中式上衣配灯笼裤,中老年广场舞则根据舞蹈内容的不同,或体现紧身健美,或展示民族风情等。

5 基于休闲方式的城市中高龄服装设计

在银发经济不断发展壮大的今天,我国中高龄人群的服装市场,仍存在服装品种少、品质差、款式陈旧、缺乏时尚感和实用性等问题。对于具有一定购买力,追求品质生活的城市中老年,选择从品质到外观均满意的服装存在一定困难。基于对城市中高龄人群休闲方式的研究调查,分析其中高龄人群休闲活动的服饰需求,提出中高龄人群的服装设计要素。

5.1 运动与休闲并重的服装风格

现今国内中高龄服装市场,大多没有明确的服装风格定位。市面上出现最多的仍是类似于居家服的宽松大码,且没有过多服装款式变化的直筒型服装。

通过对中高龄人群日常休闲方式的研究,以及对中老年各类服装着装频率的调研,城市中老年人群服装,多以运动休闲风格为主。一方面,中老年人群的休闲运动,多以日常康体的小体耗运动项目为主。满足运动需求的服饰风格设计,是中高龄服装重要趋势。另一方面,城市中老年人通过休闲项目,愉悦身心,满足日常生活和精神需求,要求服饰除了满足运动需求,还应符合休闲的目的。即从服装上满足放松身心、休闲愉悦的目的。因此运动与休闲并重的服装风格,是现今城市中高龄人群的主要着装风格方向。

5.2 易穿脱、易搭配的服装款式设计

老年人由于年岁增长,身体机能的退化,反应慢、四肢活动不够灵活,对于环境的适应性差等都是老年人很常见的问题,特别是城市中高龄人群,年轻时往往有各自的职业背景,有事业追求,自尊性较强,在情感上更渴望被关爱和平等对待。因此,对这类中高龄人群的服装设计需要最大限度地满足其功能性需求。

体现在日常着装上,易于穿脱的服装款式能够增加老年人着装的效率和舒适度,方便日常着装打理,展现自理自立能力,满足中老年人自我实现心理需求。中高龄服装可多以前开襟的

方式,替代套头衫,侧开襟或后门襟等开合方式。能够最大限度减少穿脱时肢体的活动幅度,减少手臂活动不便的中老年人群着装困难。

体现服饰审美上,易于搭配的服装款式,帮助减少对时尚审美有较高要求的中高龄人群服装搭配选择上的困难,体现着装者的服饰品位。易于搭配的服装款式,通常服装色彩款式为百搭型。男女服装针织开衫,可搭配运动感十足的T恤或POLO衫,亦可搭配衬衣或裙装。风格上,可运动,可休闲,亦可片正式,适合多数着装场合。百搭的服装款式配以合适的色彩,如接近中性色的柔和色以及无彩色能其大多数色彩搭配。

5.3 体现标志性与识别性的色彩图案

5.3.1 展现群体性

在城市中高龄人群所参与的休闲项目中,大多数为群体性活动。中老年人借由休闲活动,增加社交机会,扩大交际范围。因此,在群体休闲项目中,群体的标志性,以及个体的识别性,成为中高龄人群服饰需求的一部分。团队成员共同的色彩图案容易增强集体凝聚力,使参与团队休闲活动的中高龄人群有安全感、归属感、认同感与自豪感。

5.3.2 体现安全性

中老年服饰设计中强调具有标志性与识别性的色彩图案,能够在一定程度上提高中老年群体休闲活动的安全性。在人流量大、丛林密集的户外,具有诱目性的服装色彩图案,容易让中老年人被寻找到,如中老年旅行团,常佩戴统一颜色鲜艳帽子,方便识别以防走失。在光线不够充足的情况下,具有反光或者荧光效果的服饰图案细节,能增加出行安全。在科技通讯发达的今天,对于具有突发性身体疾病的中老年人,可通过服饰图案标志细节,迅速寻找到病人的相关信息,帮助联系家人和疾病救助。

5.4 体现实用性与便捷性的细节设计

中老年服饰应具备要实用性与便捷性。体现在服饰细节上主要从服装的细节款式时,面辅料等小部位的实用性设计。

中老年休闲活动中,一些户外活动如休闲体育项目,常涉及到随身物品无处安放。随着年龄的增大,中老年人记忆力减退,容易造成随身物品遗失的情况。可在不影响日常活动的部位(如胸前、手臂等部位),设计储物口袋,方便放置随身物品,如手机、钱包、钥匙等。还可采用拉链式闭合方式,防水防漏,同时可在口袋内部设勾襻等小细节,固定锁住钥匙等体积小而重要物件。

中老年人手指灵活程度减弱。在实际服装穿着中,过小的钮扣和扣眼会给其日常着装带来麻烦。因此,中高龄服装钮扣的设计上,可尽量采用减少钮扣的数量,并采用体积大而轻的钮扣的样式,同时扣眼即可用斜角度,代替以往的水平或垂直扣眼方式,方便中老年人扣解,或用暗扣、魔术贴、拉链、松紧带等方便操作的方式,代替繁琐的钮扣。

5.5 体现舒适性、科技性与时尚感的面料设计

随着纺织科技的不断发展,以及中老年人群服装品质的需求不断提高,中老年人群对于服装面料的要求越来越高。过去,大多数人认为天然环保是最佳面料选择,而近几年各种具备发热保暖、吸湿排汗、防雨透气、亲肤抗皱、疾病监控等科技功能的服装面料的逐步出现,并受到中老年人群的追捧。在中高龄的服饰设计时,要充分考虑辅助服装面料的舒适性、科技性,提高服装的使用功能。除此之外,还应迎合人群审美趣味与需求,选用具有时尚特征、符合流行趋势、体现质感的服装面料作为中高龄人群服装设计的重要因素。

6 结论

随着我国经济和社会的不断发展,城市中老年人对于生活品质的要求逐步提高。越来越多的中高龄人群开始有规律地参与有利于身心健康的休闲活动。以中老年人休闲方式为背景,以城市中高龄人群为设计中心,分析归纳城市中高龄服装的设计方式,为进一步开展我国中高龄服装设计提供理论依据。

参考文献:

[1] 丁志宏.城市老年人休闲活动特点分析[J].人口与发展,2009:209-212.
[2] 李敬姬.中韩老年人生活方式、休闲活动及生活质量的研究:以上海与首尔的老年人为例[D].复旦大学博士学位论文,2011.
[3] 于建新.老年服装及其市场发展研究[D].天津工业大学硕士毕业论文.
[4] 陈银辉.老年人生活行为与家具设计研究[D].南京林业大学硕士学位论文,2008.
[5] 刘丹丹.老年人生活形态综合分析与老年玩具设计事理学研究[D].陕西科技大学硕士学位论文,2012.
[6] 王露.以学科交叉和用户参与为特点的英国老龄服装设计研究模式[J].装饰,2012,5(229):82-83.
[7] 吴亚刚.中老年服装设计与开发[J].上海纺织科技,2003,2(31).
[8] 肖红,游旭群.西安地区中老年服装需求与面料设计[J],西安工程大学学报,2008,4(22).
[9] 李琼舟.丝绸在中老年服装设计中的创新运用[J],丝绸,2012,5(49):33-36.
[10] 王凤.浅谈中老年服装设计[J].江苏纺织,2005(9):38-39.

中国传统图案在中老年服装中的应用

◆ 杨 韧[1,2]① 付少海[1] 梁惠娥[2] （1. 江南大学生态纺织教育部重点实验室；2. 汉族服饰文化与数字创新实验室，江苏无锡，214122）

【摘 要】 中国传统图案是我国几千年文化的积淀，具有显著的东方文化风格和特色，它为中老年服装设计提供了极其丰富的设计素材和创作灵感，极大地促进了中老年服装设计的发展。本文对不同题材传统图案如动物图案、植物图案、人物图案、几何图案及文字图案进行分析，结合国际流行形势，归纳不同图案在中老年女装的创新运用。研究认为，将中国传统图案以现代化、国际化的方式融入中老年服装设计中，拓展了中老年服装风格，提升了中老年服装内在文化内涵，满足中老年服装个性化审美需求。

【关键词】 传统图案；图案设计；中老年服装

Application of Chinese traditional patterns in the middle aged and old people's garment

◆ YANG Ren[1,2], FU Shaohai[1], LIANG Hui'e[2] *(Jiangnan University 1. Key Laboratory of Eco-Textiles, 2. Laboratory of Han Nationality Costume Culture and Digital Innovation, Wuxi, Jiangsu 214122, China)*

Abstract: *The traditional pattern is the thousands of years of cultural accumulation, has a unique style and character of oriental culture, which provides an extremely rich material and design inspiration for middle aged and old people's garment, which greatly promoted the development of middle aged and old people's fashion design. In this paper, the different types of traditional pattern: animal pattern, plant pattern, character pattern, geometric pattern and text pattern, and connecting with the international situation, put the traditional Chinese patterns in middle aged and old people's garment design in the form of modernization, internationalization, to meet the demands of middle aged and old*

---

① 杨韧，女，江南大学纺织服装学院，硕士。E-mail: ycyangren@163.com, Lianghe@jiangnan.edu.cn

people's personalized aesthetic.

Keywords: *traditional patterns; pattern design; middle aged and old people's garment*

中国传统图案作为中国传统文化的一个重要脉络,将其运用到中老年服装设计中,不仅丰富中老年服装的设计题材,而且与中老年服装品位、时尚和中国风格的新发展趋势相符合,在满足中老年对于服装时尚化、个性化的需求的同时,提高中老年服装的设计品质与内涵。将中国传统图案的精髓融入中老年服装设计创意中,应以突出传统文化特色为主,将本土性和世界性融合、传统性和现代性继承、创新,使中老年服装设计具有浓郁东方风情和民族风格特色。

1 中国传统图案

1.1 中国传统图案概述

传统图案源于古人对自然、对图腾以及对原始神话传说的崇拜与信赖,以自然事物为基础的再创造,对客观事物的再现和模仿。从各历史时期沿传下来的传统图案风格各异:新石器时期用线舒畅,色泽单纯、明快,形式各异,造型完美,气韵生动;商周时期题材多样,手法独特,造型夸张、凝练;春秋战国时期成熟精工,平实奔放,生动活跃;秦汉时期图案逐渐开始具有叙事功能;魏晋南北朝时期佛教图案和佛教人物占主导地位;唐朝时期雍容华贵、题材丰富、结构丰满;宋元时期以写实风格为主,线条流畅,纹样简洁,典雅秀美;格调高雅、层次丰富、表现细腻、追求吉祥寓意。中国传统图案题材丰富,结构严谨,表现手法多样,吉祥意味浓厚,具有民族特色又具有不同的时代风格,具有独特的东方文化风格和特色。传统图案是古人追求美好生活而创造出来的一种艺术形式,"图必有意,意必吉祥"是其典型特征。其构成手法主要有三种,以被表现的图、形、物的名称谐音而成,给被表现的图形物以特定的大众认可的含义,以及图形之外辅以文字说明。传统图案根据题材内容分为动物图案、植物图案、人物图案、几何图案以及文字图案。

1.2 中国传统图案的提取与再设计

随着时间的推移,传统图案既不断变异,又代代相承,既有皇家之富贵,又有民间之纯朴;既有北方之粗犷,又有南方之秀美。古代纺织品、家具、书画、陶瓷、玉器、剪纸、青铜器等是获取传统图案的主要介质,被符号化的传统图案是我国传统文化的视觉表现,其内在的精神内涵是对我国传统文化的传承与弘扬。传统图案是传统、民族的风格,与现代的审美习惯有所差异,因此,在形式美的法则下,可以运用重复、发射、渐变、对比、特异,或分割、重组、添加、去繁等手法对传统图案进行提取和再设计,将中国传统图案以现代化、国际化的视觉效果呈现出来,把美好的寓意精神元素融入到现代设计之中,呈现或简洁明快,或含蓄优雅的装饰效果来满足中老年的个性化审美需求。这些传统图案经过设计师的发掘,结合中国特色以及国际流行形势,在中老年服装中自然地流露出中国的文化气韵,展现浓郁的东方风情。

2 中国传统图案在中老年服装设计中的应用

产生于独特的历史背景和文化底蕴的中国传统图案与中老年人处世哲学观念及生活态度相符合。在中老年服装设计中运用传统图案，并不是将传统图案生搬硬套到中老年服装中，要做到在了解、借鉴历史的同时不被过去所束缚，要在学习和保留传统图案精髓的同时，能够结合中老年的审美情趣及服装设计特点，使二者能够得到完美的契合与匹配。

2.1 动物图案的应用

动物图案是通过对动物形象的表现与艺术处理来表达人们对于生产、生活方面的美好愿望与期盼，常用的动物图案有龙、凤、虎、蝙蝠、蝴蝶、鱼等。龙在历史上是皇权的象征；虎寓意健壮、驱邪避鬼；鱼图案则表示子孙绵延；传统的动物图案很多很多。不同的动物类型，给人的感官体验也是不同的，体积庞大拥有锋牙利爪的动物，大多是力量的象征，具有权力、制约与震慑作用，大多运用在男装上；小巧可爱、美丽的动物反映对自然的热爱，对生活的满足或对爱情的祝福，多运用于女装中。历经了相当漫长发展过程的动物图案，由于人们生活方式，审美观念、阶级理念以及民族文化等的改变，其形式与内涵受到了一定程度的影响。在现代社会生活中，动物图案已不再像古时那样被赋予多种含义，更多的只是作为装饰运用于服装中。中老年服装设计在选择动物图案时，首先需要考虑中老年的审美情趣以及接受程度；其次，根据现代图案装饰进行再设计，使之适合中老年人群。图1a 是将动物图案"鱼"为设计元素运用到中老年服装的效果图，"鱼"通"余"，寓意祈祷富饶。图1b 图中鱼图案运用于中老年的下装，上装的莲花图案与之呼应。

a 传统动物图案—鱼　　　　b 服装效果图

图1　传统动物图案在中老年服装设计中的应用

a Traditional animal pattern　　　b Effect chart of garment

Fig.1　Application of traditional animal Pattern in the middle aged and old people's garment

2.2 植物图案中的应用

植物图案是源于原生态的艺术形态,具有极强的人文意义。早期植物图案数量很少,为其他图案的配衬,魏晋南北朝起植物图案开始流行,从唐后期以来,植物花卉图案成为装饰图案的主流。植物图案在产生、发展及使用过程中被赋予吉祥寓意,体现了人与自然彼此交融,相互沟通的格局。植物图案给人以自然、亲切、美好的形象,在中老年服装中得到广泛的应用,将植物图案通过模仿、联想、组合、夸张、类比等手段进行再设计;或通过对植物图案形象的变异模仿;或提取它们特定的民俗意义;或与其他动、植物图案相结合,使之符合现代审美情趣,得到中老年的喜爱。图 2 是将植物图案运用于中老年服装中的效果图。a 是江南大学传习馆中的一个纺织品,图中的刺绣是动物图案与植物图案的结合。b 对其进行模仿并组合形成对称图案运用中老年的晚礼服上。

a 传统植物图案(江南大学传习馆)　　　　　b 服装效果图

图2　传统植物图案在中老年服装设计中的应用

a Traditional plant pattern (Jiangnan University ECS)　　　b Effect chart of garment

Fig.2　Application of traditional plant pattern in the middle aged and old people's garment

2.3 人物图案的应用

人物图案生动丰富,以各种形式出现在各种载体上,有抽象和写实两种形态。人物图案一般有辅助图案的烘托,如花鸟、植物、动物等,这些辅助图案突出人物特征,构成人物环境,形成整体风格,使主题更加鲜明。人物图案题材大多来自人们喜闻乐见的神话传说,如八仙过海、福禄寿喜仙人、西王母祝寿等;历史题材,如竹林七贤、昭君出塞、十八学士等;世俗生活,如百子图、庭院仕女、采莲图等。人物图案反映出本民族趋吉避凶、祈愿祝福的美好愿望,一直备受人们的欢迎,有着其他图案所无法取代的魅力及独特的文化内涵和美学特征。并不是所有的人物图案都适合用于服装,更不是能够用于服装图案的人物图案都适用于中老年服装,在服装中运用人物图案不如其他图案来的广。选用人物图案时,首先要注意人物图案的题材及形式美感,其次是图案的趣味性。图 3 是将人物图案运用于中老年服装中的效果图。a 是"十美踢球图"的局部图。b 图从中选取了中间正在踢球的人物,图中人物极具动感,趣味十足,将其运用到中老年服装中,既有装饰性又有趣味性。

a 传统人物图案—"十美踢球图"　　　b 服装效果图

图3　人物图案在中老年服装设计中的应用

a Traditional statues pattern-"Ten beauties playing football"　　b Effect chart of garment

Fig.3　Application of traditional statues pattern in the middle aged and old people's garment

2.4 几何图案的应用

几何图案是中国图案史上最早进入装饰领域的装饰图案,也是在我国装饰史上运用最广泛、最活跃的装饰图案,它来源于生产和生活,人们通过自己的抽象思维,对自然界存在的客观事物及图形进行的有意识地重新排列、组合、变形而形成的具有高度概括性、简洁性的纹饰。几何图案变化自由,视觉效果好,并且绘制简易方便,有较强的规律性可循。将象征长寿、生长、富有、仁德的万字纹、龟甲纹、方胜纹、方纹的几何纹样应用在现代服装图案设计上,既古色古香,又极具创意。几何图案历史久远,开创了人们对美的形式的追求,其结构美、生命张力美及空间美,展示了它的独特艺术魅力。随着现代审美意识的变化发展,几何图案在中老年服饰装饰中的应用空间越来越大。图4几何图案在中老年服装中运用的效果图。a 是从传统几何图案演变而来的几何图案。b 为将它运用于中老年礼服的效果图,经过缩小的几何图案显得更加密集繁复。

a 传统几何图案　　　b 服装效果图

图4　几何图案在中老年服装设计中的应用

a Traditional geometric pattern　　b Effect chart of garment

Fig.4　Application of traditional geometric pattern in the middle aged and old people's garment

2.5 文字图案的应用

文字是人类思想感情交流的必然产物,是人类文化的重要组成部分。文字图案具有特殊文化意韵和魅力,表现出来的文学艺术欣赏,不仅使其本身具有审美价值,同时,通过夸张变形等手法,使得其与服饰图案更为贴近,并将它作为一种特殊的装饰图案运用于服饰中,显得古朴别致,具有很强的形式感和装饰性。人们会将能够体现吉祥文化的文字用于服饰中,如"福""禄""寿""喜"等;出于对宗教的信仰和崇拜之情,也会将宗教的符号运用于服饰中,如"卍"字纹;此外,汉字、少数民族文字、佛教的梵文、阿拉伯文,也会因其独特的装饰魅力而被运用在服饰中,其内容有诗、词、歌、赋等。将文字纹作为装饰运用到中老年服装上,首先,作为一个独特的装饰语言,它为中老年服装设计提供新鲜元素,开辟一条崭新思路;其次,文字图案本身的艺术性及形式感,提升了中老年服装的设计品质;最后,作为人类文化的重要表现符号,为中老年服装增加了其他图案所不具备的文化艺术气息。图5是文字图案在中老年服装中运用的效果图。a图将文字以符号的形式进行组合排列,组成的图案既有形式美感,又有美好的寓意。b为将文字图案运用于中老年服装中的效果图,搭配飘逸的面料、宽松的款式,可以遮掩中老年腰腹部的赘肉,弥补体型上的不足,同时具有现代感,突出中国传统的文字文化。

a 传统文字图案　　　　　　　　b 服装效果图

图5　文字图案在中老年服装设计中的应用

a Traditional text pattern　　　　　　b Effect chart of garment

Fig.5　Application of traditional text pattern in the middle aged and old people's garment

3　结束语

传统图案装饰意味强,同时还能达到情景交融、含蓄隽永的艺术效果,而且具有独特的民族情感和民族精神,是任何装饰图案所不能取代的。将一典型的民族文化运用到中老年服装设计中,不仅使服装体现设计的风格性,更夹带了民族精神和民族审美情绪,促进了民族文化的弘扬与传承。中老年服装在现阶段的发展并不令人欣喜,随着越来越多人的重视,提升中老年服装品质,使之时尚化、个性化是中老年与设计师的共同目标。传统图案为中老年服装设计提供了极其丰富设计素材和创作灵感,对中老年服装设计的发展具有很大的促进作用,拓展了中老年服装风格,提升了中老年服装内在文化内涵,同时,中国传统图案及文化也得到弘扬、发展与再创造,具有深远的意义。

参考文献:

[1] 李晖,宗明明.中国传统文化与老年服装创新设计探微[J].大众文艺,2012(02):101-102.

[2] 王传龙.中国传统图案与现代设计[D].重庆大学硕士学位论文,2009:2-11.

[3] 彭景.中国传统图案在现代服装设计中的运用[J].艺术百家,2013(zl):149-151.

[4] 张毅.吉祥图案在汉族民间传统服饰中的应用[J].纺织学报,2013(05):107-110.

[5] 孙成成.清代宫廷服饰纹样在高级定制服装中的应用研究:以动物纹样为例[D].北京服装学院硕士学位论文.2012:4-6.

[6] 李金侠.近代民间服饰上植物纹样的美学意义[D].江南大学硕士学位论文.2009:4-7.

[7] 李文倩.纺织品中的人物纹样研究[J].文艺生活·文海艺苑,2011(5):101,104.

[8] 王独伊.解读人物图案在纺织品设计中的文化内涵与流行成因[J].大众文艺,2009(19):121.

[9] 梁惠娥,胥筝筝.中国传统服饰中的几何纹样及其艺术魅力[J].艺术百家,2010(zl):185-187,184.

[10] 崔婧婧.中国传统服饰中几何纹元素在现代服装设计中的应用[J].文物世界,2015(1):33-37,6.

[11] 王涛.浅谈几何纹在中国传统服饰中的应用艺术[J].文艺生活·文艺理论,2012(6):83.

[12] 李非非,崔荣荣,牛犁.民间服饰中文字纹样形成的影响因素[J].纺织导报,2013(9):94-95.

浅析明代补子图案在中老年家居服设计中的应用

◆ 董稚雅[1][①]　梁惠娥[1,2]　（江南大学汉族服饰文化与数字创新实验室，江苏无锡，214122）

【摘　要】　补服是明代的品官之服，官员依照品级职司的不同，在服装的前胸和后背处缝缀相应的刺绣织品，即"补子"，用以区分官品等级。补子图案细腻精美，具有极高的艺术价值和丰富的文化价值。本文通过分析明代补子图案的造型及寓意，探讨其在中老年家居服设计中的应用方法，以期为中老年服装设计提供新的设计思路，探寻我国传统服饰元素与现代服装设计的契合点。

【关键词】　补子图案；中老年服装；家居服

Preliminary analysis about the application of Buzi pattern of Ming Dynasty in the design of middle-aged and senior people's leisurewear

◆ DONG Zhiya[1]　LIANG Hui'e　*(1. Laboratory of Han Nationality Costume Culture and Digital Innovation of Jiangnan University, Wuxi, Jiangsu 214122, China)(E-mail: milanmilandream@hotmail.com, lianghe@jiangnan.edu.cn)*

Abstract: *Bufu is a sort of official clothing in Ming Dynasty, which was decorated with embroidery fabrics on the chest and back according to different ranks and duties of officials. The embroidery fabric is called Buzi, which has a very high artistic and rich cultural value. By analyzing the characters and inner meaning of Buzi in Ming dynasty and discussing the methods of its application in the middle-aged and senior people's leisurewear design, in order to provide new design ideas for middle-aged and senior people's clothing and explore the correspondence of Chinese traditional costume elements and modern fashion design.*

Keywords: *Buzi patterns; middle aged and senior people's clothing; leisurewear*

---

① 董稚雅，女，江南大学汉族服饰文化与数字创新实验室。E-mail: milanmilandream@hotmail.com, lianghe@jiangnan.edu.cn

目前，我国中老年人口数量正逐年上升，老年人作为一个庞大的消费群，根据其生理心理特点而做针对性设计的老龄设计理念也逐渐进入包括纺织、服装业在内的各类生活必需品生产行业。在我国一些经济发达的地区，城市文明程度与居民生活水平较高，越来越多的中老年人希望与时俱进，追求时尚健康的生活方式，保持或提高自己的社交活跃性，这在一定程度上扩大了中老年人对于服装多样性的需求。现代家居服在改革开放后西方文化的传播和我国社会经济发展的推动下，作为一个独立的服装品类逐渐被国人所接受。得益于人们生活水平的提高，中老年人的消费观念也发生了变化，他们一改过去艰苦朴素的生活方式，开始追求积极健康的生活方式，这在一定程度上促进了家居服在中老年消费者中的推广。然而我国中老年家居服市场仍存在一些问题，需要设计者结合消费者心理与本土服饰文化，不断创新设计理念与设计思路，设计出具有特色的满足中老年消费者需求的家居服产品。

1 我国中老年家居服市场现状

家居服是指在家中穿着的服装，按功能可分为睡衣、浴袍、会客服、家居休闲运动服等。现代家居服的概念起源于西方。20世纪80年代，得益于改革开放后西方文化的传播，家居服作为一个独立的服装品类逐渐被国人所接受，在家中穿着舒适便利的家居服也已成为一种时尚健康的生活方式在我国逐渐普及。现有的家居服产品种类繁多，按其风格，可分为休闲、民族、甜美、复古等；按照款式，可分为袍式、衬衫式、T恤式、吊带式等；按穿着者性别和年龄，可分为男式、女式、儿童、中老年等。

随着许多知名家居服企业的兴起，我国家居服产业正逐步完善。然而现有的家居服产品普遍受西方服饰文化及审美趋势的影响，服装款式雷同，缺乏本土特色。就中老年家居服而言，通过对国内家居服市场的走访调研可发现，市面上现有的家居服产品多针对年轻人而设计，为中老年人提供的家居服产品较少。一些品牌虽推出了中老年家居服产品，但款式保守、单一，缺乏创新，结构设计上只是一味地加肥加大，不能满足不同体型中老年消费者的需求；色彩和纹样的设计过于沉重、呆板，没有起到修饰形体、美化形象以及提升气质的作用；结合中老年人生理特点的功能性家居服设计少之又少，产品的品质、功能和舒适性得不到很好的保证。

社会经济的发展和人们生活水平的提高促使一些受教育程度较高的中老年人希望与时俱进，保持积极乐观的生活态度并追求更高的生活质量。他们追求健康时尚的生活方式，并十分注重自身形象。在居家生活中，不少老年人消费观念的转变使他们一改往日将旧衣服或内穿的线衣当家居服穿着的艰苦朴素的生活习惯，转为注重家居服品质、舒适性、功能性以及服装对个人形象的美化和修饰作用。中老年消费者产品需求的不断提高，要求设计者充分了解中老年人的着装心理，充分体现中老年人特有的气质和风度。随着中老年人对戏曲、书法、国画等我国传统艺术文化学习热潮的兴起，将我国传统艺术文化与时尚结合，设计出具有本土特色的服装，为中老年服装设计提供了新的理念与思路。

2 明代补子图案的形与意

明代是我国古代礼治文化的顶峰。这一时期的统治者为巩固中央集权,维护等级森严的封建统治,制定并颁布了用以区分官员职司品级的补服制度,并为清代所沿用。补服是一种品官常服,官员依照品级职司的不同,在服装的前胸和后背处缝缀不同图案的刺绣织品,即"补子", 用以区分官员品阶和类别。文官饰以飞禽,武官饰以猛兽。补子图案细腻精美,具有极高的艺术价值和丰富的文化价值。

2.1 明代补子图案的形制特征

将补子作为服装的一部分,以动物图案区分官阶的做法早在唐代就已出现。据《旧唐书》记载,武则天时期召集官员赐予"绣袍",在官服上以成对的禽或兽为纹饰,"左右监门卫将军等饰以对狮,左右卫饰以麒麟,左右武威卫饰以对虎……宰相饰以凤池,尚书饰以对雁"。[1]由此可见,这种图案与明代补子图案极为相似,文官绣以飞禽,武官绣以猛兽。虽在唐代已将补子作为一种官员身份的标识,但当时的法律并未将其制度化,真正代表官品等级的补服制度则定型于明代。

《明史·舆服志》中对文武官常服中补子图案的使用有如下记载:"(洪武)二十四年定,公、侯、附马、伯服,绣麒麟、白泽。文官一品仙鹤,二品锦鸡,三品孔雀,四品云雁,五品白鹇,六品鹭鸶,七品鸂鶒,八品黄鹂,九品鹌鹑;杂职练鹊;风宪官獬豸。武官一品、二品狮子,三品、四品虎豹,五品熊罴,六品、七品彪,八品犀牛,九品海马。"[2]同时,《大明会典》中对明代品官补子图案的使用也有详细记载,并附图以诏令天下(如图1至图19所示)。由此表明,补服制度及补子图案已从日常服饰方面使官员明确各自地位,昭示自己的身份。

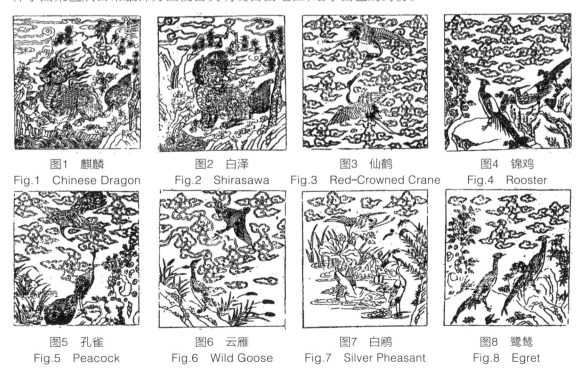

图1 麒麟　　　　图2 白泽　　　　图3 仙鹤　　　　图4 锦鸡
Fig.1 Chinese Dragon　Fig.2 Shirasawa　Fig.3 Red-Crowned Crane　Fig.4 Rooster

图5 孔雀　　　　图6 云雁　　　　图7 白鹇　　　　图8 鹭鸶
Fig.5 Peacock　Fig.6 Wild Goose　Fig.7 Silver Pheasant　Fig.8 Egret

图9 溪鹈　Fig.9 Purple Mandarin Duck
图10 黄鹂　Fig.10 Oriole
图11 鹌鹑　Fig.11 Quail
图12 练鹊　Fig.12 Magpie
图13 獬豸　Fig.13 Haetae
图14 狮子　Fig.14 Lion
图15 虎豹　Fig.15 Tiger & Leopard
图16 熊罴　Fig.16 Bear
图17 彪　Fig.17 Young Tiger
图18 犀牛　Fig.18 Rhinoceros
图19 海马　Fig.19 Sea Horse

就补子形制而言，明代补子均为方形，边长一般在 32～40 厘米之间[3]，四周多为光边。补子多以红色等素色为底，金线绣花。有的补子通过缂丝或是刺绣完成再缝缀到服装上的，也有根据服装整体款式和尺寸设计安排花样，将补子纹样与衣料一同织成的。明代补子缀于大襟袍上，胸前与背后的补子都是完整的，但也有一些补子是中间破开，沿中心线被分为两块。[4]

2.2 明代补子图案的文化寓意

由于古代科学技术不发达，人类缺乏对自然界诸如风、雨、雷、电等自然变化的认识，因此我们的祖先认为一切自然变化都是神的意志和力量。在随后的生产劳动中，他们依照自己心目中的英雄形象，将变化莫测的自然现象与形态各异的自然物形象化、人格化，创造出许多神话故事，并衍生出了对动物、植物、日、月、星辰等自然物的图腾崇拜。我国古代服饰中龙、凤等图案的使用便是古人图腾崇拜的体现。自然界物种繁多,无论是凶禽猛兽还是花鸟虫鱼,人们都赋予其特定的含义,寄托了美好的愿望与希冀。明代补子图案作为我国古代服饰图案中极富特色的一种,不仅被赋予了特定的寓意,也是我国古代礼治文化的体现,具有深刻的文化内涵。

明代补子图案构图巧妙,文官皆用双禽比翼而飞,相映成趣,武官用单兽,或蹲或立,威风凛凛。补子图案以双禽或单兽为视觉中心,四周以云纹、山石等作为装饰,整体造型饱满方正,气度威严。至于文官用禽,武官用兽的原因,明代《大学衍义补》中言道:"文武官一品至九品,皆有应服花样,文官用飞鸟,像其文采,武官用走兽,像其猛鸷也。"[5]无论是俊逸的仙鹤还是威严的雄狮,这些鲜活的动物形象被充分拟人化,用以象征和指代封建阶级社会所推崇的道德操守和意志品格。

在禽鸟纹补子中,一品文官员饰以仙鹤,具有仙风道骨之风雅,出类拔萃之高贵。鹤前冠以"仙"字,象征神仙之鸟,寓意官位显达,不同于世俗之民。此外,仙鹤在我国传统文化中还具有长寿安康、高贵优雅的寓意。文官一品补子集仙鹤、云、水、花草于一体,图案中的仙鹤双翅舒展,翱翔于云间,尽显其灵性与超逸之美。二品至九品文官补子图案分别为锦鸡、孔雀、云雁、白鹇、鹭鸶、鸂鶒、黄鹂、鹌鹑。从外形而言,这些禽鸟都有着优美的体态与多彩的羽毛,富有美感。与仙鹤一样,它们也被赋予了特定的文化内涵,是吉祥、富贵、忠义、优雅等美好寓意的象征。

在兽纹补子中,"公、侯、附马、伯服,绣麒麟、白泽"。麒麟是古人虚构的生物,集龙、马、鹿、虎、鱼等生物的特点于一身,因其样貌凶猛,具有降魔祛邪的功用,遂被古人看作是招福纳祥的瑞兽。麒麟在古代传说中的地位仅次于龙,是祥瑞的化身,其造型被广泛应用于建筑、雕塑等领域。在补子图案中使用麒麟,表示穿着者地位尊贵。白泽也是古代传说中的神兽。据宋代张君房纂《云笈七签·轩辕本纪》记述:"帝巡狩东至海,登桓山,于海滨得白泽神兽,能言,达于万物之情。因问天下鬼神之事,自古精气为物,游魂为变者,凡万一千五百二十种,白泽言之,帝令以图写之以示天下,帝乃作《祝邪之文》以祝之。"[6]因此后人将白泽看作祛邪避害之物。补子图案上的白泽同象征着高贵与威严。武官补子图案一、二品为狮,三品虎,四品豹,五品熊罴,六、七品彪,八品犀牛,九品海马。诸如狮、虎、豹、熊等动物,皆具有凶猛强悍的特征,寓意穿着者骁勇善战、威武彪悍。此外,这些动物也是古人用以镇宅辟邪的瑞兽。

3 明代补子图案在中老年家居服设计中的应用

明代补子图案构图巧妙、色彩艳丽、造型美观、工艺精湛,是我国古代服饰图案中的精品,具有极高的艺术价值,对现代服装设计有着很好的借鉴意义。同时,补子图案具有深刻的文化内涵,它寄托了古人追求功名富贵、平安幸福、福寿安康等人生期盼,其内在的文化寓意一定程度上与中老年人的精神诉求相契合,更易于在中老年服装中进行推广使用。因此,选择明代补子图案应用于中老年家居服设计中,有助于弘扬我国传统服饰文化,表现中老年人独特的气质精神和审美意蕴。

3.1 明代补子图案在中老年家居服设计中的应用价值

明代补子不仅用于品官常服,其居家燕服,即官员的家居便服,也缝缀补子。因此,将补子图案应用在中老年家居服中也是对我国古代家居服饰文化的传承。据《明史·舆服志》记载:"(嘉靖)七年既定燕居法服之制,阁臣张璁因言,'品官燕居之服未有明制,诡异之徒,竟为奇服以乱典章。乞更法古玄端,别为简易之制,昭布天下,使贵贱有等'。帝因复制忠静冠服图颁礼部,敕谕之曰,'祖宗稽古定制,品官朝祭之服,各有等差。第常人之情,多谨于明定。朕因酌古玄端之制,更名忠静,庶几乎进思尽忠,退思补过焉。朕已著为图说,如式制造'……忠静服仿古玄端服,色用深青,以纻丝纱罗为之。三品以上云,四品一下素,缘以蓝青,前后饰本等花色补子。深衣用玉色。素带,如古大夫之带制,青表绿缘边并里。素履,青绿绦结。白袜。"[7]由此可见,明代对上至帝王下至文武百官的燕服,即家居服都有明确规定。燕服胸背饰以官阶对应补子,说明明代补子图案不仅被用作区分官员等级的象征应用于常服之中,也可被视作家居燕服的装饰图案,为补子图案在现代家居服设计应用提供了灵感来源与可靠依据。将补子图案应用到中老年家居服设计中,不仅是对现有家居服设计思维的创新,同时也有助于我国古代家居服文化的梳理,对于弘扬和传承我国古代服饰文化具有重要意义。

补子图案寓意富贵安康,具有丰富的文化内涵,是我国古代吉祥文化的体现,在一定程度上与中老年人的精神诉求相契合,表现中老年人独特的气质精神和审美意蕴。如仙鹤,在古代被视为吉祥灵物,象征高雅纯洁、长寿安康;狮、虎、豹等均为古代神兽,具有祛邪避害的功用,象征顽强的生命力;主体图案外的云纹、海水江崖等装饰性图案,也具有吉祥如意、福山寿水等美好寓意。这些图案的寓意与老年人希望永葆青春、健康长寿的心理诉求及潇洒大方、优雅稳重的审美诉求相一致,较年轻人服装更适合于在中老年服装中使用。从色彩角度而言,补子图案以红色等素色为底,颜色鲜明端庄,但又不至于过分"争奇斗艳",既可以表现中老年人成熟稳重的气质,又能彰显其富有活力、积极健康的形象。

3.2 明代补子图案在中老年家居服设计中的应用方法

图20 明代一品文官仙鹤补子
Fig.20 Buzi of First Rank Civil Official in Ming Dyasty

图21 明代一品文官仙鹤补子重绘图
Fig.21 Buzi of First Rank Civil Official in Ming Dyasty Illustration

明代补子图案在中老年家居服中的应用方法包括直接应用、打散重组、抽象提炼。本节以明代文官一品仙鹤补子(如图20、图21)为例,对上述各方法进行说明,仅考虑图案形态及构成的改变与应用,图案色彩的变化应用不在本文讨论范围。

3.2.1 直接应用

直接应用,即不改变图案自身的形态及构成,依据服装的款式,仅改变图案整体比例大小,将其以印花、刺绣等工艺手法表现在服装中。明代补子图案由禽鸟、猛兽等主体图案及云纹、山石、海水等背景装饰图案构成,内容丰富,构图巧妙,可将其作为定位图案直接应用于服装上。图22将补子图案作为印花直接印于睡袍背部,图23将其作为上衣口袋的装饰图案,二者都是将补子图案直接应用于中老年家居服设计中的示例。

图22　直接应用：背部印花
Fig.22　Direct Use: Print on Back

图23　直接应用:口袋印花
Fig.23　Direct Use: Print on Pocket

3.2.2 打散重组

打散重组,是把原有图案内部各构成元素进行拆解再重组,构成新的图案。该方法可在保留整体图案各组成部分原始形态的基础上,对各独立元素进行拆分,选取其中全部或部分具有特色元素,通过变换图案位置或增减图案数量的方法,对其进行重新排列组合,以构成新的图案,使图案整体适用于装饰对象。图24是通过将原有补子图案中的仙鹤、祥云、花草等图案重新排列组合,再设计而成的单位纹样平铺图案,图25是其在服装上的应用效果。

图24　打散重组再设计图案
Fig.24　Scattering, Reorganizing and Redesign

图25　重组图案平铺应用
Fig.25　Application of Redesigned Tiling Pattern

3.2.3 抽象提炼

抽象提炼，即在把握图案原形的基础上，对原始图案局部进行提炼，将其简化概括为较为抽象的新图案。可对原始图案进行适当的夸张变形，将一些复杂的细节装饰或图案色彩进行省略概括，使图案更加适用于装饰对象。图26是通过对原有补子中的仙鹤图案进行提取，舍弃其细节与多余色彩，只保留整体轮廓，将提取后图案的大小进行调整并重新排列组合后，形成的抽象单位纹样平铺图案。图27是其在服装上的应用效果。

图26　抽象图案提取
Fig.26　Extraction of Abstract Pattern

图27　抽象图案平铺应用
Fig.27　Application of Abstract Tiling Pattern

4　结语

明代补子图案造型独特，寓意深刻，具有极高的审美价值和文化价值，是我国古代服饰文化的精髓。通过对补子图案造型特征的分析以及内在寓意的挖掘，探究其在中老年家居服中的使用方法，为我国传统服饰图案在中老年服装设计中的应用提供了范例与灵感，对于弘扬我国传统服饰文化，提升中老年服装内在文化内涵都有着重要意义。同时，将明代补子图案应用于中老年家居服设计，也是对现代家居服设计理念与思路的创新，有助于设计具有本土特色的家居服产品，促进我国家居服产业的发展与完善。

参考文献：

[1]（后晋）沈昫.旧唐书（卷45）[M].北京：中华书局，1997.

[2]（清）张廷玉.明史 [M].北京：中华书局，1974.

[3] 赵连赏.明清官员的补服[J].文史知识，2006，07：67-75.

[4] 王渊.补服形制研究[D].东华大学博士论文，2011.

[5]（明）邱浚.大学衍义补[M].北京：京华出版社，1999.

[6]（宋）张君房，蒋力生校注.云笈七签[M].北京：华夏出版社，1996.

[7]（清）张廷玉.明史 [M].北京：中华书局，1974.

女书文字的审美意蕴及其在中高龄群体服装中的应用

◆ 贾蕾蕾[①] 梁惠娥[1,2] （江南大学纺织服装学院，传统服饰文化与数字化创新实验室，江苏无锡，214122）

【摘　要】 女书是世界上唯一的女性文字，其蕴含着丰富的文化内涵和美学价值。研究女书文字的审美意蕴，对女书文字的造型与结构进行分析，并从设计学的角度对女书的审美特性加以阐述，提取出女书的审美意蕴并挖掘其艺术特征，将其融入到中高龄群体服装设计中。女书具有丰富的艺术形式与深厚的文化底蕴。通过借鉴女书的艺术特征与审美意蕴，结合现代审美需求，对其在中高龄群体服装中创新运用的方法进行探索尝试。既能强调民族文化内涵，又能弘扬中国传统文化，力求开拓女书在中高龄群体服装设计中传承与创新之路。

【关键词】　女书；审美意蕴；艺术特征；中高龄服装

The aesthetic implication of Nvshu and its application in the design of elderly population the clothing

◆ JIA lei-lei　LIANG hui-e *(Laboratory of Han Nationality Costume Culture and Digital Innovation of Jiangnan University, Wuxi, Jiangsu 214122, China) (E-mail: 15261663989@163.com, lianghe@jiangnan.edu.cn)*

Abstract: *Nvshu is the world's only female script, which contains the cultural connotation and aesthetic value. Aesthetic implication of the study of women's script, the shape and structure of women's script analysis, and the aesthetic characteristics of the female scripts from the design point of view expounded, aesthetic implication of extract of Nvshu and mining its artistic features, into the older clothing design. Nvshu has rich artistic form and rich cultural heritage.*

[①] 贾蕾蕾，女，江南大学纺织服装学院，传统服饰文化与数字化创新实验室，E-mail: lianghe@jiangnan.edu.cn

By drawing on the Nvshu artistic features and aesthetic implication, combined with modern aesthetic needs, to the elderly clothing innovative use of method explored and attempted. Can emphasize the connotation of national culture, and promote traditional Chinese culture, and strive to explore women's script in the inheritance and innovation in older clothing design.

Keywords: *Nushu; aesthetic implication; artistic features; elderly clothing*

随着我国中高龄人口的进一步增长以及人民生活水平的不断提高,中高龄人群的消费产业也蓬勃发展,其中服装产业也相应地提升。近几年,弘扬中国传统服饰文化的势头一浪高过一浪,国内外设计师重拾传统,运用传统文化元素突破创新,用于当下,值得肯定和鼓励。[1]女书作为中国传统文化,不仅是人类唯一的女性文字,世界文化史上的奇葩,而且是迄今活着的世界性古老文字,具有传统性和古老艺术性的特征,其自身所包含的文化底蕴和美学价值对于现代设计来说是一笔宝贵的财富。本文尝试运用女书作为设计元素,以给当今中高龄服装设计者提供丰富的灵感源泉与设计思维。

1 女书概述

20世纪80年代,女书作为世界上唯一的女性文字被专家学者发现,进入人们的视野,很快引起学者的浓厚兴趣与广泛的关注。研究发现,女书流传以湖南江永、道县、江华、东安和广西富川、钟山等县为中心,辐射周边邻近十余县等地区。千百年来,它靠母传女、老传少,代代相传,形成了人类历史上一个神秘又独特的文化现象。女书既是由女性创造、在女性中流传和使用的一种特殊文字符号体系,又是女性用来描绘女性生活的一种特殊民间文学。

女书是一种神秘古老的中华文化,随着历史的发展,拨开女书文化那神秘的面纱,大致可将女书的概念分为狭义和广义。狭义的概念是指女书作为一种独特性别的文字,包括其字体的形态、语音、笔画结构、组合方式、使用功能等,它仅限于妇女交流、使用,区别于其他任何文字;广义的概念是指女书是一种文化,包括产生这种文化的人文地理、女书流传区域各种风俗习惯以及女书文字写成的作品和写有这种文字的物体。[2]

2 女书文字的审美意蕴及艺术特征

2.1 女书文字的审美意蕴

女书文字其形体呈长斜体菱形,柔中带刚,笔画纤细飞扬,轻盈飘扬,结构精巧美妙,排列工整,充满了灵活的律动感和独特的审美意蕴。表现在柔性美、古典美、动态美、和谐美四个方面。(如表1所示)它们看上去或典雅或刚健,或圆润或犀利,形态各异,极具韵味,让人久久回味。

表1 女书文字的审美意蕴
Tab.1 The aesthetic implication of Nvshu

审美意蕴		特 点
柔性美		纤细修长、婀娜多姿的女字，轻盈飘逸，包含了阴柔之美，散发着女性永恒的光芒。每笔每划都有浓烈的女性韵味，点精巧轻盈，弧线千姿百态。笔式倾斜犹如女性在宣泄内心情感，每笔都显得秀丽纤雅，充分显示了女性的个性特征，具有明显的柔性美
古典美		女书是古老文字，只有点、竖、斜、弧四种笔画，结构只有左右上下两种方式，借鉴了古汉字中篆文和甲骨文瘦硬纤细的特点。笔画不多，但用最简单的形式创造出具有建筑美的造型、音乐韵律的节奏，体现了中国书法的古典美学思想
动态美		动态美首先是通过笔画形态变化体现的，其长短、粗细、浓淡、弯度的变化，如河水流动产生动态美一样。其次是通过体态变化来体现，女书在书法作品中的变化丰富，可修长也可粗矮，可前倾也可后仰，这丰富的变化为造就女书的动态美创造了条件
和谐美		女书的和谐美是通过笔画线条的呼应和篇章布局的变化来体现的。其篇章布局十分丰富多彩，采用虚实、疏密、穿插等艺术手段来处理，达到一种和谐美的境界

2.2 女书文字的艺术特征

女书除了具有多方面的学术价值以为，还有那以它纤细优雅的造型、巧妙有序的结构展示出一种独特的艺术美，这便是女书文字独特的艺术特征。

2.2.1 纤细优雅的造型特征

女书文字书写呈长菱形，字体秀丽娟细，造型奇特，也被称为"蚊形字"[3]。由于女书除了日常用作书写以外，也可以当成花纹编在衣服或布带上，所以字形或多或少也有所迁就，变成弯弯的形状。从形体和外观上看，女书与汉字有着深刻的渊源。女书从整体上看，是一种由右向左略有倾斜的长菱形的字体。它的行款方向是由上至下，由右向左，没有标点，排列十分整齐，笔画的线条纤细一致，笔势犀利，既有小篆体匀称的特点，又有甲骨文劲挺的姿态。与汉字相比，最大的不同是汉字呈方形，上下左右结构，组合对称，显得厚实稳重。女书则呈菱形，只有点、竖、斜、弧四种笔画，左右错开排列，左在下，右在上，上下保持在斜菱形的范围内，显得修长秀丽。

2.2.2 巧妙有序的结构特征

女书从整体上看呈菱形框架，是一种自上而下略向右倾的斜体字，类似"多"字体势。一般右上角为全字的最高点,左下角为全字的最低点。(如图1[4])在汉字中,"多"字体布局结构被认为较难安排,而女书却巧妙地就势组成单体、上下、左右、双合、夹心、包围等参差错落的多种结构形式。这些形式中隐含着一种对称性,如右上与左下保持均衡对称,倾斜之中不失中心,给人以一种特殊的美感。从文字构成上来说,由笔画所组成的构字单位,称为部件。[5]由一个部件组成的字是独体字。独体字笔画较少,字数也较少。由几个部件组成的字是合体字。多数合体字是由来两个不同的部件组成的,其中右边靠边靠上的部件形体比较大,是字的主体,变异的比较少;左边靠下的部件形体比较小,是字的配件,变异比较多。这种右上大左下小的合体字数量最多。少数合体字是由两个相同的部件组成的,二者形体一样大,但是方向有的相同,有的相反。这种由相同部件组成的合体字称双体字,双体字数量较少。此外还有由三个部件或四个部件组成的合体字,但数量有限。

图1　女书文字
Fig.1　Nvshu

3　中高龄服装设计的市场现状

　　通过观察国内各大综合型商场与专营服饰店可以发现,上市的各类服装面料适合于年轻人的居多,而为中高龄消费人群提供的比较少。当下不少中高龄人群已改往昔只追求面料牢度和价廉的观念,转为注重面料的品质、功能和舒适性,同时对修饰和美化个人形象也有更高的要求[6]。当今,追求生活质量的提高,彰显个性已成趋势,中高龄人群也是如此。中高龄消费者对老年服装不满的原因主要有款式呆板、质量差、颜色单调、体型不适、缺乏活力。

　　第一,中高龄服装款式呆板、陈旧是最突出的问题。市场中的服装款式几乎每年都一个样,缺乏变化,不新颖,无法吸引中高龄消费者的眼球。第二,很多中高龄服装质量差,主要体现在面料上。面料差的服装会给人带来不适之感,面料的舒适度是决定中高龄人群消费的重要因素之一。第三,颜色单调、暗淡,总体缺乏活力。目前市场上服装的颜色大部分偏黑、蓝、灰这些暗调子,而在调研中发现,除了一些传统的色彩,中高龄消费者其实更喜欢穿亮一点的颜色,可以使他们显得年轻愉快。第四,市场中的中高龄服装与大多数中高龄人群的体形不适,主要表现在胸、腰、腹、臀和上臂这几个部位。而设计师在制作中高龄服装时并没结合消费者人体形态的特殊性,使消费者难以买到合体的服装。

　　随着经济发展和消费水平的提高,中高龄人群对服装的需求在逐渐发生变化。根据市场调研分析,笔者认为中高龄服装的设计应结合实际,融入传统文化元素,为求满足中高龄消费者的需求。

　　首先,在中高龄人群的传统衣着习惯背景下,融入设计元素,使中高龄服装受消费者青睐。在设计过程中,应加入适量的时尚元素,同时也不能过于花俏,不脱离中高龄人群传统的衣着习惯。在色彩上趋向多元化,中高龄人群对服装色彩的需求由单调的黑、白、蓝、灰色逐步转向明快多彩色系。第二,改进中高龄服装版型,研发适合中高龄人群体形的服装。现实中的中高龄服装并不受消费者的喜爱,除了款式、色彩等因素外,还表现在服装胸、腰、腹、臀和上臂这几个部位处理不好。因此为了适应中高龄人群的特殊体形,现有的成衣结构制作应在原有的版型基础上进行改进和创新。第三,结合中高龄消费者的特点,面料是很多消费者购买服装时选择的一个重要因素,他们对面料的舒适性要求较高。

4 女书文字在中高龄群体服装中的应用

随着时代的变迁、科技的发展以及现代人们的生活方式与思想观念的变化,服装中单纯的女书文字形式已无法满足中高龄消费者需求。那么,女书文化应如何融入到中高龄服装设计中?落实到载体上可通过形制相似构建、功能应用拼接、文化情感营造三个方面来实现。

4.1 形制相似构建

形制相似构建再现这一方面主要是根据物本身的局部拼接、元素的解构重组等方式从而形成新的设计。女书文化的源头根植于民俗文化,并依托于民俗而存在,在代代相传中,虽有丰富的文化内涵,但表现形式往往程式化。女书文化的传统审美与现代审美结合的重要途径是运用现代设计手段,采用形制相似构建的方法把传统女书文化融入现代设计语言中,获得适合现代市场需求的新血液。如图2所示,女书文字在中高龄群体服装中的应用设计1,采用了形制相似构建的方法进行设计。将女书"父"字应用于中高龄群体男士服装中,首先提取女书"父"的形态;然后根据消费者对服装款式的爱好设计了一款中式上衣,在服装的分割线上巧妙地运用女书"父"字形态;最后对服装进行色彩设计,采用较为年轻的色彩,摒弃黑、白、灰色,让中高龄群体的消费者通过服装感觉自己的年轻活力。

图2 女书文字在中高龄群体服装中的应用设计1
Fig.2 Design and application of Nvshu in the elderly population in clothing 1

4.2 功能应用拼接

功能应用拼接再现这一方面是指传统造物应用于现代使用情境中,或传统使用情境应用于现代造物中。这不单单是功能与形式的转化,更是物境的转化。将女书文化融于中高龄群体服装设计中时,要按照美的构成规律将形状、材质、结构、色彩等合理布局,使传统女书文化通过其视觉表现形式满足受众的精神需求。女书一般在宗教祭祀、读唱娱乐、结拜姐妹、信件来往、诉苦写传、记事记史、改写汉字韵文这七个方面进行使用,可看出女书的功能是当地流传区域妇女的交流与娱乐的载体。服装也有传递讯息的功能,结合女书将其功能应用于服装中,传达与交流的情感便会自然地流露到服装中。这种将女书文字应用在服装中的方式,可以带给人们很强的视觉冲击力,使人们可以更加直观地接触和认识女书文字,这不仅是一种文字的传承,更是一种带有民族特色的设计。如图3所示,女书文字在中高龄群体服装中的应用设计2,采用了功能应用拼接的方法进行设计。将女书扇面五言文应用于中高龄群体针织男士服装中,利用针织提花技术将女书扇面五言文直接应用在服装中,其局限性是只能采用三种颜色以内色彩。这一设计不仅将传统文化融入到服装中,也为中高龄群体服装加入新鲜血液,将服装代替了扇子这一载体,使人们只管认识女书,传递感情。

图3 女书文字在中高龄群体服装中的应用设计2
Fig.3 Design and application of Nvshu in the elderly population in clothing 2

4.3 文化情感营造

文化情感营造再现这一设计,缺乏前两者的巧妙,在现代设计中多表现为一种文化创意产业的再生和价值再造。[7]现如今很多女书元素通过现代技术,印刷或制造与不同材质的物品上,成为文化的表征产品。也就是现在常用的借形为饰,多以装饰为手段,将女书元素直接运用到设计中。在女书作品中,女书文字常以相同文字组合和多种文字组合出现。在织带、把八宝被上,女书字符的组合形式以交错组合、上下组合较为常见。如图3所示,女书文字在中高龄群体服装中的应用设计3,采用了文化情感营造的方法进行设计。借鉴传统女书的组合形式,在女书文字的创新中,将"草"字进行不同的组合应用,不仅丰富了

图4 女书文字在中高龄群体服装中的应用设计3
Fig.4 Design and application of Nvshu in the elderly population in clothing 3

女书文字的视觉效果,也为中高龄群体服装款式增添了独特的装饰性。女书以装饰图案的形式融合在服装设计中的案例相对较多。

5 结语

本文基于女书文字审美意蕴与艺术特征的分析,以及对中高龄服装现状的分析,将女书文字的审美意蕴提炼创新为一种独特的设计语言,探索了女书在中高龄服装中的应用。现今对女书这一传统文化的吸收绝不是形式上的模仿,而是要吸收其浓厚的民族气息,寻找和挖掘能够体现当代艺术精神的元素,探寻出融合于中高龄服装设计的方法。在服装设计学习的过程中,应多学习传统文化的精髓,学习、继承、改进、发展,并给予它新的形式和新的变化,要用新的思路去分析现代社会中中高龄消费者的观念,重视中高龄消费者不断提高的审美需求。进一步满足中高龄服装消费者不同的消费需求。这不仅弘扬我国传统文化,而且对提升中高龄服装的内在文化具有重要意义。

参考文献:

[1] 刘水,贾蕾蕾,梁惠娥.云肩形色之意及其在现代女装中的创新运用[J].丝绸,2015,06:42-47.
[2] 谢明尧,李庆福,欧阳红艳,贺夏蓉.女书习俗[M].长沙:湖南人民出版社,2008.
[3] 李庆福.女书文化研究[M].北京:人民出版社,2009.
[4] 赵丽明,宫哲兵.女书——一个惊人的发现[M].武汉:华中师范大学出版社,1990.
[5] 郎丽.基于女书的文字设计研究[D].中央美术学院,2007.
[6] 贺婷.浅析我国老年服装市场的现状及对策[J].四川教育学院学报,2007,08.
[7] 熊微.传统文化融于现代设计的结构层次与转化方法[J].创意与设计,2014.05:52-58.

二维码个性化设计在中高龄服饰中的应用

◆ 夏 俐[①] （上海视觉艺术学院，上海，201620）

【摘　要】 随着信息社会的发展、高科技技术的不断更新，二维码也渐渐进入人们的视野并且成为一种新的潮流与趋势。二维码的应用得到发展的同时，其在中高龄服饰设计领域里的创意空间也不断被发掘，由原有的单一色彩到多样色彩的搭配、由二维平面到三维立体的过渡、由单一材质到综合材料的应用等，无不体现了二维码在图形创意空间里的潜在个性魅力。通过对二维码的解析以及在中高龄服装设计领域里的应用分析与研究，探索二维码带来的全新感受。

【关键词】 二维码；个性化设计；中高龄服饰

Personalized design of QR code in application of Fashing Design of mature people

◆ Xiali *(Shanghai Institute of Visual Art，Shanghai，201620)*

Abstract: With the development of information society and technology constantly updated, QR code also gradually into people's horizons and become a new trend . At the same time of its developing, the application of QR code in the field of fashing design of mature people creative space has been discovered, diverse from the original single color to colorific collocation, the transition from the two-dimensional surface to the three-dimensional surface, from single material to the application of composite materials, etc., which embody the QR code in graphically creative space potential character charm. By parsing the QR code, and the application of in the field of fashion design of mature people analysis and research, explore the QR code to bring a new feeling.

Keywords: *QR code; Personalized design; fashing of mature people*

[①] 夏俐，女，上海视觉艺术学院时尚设计学院讲师，硕士。E-mail: lilisunny520@163.com

引言

在科技日益更新的时代,随着智能手机的快速普及,我们已经生活在二维码的潮流中,无论在报纸、电视、互联网上,还是在咖啡厅、超市、地铁、商场里都随处可见二维码的身影,比如扫描食物上的二维码就可知道它的营养成分;扫描衣服上的二维码就可以了解它的材质产地;扫描图书杂志上的二维码就可以了解它的大体内容;扫描名片上的二维码就可以进入对方公司。二维码是用特定的几何图形按一定规律在平面(二维方向)进行分布的黑白相间的图形,用来记录数据符号信息。它作为"便携式数据库"已经被广泛地应用到生活的方方面面,将二维码更好地应用于中高龄服饰设计中,使二维码设计个性化与服饰风格完美地融为一体,实现中高龄服饰设计功能性与美观性完美结合将是一个必然的趋势。

1 二维码

二维码的起源是日本,原本是 Denso Wave 公司为了追踪汽车零部件而设计的一种条码。它是用某种特定的几何图形按一定规律在平面(二维方向上)分布的黑白相间的图形记录数据符号信息的,在代码编制上巧妙地利用构成计算机内部逻辑基础的"0""1"比特流的概念,使用若干个与二进制相对应的几何形体来表示文字数值信息,通过图像输入设备或光电扫描设备自动识读以实现信息自动处理(图1)。二维码是比一维码更为先进的条码格式:一维码只能在一个方向上记录信息,而二维码在水平和垂直两个方向都可以记录信息;一维码只能由数字和字母组成,而二维码能存储汉字、数字和图片等信息。二维码有信息容量大、密度高、可以表示包括中文、英文、数字在内的多种文字、声音、图像信息、保密性强、追踪性高、抗损性强、备援性大、成本便宜、易识别互动性强、体验性好等特点,关键点是能够更好地与智能手机等移动终端有机结合。

图1 二维码的结构

近期,在加拿大的一个农场里,克雷一家在一位设计师和技术工人的帮助下创作出占地31.2万英尺(约2.9万平方米)的二维码图形,这是迄今为止最大的二维码,现在该二维码图形已被列入吉尼斯世界纪录。据说当有飞机经过农场上空时,乘客在飞机上看到该二维码,并用手机拍摄扫描后可以连接到农场的网站;另外,还有一些夸张立体的空间造型二维码,这些独特的二维码都不得不让人为之赞叹(图2)。

图2 最大的二维码和最立体的二维码

2 中高龄服饰中使用二维码的主要功能

在中高龄服饰设计的功能性设计上置入二维码,主要是因为二维码中能够容纳非常多的内容,包含穿着者的个人详细信息、健康状况、主要疾病治疗情况、常用药及药物过敏情况以及家庭地址、紧急联系人电话等等,有了这些内容在穿着者需要治疗时只要让医生扫二维码就可以使医生快捷准确地了解病情,对症医治。

3 中高龄服饰中二维码的个性设计

二维码作为一个平面图形,色彩呈现单一的黑白效果,外形也是简单的正方形,给人单调的感觉,缺乏视觉上的美感和个性,因此完全可以通过平面设计改变它单一的形象,赋予它创意的内涵,让它在传递信息的同时也显现独特的个性设计。基于二维码业务发展的现状,二维码的创意图形设计浮出水面,国内也出现了像码客帝国、二维工坊这样的二维码服务平台。

由于二维码的尺寸可变,条码符号形状、尺寸大小比例可变;并且二维码通常有三个定位点,这三个定位点提供读码机辨识,因为有这些定位点,所以二维条码不管是从何种方向读取都可以被辨识,在中高龄服饰中二维码的个性设计上就有很大的空间,使二维码个性化设计在图形和色彩上与服装风格协调并呈现显装饰美感。

改变二维码图形的外观,但依然保持它能够被读码软件识别并准确读取信息。主要可以从以下四个方面着手设计:

1. 定位点与矩形点的形状。这些点不仅可以是生码软件生成的矩形,也可以被设计圆角矩形、正圆点、椭圆点甚至是不规则的点。

2. 色彩。传统的二维码采用的是黑白单色,而扫码软件同样是根据色彩的色阶(而不是色相)来进行信息的读取,所以只要在设计时让色彩保持在一定的明度范围内(测算这个范围最有效的办法就是运用不同的读码软件对应用了不同色阶组合的同一个彩色二维码进行解码测试），不同色相的色彩组成的二维码图形一样可以被顺利读取（图3）。

图3　二维码色彩保持一定的明度对比

3. 整体长宽比例。传统的二维码都是正方形，但由于二维码容错机制的存在，我们不仅可以改变这个正方形的长宽比例（经过实验表明，当长宽比例小于1∶1.5时识别率基本接近100%，而当长宽比大于1∶2时，识别率就会迅速降低），甚至可以对其进行小幅度的透视与扭曲，从而设计出带有透视感的二维码或者表现旗帜飘动、水波扩散这样的创意效果。

4. 空隙与可以去除的点。一个二维码在矩形构成的信息点组合中会包含大量白色的空隙，这些空隙可以加以利用，放置其他的图形，只要这些图形在色阶或者形状上不影响到有效信息点的解读（图4）；同样，由于容错机制的存在，有些二维码的矩形点是可以去除的，这为图形的设计提供了更大的空间。

图4　利用空隙和可去除点创意的二维码

3.1　结合服装风格的二维码图形设计

中高龄服装款式类别分为休闲、运动和正装，服装的风格分为民族风、极简风、休闲风、运动风等等，二维码的创意设计应当与服装风格和谐一致，在相同的风格内进行设计，并对服装

的整体风格起到渲染和强调作用。例如从绘画艺术中汲取营养和创作灵感,运用抽象画家蒙德里安的绘画作品的独特风格呈现的色彩和装饰感冲击着我们的视觉,由此衍生的二维码图形设计使人在穿着的同时又叹服服装与绘画的那种完美结合(图5)。最常用的一种方法是内外结合法,即将二维码内部可更改的空间与外部自由设计空间相结合来进行图形设计,这种方法因为空间的灵活性与多样性,可以因地制宜地最大限度地实现创意的表达,在进行二维码创意图形设计时都必须重视色彩优先原则。色彩是图形设计的重要组成部分,运用合理的色彩搭配可以增强图形的表现力,吸引更多的关注。设计师应当从整体服装的色彩元素中寻找创意二维码图形的色彩。

图5 抽象画风格的二维码创意设计

3.2 个性化的色彩

无论是时尚夸张的色彩风格还是端庄典雅的色彩风格,设计师通过对其颜色的改变,不仅能使二维码变成彩色或渐变色,而且可以让其变得更加丰富多彩,如明度较强的色彩使用、对比度色彩使用、同色系色彩使用等。在色彩分布形式上可以采用平面构成的样式,例如上下分割式、对角分割式、同心圆式等,让其色彩的划分更加具有设计感和规律性(图3)。同样色彩也可以是无序的,更多的色彩不仅能使二维码的辨识度更高,还可以吸引更多的扫描率。另外,要尽量使用浅色背景深色条码,因为目前国内很多解码器无法识别色彩对比度较低的二维码。因此,在二维码色彩设计中的用色,设计师需要巧妙地运用色彩关系和空间构成关系,在限定的空间里,发挥出色彩特有的个性,为二维码的创意图形设计锦上添花。

3.3 美观的线条

如果单纯地改变颜色还不能与服装风格协调,那么在二维码中对于线条的设计就可以更加大胆。二维码最初是为工业应用而设计的,线条硬朗规范。但在日常生活中,这样的线条未免显得不够美观。因此,可以尝试加入一定的弧度,使线条变得有趣更具有个性风格(图6)。

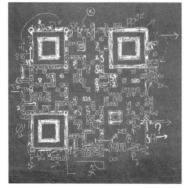 

图6 个性线条的二维码

虽然二维码只是由一些凌乱的小方块组成,但是从图形设计的角度观察,这些不规则的小方块也是极富设计感的。这种线条不是人为刻意形成的,而是电脑根据文字信息的输入随机产生的,即便是 30% 的冗余代码,在每次形成之后都是不一样的。设计师也可尝试让变化有规律的中国传统民间装饰性图案与不规则的黑色小方块相互结合,则更富有创新性。设想以装饰图案规律为基础,以二维码不规则方块为设计元素,配以色彩作为点缀,最终形成的装饰图案既有现代装饰意味,也具有自然风格的流露(图7)。设想二维码方块线条元素可单独进行创作,因为二维码的黑色方块是随机形成的,每次形成后的造型千变万化,这就为设计师提供了更为广阔的联想和设计空间,有的像动物,有的像植物,有的甚至像人物。

图7　二维码与装饰图案的结合

3.3　二维码在服装上的处理工艺

　　二维码制作工艺上,也要针对不同的服装面料采用不同的制作工艺,只是在具体的服装上面要结合面料去选择适合的工艺手法,这样才能使二维码与服装完美结合。二维码平面的处理手法可以通过手绘、热转印、丝网印、数码印、扎蜡染实现;立体二维码是指在服装上的装饰具有立体感或是有浮雕效果。其表现特征是具有三维空间的立体造型,处理手法可以通过编织、刺绣、镂空面料图案、盘花以及结合面料再造的肌理纹理呈现千变万化的丰富美感(图8)。

  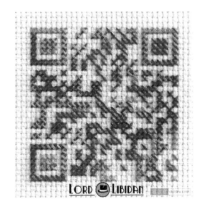

图8　二维码在服装上的处理工艺

3.4 二维码在服装上的应用面积

在二维码的应用面积上可大可小,部分中高龄人群有独特热情奔放的个性,注重装饰感的夸张手法,在二维码的应用上可以结合整体服装的廓型,在服装上做大面积的构成处理以产生类似平面构成及空间混合的效果,这样在初见时并不便察觉,只有在一定距离和特定角度扫一扫时才会显现二维码(图9)。

图9 二维码在服装上的应用

图10 二维码在服饰中应用

当然,也可以根据需要在应用手法上采取以点带面的设计方法,通常可以在领口、门襟、下摆、袋口、袖口等处作细微的装饰应用,当然也可以应用在服饰配件上,例如在领带领夹、耳环、鞋帽、钮扣、首饰、腰带夹头、手套、阳伞等,这样较小面积应用二维码图形也更加有利于个人信息的安全保护(图10)。

4 结论

二维码创意图形的设计工作从生码软件将信息转换为二维码起步,然后设计者运用扫码软件进行读码,寻找出有效的信息点和可以容错的空间范围;接下来根据信息特征和客户要求进行创意,之后将创意点运用平面设计的手法,融入到有效的信息点和可以容错的空间范围中,完成二维码图形的设计。这种创作有别于普通的平面图形设计,它受到二维码容错范围的限制,但也正是由于这种限制的存在,二维码的创意图形设计才变得更有挑战性和趣味性。值得注意的是,由于艺术创意带来的图形的不确定性和不规则性,所有经过创意图形设计的二维码都需要经过多个不同扫码软件的测试,确认信息读取的精确后才算最终设计完稿,结合中高龄服饰设计个性二维码是具有挑战性的。

参考文献：

［1］ 原研哉.设计中的设计［M］.济南:山东人民出版社,2007.

［2］ 王丽娟.告别黑白小方框,二维码也能有设计感［OL］.http://wo.poco.cn/10855695/post/id/2646716.

［3］ 潘雪峰.柏琳日本手机二维码业务发展的启示［J］.—中国新通信,2007（6）.

［4］ 王彦博.个性时尚二维码在线制作［J］.电脑知识与技术-经验技巧,2013（5）.

服"适"生活
智时代下的中高龄服装创新设计与服务的思索

◆ 刘 坤[①]　（上海视觉艺术学院，上海，201620）

【摘　要】　社会变迁，时代发展迅速，生活中的智能化方式也越来越多，这不仅表现在各种家用电器上，医用、生活、行动、学习各个领域都在比对着智能化的程度，可谓是一个对比智慧的智能时代，我更加觉得"智时代"这个词可能更贴近也更适合这样一个现代化的生活状态。"智时代"不仅仅意味着大数据、3D打印、智能手机、APP和电商，如不具备相应的思维方式和认知模式，就无法利用这些新兴技术实现设计的创新与可持续性发展。服装设计作为一个不断跨界的边缘艺术类设计，在应用领域的局限性让它不会在这样一个时代独占鳌头，但它是应用技术与时尚设计巧妙结合的"轻设计"类型，也就更能附着于新兴技术上从而拓展出一个新的设计层面。尤其在中高龄人群中的服装设计应用，更可以结合智能化的创新点服务于中高龄人群。

【关键词】　城市中高龄人群；智能；服装创新思维设计要点；设计服务

Discussion on Fashion Design and Service for Mature People in the Era of Intelligence

◆ Liu Kun　*(Shanghai Institute of Visual Art, Shanghai, 201620)*

Abstract: *Social change, the rapid development of the times, more and more intelligent ways of life. This is not only reflected in all kinds of household appliances, medical, life, action and learning in various fields are in than in front of the intelligent degree, can be described as a contrast of the wisdom of the era of intelligence, I am more and think "Chi era" the word may be closer to and more suitable for such a modern life. "Chi era" means more than just data, 3D printing, smart phones, app and business, if you do not have corresponding thinking mode and cognitive models, we can not use these*

[①] 刘坤，女，上海视觉艺术学院时尚设计学院，服装设计师。E-mail: hanhan030705@qq.com

emerging technologies to achieve design innovation and sustainable development. Costume design as a continue to cross of edge of art design, in the field of application of the limitation of make it not in such an era champion, but it is "light" type of ingenious combination of technology and fashion design, also can be attached on the emerging technologies to develop a new level of design. Especially in the middle–aged and elderly people in the clothing design and application, more can be combined with intelligent innovation service in the elderly population.

Key words: *citymiddle aged; Intelligence; Clothing creative thinking design; Design servic*

引言

"中高龄服装设计"尤其是城市中高龄服装设计,其出发点应该结合目前城市生活的智能化特色,城中的老人生活环境拥挤而快速,越来越多的中高龄人士加入到"低头族""麻将休闲族""老年大学族""爷爷奶奶族",设计师要通过对于穿着者的关注与研究、独特的认知方法,摆脱传统工业式服装设计思路,在智能创新设计与服务上切入,希望能找到更好的中高龄服装的设计思路,长久地发展且拓展下去。

1 中高龄城市人群的着装需求

中高龄人群随着年龄的增长,身高在逐渐缩短,驼背者也较多,尤其是在65岁以上这种情况更加明显。中高龄人群稳重而实在,更多的是对子女、对第三代人的倾情付出,对流行已经不太关注,造型上喜欢沉稳优雅的风格,严谨而略带保守,中高龄人群已经渐渐推出社会舞台,追求安详宁静的生活,对流行事物不感兴趣,造型要求宽松舒适,零部件以简单实用为主[1],因此设计师更多的设计思路会集中在修正体态上,目前市场上所出现的设计形式也居于宽松、舒适、得体以及与之相悖的一个设计方向——花哨的老来俏设计两个方向。

1.1 对中高龄女装品牌设计的调研

通过对淘宝网中高龄女装品牌"乐卡索"的调研,了解到目前中高龄女装的分类相对简单,可以说是"简陋",但就是这样一个状态,乐卡索品牌却创造了淘宝网中高龄女装的销售排行榜第二名的成绩,它的经验来自于哪里?对于同样奋斗在实体行业的我来说是一个强大的吸引。乐卡索的女装分类相当简单:

1.1.1 日韩风格女装

日韩风格的女装包括连衣裙、套装、衬衣,颜色多素雅,款式设计多以修身款式为多,适合人群:体态较好的气质女性、职业女性;面料多以化纤合成面料为主;价格在350~550元之间,是品牌的形象高端产品层,销量一般。

1.1.2 中式风格女装

中式风格女装多是以出席场合为主的款式,已经不再是传统的旗袍,而是以有中国元素的改良款式为主。大廓型款式可以配以中国式刺绣、盘扣、立领、滚边和镶边,这样的手法可以让一个普通的服装变得正式又有韵味,适合出席聚会、婚礼、生日等必要的场合,在体型上也不会受到限制,因为中国的特色美德就是"大融"。中高龄人群经历丰富,花白的头发配上这样的款式正是从容大体的气质。此类款式面料如真丝、全麻、混合丝绵等,稍有考究,价格在155~355元之间,销售的高端出现在春节、妇女节、五一节以及国庆节,明显是婚庆、节假日的招牌款式。

1.1.3 日常印花系列女装

这类女装的特色就在于花色,选择的面料多是花得不能再花的花团簇拥的面料,由于一个面料分四个颜色倾向,款式以套头款、开衫款、半开领款三个,每个款式从S码做到XXXXL,选择面之宽可想而知。此类面料有两种:丝棉面料和polo衫面料,价格在45~199元之间,是品牌的销售生力军。

由此看来,中高龄的女装销售市场是有一定的特点的,是很具有中国特色的销售。中高龄服装目前的设计可谓是中高龄"ZARA"式的销售战略,是"短""平""快"的策略。"短":款式中大量的短款式衬衣、T恤衫为多,其二是设计信息的传播途径短,从消费者到设计师之间不存在任何互通的交流,我设计什么你认为合适就购买,仅此而已;"平":款式设计平铺直叙,没有什么大的设计点;"快"就是表现在纸板制作快、流水线制作快、销售快、全国流通快。中高龄品牌乐卡索虽然没有做过什么千人万人的调研,确是从自己身边的母亲开始入手,抓住了她们的生活范围的要求,做出了这样看上去简单却很有生活根源的设计,这也是中国绝大多数中老年人群的生活,有了生活的基础,才能被这么多的人群购买,创造了在淘宝网的品牌销售业绩。基于对实体品牌的调研,反映出中高龄人群的主动性服装着装要求。

1.2 中高龄人群的主动性服装着装要求

1.2.1 符合中高龄人群生理特点的保健类服装

此类服装要求符合中高龄人群的生理特点,要有医疗、健身、减龄、标识、保险等功能,这类服装是创新设计的入手处,是时尚服装设计跨入实用智能行列的切合点。从面料来说是功能性面料的开发,比如说含有中草药的面料、夜光显示面料、冬季室内室外可调温面料;加入智能科技技术含量的设计,比如心跳技术测试、位置显示设置、二维码扫描设置等[2]。

1.2.2 符合中高龄人群生活状态的生活类服装

作为中高龄服装品牌设计要从此类人群的生活实际状态入手,做符合他们心理需求的服装才能销售的出去。即在家有家居服,这个家居服能穿到小区,能穿到菜场,总之能够在不正式的场合穿着,要舒适、要随身、要稳重、要易打理;生活类服装第二种是出席场合的服装,要有气质、有档次、有设计、有时尚感,总之出得厅堂,展现职场未尽的情怀。

2 中高龄服装设计的服务方向

中高龄服装设计的服务方向可以从功能化、智能化和设计本身的定制化发展两个方面来考虑,前者是基于应用、实用的角度关爱中高龄人群生活,后者强调的是从服装设计本身的高时尚角度出发。

2.1 中高龄服装设计的智能化创新设计服务

2.1.1 智时代创新设计服务之一:医疗保健服务

中高龄服装设计加入"智时代"的理念,就不再属于服装行业本身的事情,它是属于整个社会的物品,社会价值远远大于服装本身。例如加入心跳技术显示的服装、加入血压显示的服装、加入中草药康复功能的服装,这类的服装带有智能设备,就不再是普通家居服装可以达到的,他的应用领域可以是从医疗领域渗透到日常生活,此类服装的设计要从医院、保健院、理疗中心等地方合作开发,加大中高龄人群对此类服装的认知度、试穿度、可信度,才能加快此类服装的流传速度。

2.1.2 智时代创新设计服务之二:运动健身服务

越来越多的中高龄人群意识到健身的重要性,但怎么健身、时间长短、自身健康状况如何在健身过程中得到控制和监测,让锻炼身体的中高龄人群在锻炼前以及锻炼过程中不断自我调整,在智时代服装的运动性能上加入监测功能就显得尤为重要。在中高龄人群运动服装中设置环境监测功能,即外围环境温度、湿度、气压、空气质量等数据,提醒着装者是否适合锻炼;在运动服装中加入锻炼强度测试,可以设置时间长短的功能性运动服装,以及位置显示功能,让运动者不论走到或者跑到哪里都能够受到家人的监测,提高交互信息的速度。

2.1.3 智时代创新设计服务之三:行为保险服务

单独外出的中高龄,在行为上要受到保险服务,例如在昏厥过程中路人可以通过二维码扫描获得家人联络信息;在服装的警示色彩上设置国家标准,可以参照台风预警色彩,例如70岁老人黄色预警,75岁老人橙色预警,75岁以上老人红色预警,让路人轻松知道有这种标志的老人属于哪一类年龄层,从而让座、现行等优惠可以更快更直接地让老人享受到。在晚间独行的老人除了家人必要的陪同外,实在有独行要求的老人要有位置显示、荧光显示等功能,确保过马路、坐车子的安全系数。

2.2 中高龄服装设计的高级定制服务

中高龄服装的设计可以分离出高级定制单项,以求满足部分仍然在职场的职业金领人群。而高级定制在我看来就是——做的是心气儿、买的是艺术、玩儿的是形式、卖的是服务。服装设计的高级定制服务不以市场追求为追求,不以时间来换算成本,一切以客人的意愿为依托,一切为了最终效果而让步。中高龄服装的定制就像一瓶香水,有着丰富的前调、中调和余味,绝不单单是一种气息。如今中高龄的高级定制可以发展成新奇、创意、个性的部分而不受"优雅"与"沉稳"的禁锢,做独树一帜的中高龄服装,走的是设计定制的限量版路线,从量体裁衣到制作的特殊需求,能给着装者提供最舒适的穿衣体验。

2.2.1 高级定制服务的面料定制

中高龄人群从着装来说,最多的聚焦会在面料上,首先以触感决定购买与否的人群有很多,所以面料定制将是高级定制的一个开始。在面料定制上包括面料的性能、面料的花色图案、面料的成分,以及面料的后期改造与制作,面料的功能性能的升级再造,高级定制的存在就是在解决市场量化产品达不到的效果,面料的定制就是高级定制的起步。

2.2.2 高级定制服务的身材定制

中高龄人群的特体有很多,大肚、驼背、臂粗等等,不是流水线上的版型纠正的范围,高级定制的量体裁衣,在设计师丈量的过程中就已经开始了加加减减的量身设计,能够很好地规避身材的缺陷。

2.2.3 高级定制的款式设计定制服务

中高龄人群考虑到自身的身体状况有些款式设计要符合穿着者的着装特殊要求。比如:有些人怕冷,所有款式要高龄设计;有些人因为心脏原因要透气,所以衣服要低领、V字领或者半开领;有些人由于关节炎需要护肘设计;有些人由于腰椎不好,要宽下摆设计……高级定制可以满足穿衣者的所有需要,你想要的款式只要面料性能允许都能达得到。

2.2.4 高级定制的服务附加值

中高龄人群的特点是一件衣服要"穿回本"高级定制的附加值领域可以加入服装维护和"redress"的理念,以求满足这类人群的需要。服装维护是指定制服装的修改、洗烫维护;"redress"的理念是根据人群的体态变化做出二次修改与设计,例如加宽加大、袖长、袖窿的修改,缝缀点缀的要求,让一件衣服达到穿着要求。这样的附加值是一般销售品牌没有的。这就是设计的后期服务。

对于中高龄服装的定制设计思路,可以从各个角度来考虑设计,有时从造型,有时从色彩,有时从表现方法和着装方式,有时有意无视原型,有时根据原型,但又故意打破这个原型,总之是反思维的,也就是逆向思维,它是在服装设计中能够进行大胆创新的一种思维方式,是在正向思维不能达到目的或不够理想时的一种尝试。逆向思维是培养创新精神和创造性能力的基础。"倒着想"或"反过来想一想",中高龄人群也许并不像我们理解的那样,他们可能不愿意这样老去,所以设计师可以无视他们的年龄,忽视他们的体态,做出与众不同的设计感觉,也许中高龄人群可以以这样的"榜样"的力量找到夕阳人群的色彩,获得更加精彩的人生乐趣。

3 结论

服装设计师的设计思维容易受到其他行业领域的影响甚至桎梏,"创新设计"将会是引领我们进入新时代的设计思维方式。智时代下对制造、服务、设计的呈现方式的思考,将转化成中高龄服装设计的发展优势,解析"智造"时代的特色,分享智时代的创新设计,转换中高龄服装设计的传统产业思路。面对百变的穿着者需求与动态,设计师需要更加系统的方法知道如何让"对手"变"战友",觉察与并解决穿着者的问题,进而获得更多穿着者的支持,创造更高的满意度。客人寻找一件适合自己的服装的过程就好比"顾客旅程地图",通过定性与定量调研,捕捉顾客实际的体验历程后,再辅以可视化的呈现,服务提供者也就是设计师可以清楚了解顾客的实际经验,也能系统性地从中获得顾客搜寻的最终结果,从而使服装设计策略更加准确,并要想尽方法促使设计师与客人之间的沟通更加顺畅。这是设计之后的服务必备的途径。对于中高龄人群而言,说服这类人群试图接受新的设计理念只有主动服务才是硬道理,否则中高龄服装设计就只剩下硬性的实用设计头衔了。

参考文献:

[1] 刘晓刚.女装设计[M].上海:东华大学出版社,2008.
[2] 刘坤.寻找银色光彩:2014中高龄时尚服饰研究[M].上海:东华大学出版社,2014.

浅谈流行元素在中高龄服饰定制设计中的运用

◆ 李昌慧[①]　（L UNION-慧设计 高级定制工作室，上海，20000）

【摘　要】　随着我国经济日益发展，人民的消费水准也随之提高，各种高级定制服务开始盛行。目前，大多数高级定制服装工作室把客户群锁定为结婚人群，为此，设计更新的产品也是针对年轻客户群的审美情趣，从而缺少了针对中高龄人群的高端设计产品，同时也缺乏中高龄人群服装的设计经验。中高龄人群由于体型、肤色、精神面貌和审美等特质有别于年轻团体，最新的流行元素无法直接运用在她们的服饰中。所以更需要对他们提供专业、有针对性地设计服务，把最新的时尚元素改良并融入到他们的服饰中，从而解决中高龄人群买衣难、消费难、服装款式千篇一律、脱离时尚潮流等问题。

【关键词】　中高龄；流行色；流行元素；旗袍；高级量身定制

The Application of popular elements in the designation of mature people

◆ Li ChangHui　(L UNION fashion design studio, Shanghai, P.R.China, 20000) (E-mail: 307165774@qq.com)

Abstract: With the development of China's economy and the improvement of people's consumption, all kinds of Haute Couture begin to prevail. At present, the married people are main target customers in the most of Haute Couture studios. Therefore, general renewal designations of costumes are based on young customers' aesthetic temperament and interest. That's why we're lacking of high-end costume and design experiences for middle aged customers. We all know that there's a big difference between middle aged customers and young people in shape, complexion, spiritual outlook and taste facts. Thus the latest popular elements can not be directly integrated in middle aged costumes. In conclusion, they need more professional and

---

[①] 李昌慧，女，Lunion-慧设计工作室，设计师。E-mail: 307165774@qq.com

tailored services. L Union will integrate the reformed fashion elements into their costumes, so as to solve their difficulties in costume consumption, such as stereotyped dress pattern, far from fashion trends and other issues.

Keywords: *mature people; popular color; popular elements; Cheongsam; Haute Couture*

引言

在经济全球化的同时，全球老龄化也越发明显，特别是经济发达的一线城市。中高龄人群量身设计定制服装的市场潜力巨大，然而目前国内的服装定制工作室并没有把工作重心放在研发中高龄客户的产品上，针对中高龄人群的设计存在不够新颖、不够舒适等多种问题，使得中高龄消费者流失。

1 流行元素在服装设计中的重要性

流行是一种普遍的社会心理现象，指社会上新近出现的或某权威性人物倡导的事物、观念、行为方式等被人们接受、采用，进而迅速推广以致消失的过程又称时尚。流行元素在商业和艺术领域应用广泛，引领各种产品和艺术的表现形式。流行的特征是入时性、突出个人、消费性、周期性。这几个特性与服装产品是非常吻合的。服装设计是科学技术和艺术的搭配焦点，涉及美学、人文、社会流行等各种要素。运用了流行元素的服装能反映出特定时期的政治、经济、文化和艺术的状况。

2 流行元素在中高龄人群服装设计中的运用

随着互联网的普及使用，各种流行信息成为传播最快最强的一股力量，从而使如今的中高龄人群和流行青年一样关注时尚。然而，无论潮流如何风起云涌、变化多端，中高龄服装的设计总是相对缺乏新意，其款式常年不变，缺乏内涵与时尚，不能满足中高龄人群对服装的需求、对流行的渴望。因此，设计师应当针对中高龄人群的需求，将流行元素适当转变、修改使之符合中高龄人群的特质后再融入到他们的服装设计中，使其不仅能够掩饰中高龄人群身材的缺陷，而且能够满足中高龄人群对时代潮流的向往，对服装内涵的深层次追求。

2.1 流行色在中高龄服装中的运用

中高龄妇女在服装颜色的选择上非常被动，且被极端分为两派——黑白灰素色派和眼花缭乱彩色派。当季流行色几乎与她们无缘也无关。但只要选用适合的当季流行色配色适合的款式，中高龄人群也能跟着国际流行趋势走起来，而且效果并不输给年轻人。

2.1.1 勃艮第酒红丝绒素旗袍

图1 勃艮第酒红色丝绒旗袍（Burgundy velvet cheongsam）

勃艮第酒红色与栗色相似，是2015年秋冬的时尚主打色之一，它具有基础黑色的优雅时髦，又能突显温柔性感的女性特质，而且稳重端庄不浮夸，非常适合中高龄女性穿着。

图1旗袍就是用勃艮第酒红色丝绒制作的素色旗袍。

丝绒质地柔软舒适，有一定的保暖性和很强的抗皱特性，非常适合中高龄人群穿着，比起缎面真丝更保温，光泽更高雅，长时间坐姿之后也不会在腹部臀部等部位起褶。腹部臀部是中高龄人群比较容易囤积脂肪的部位，服装在这些部位形成褶皱的话更容易突显和放大此部位的臃肿，是体型臃肿人群比较忌讳的现象，反光性强的面料如缎面真丝会加剧这样的视觉效果，而丝绒面料却不会。

纯色勃艮第酒红色丝绒面料视觉效果优雅大气，不俗不媚，能很好地衬托出穿着者的肤色白皙，气色健康。配以金色的滚边制作成旗袍，增加了旗袍的精致度，同时可以更加提亮穿着者肤色。最后在领口搭配珠宝造型金属扣画龙点睛，提升了整件旗袍的精致和华丽度，使它与一般的素色丝绒旗袍有本质区别，但又不至于太浮夸。

2.1.2 金色涂层真丝乔其纱中式连衣裙

金属色已经连续多年活跃于流行舞台，作为冰冷金属的特有色彩却通过高新科技手段被添加在面料上，给大众带来强烈的视觉冲击以及质感对比的猎奇感，特别是金色和银色因为其特有的金属质感和高贵气场而风靡全球，是代表着世界潮流中最为前卫的一抹色彩。然而，前卫的颜色配合前卫的款式，这几乎代表着只适合勇于活出自我的那部分年轻人，而与中高龄人群无关。

图2为假开襟中式连衣裙，使用黑底金色涂层真丝乔其纱制作。

中高龄女性的确很难穿着大面积强烈金属色的服装，虽然金色银色可以提亮穿着者肤色，但是太过强烈的金属色容易喧宾夺主，反而显得穿着者肤色黯淡，气色不佳；华丽的镜面光泽也会夸大身体的曲线，使中高龄穿着者的身体曲线问题一览无余。同时，太过于张扬、浮夸、耀眼的颜色特性也与中高龄人群的内敛沉稳特性相左。

上图连衣裙选用的面料是黑色底色的真丝乔其纱，表面喷有金色的图层，面料整体颜色高雅，地质轻盈，手感柔软。而且此面料在不同的光线折射下会呈现金色和黑色两种色彩，动态时颜色转换更具有多变性与活力感。制作成中式开襟造型的连衣裙，穿

图2 金色真丝乔其纱中式连衣裙（Golden silk chiffon Chinese style dress）

着舒适,视觉效果现代却富有历史底蕴,所以深受中高龄客户的喜爱。

2.2 多种袖型在中老年服装中的运用

中高龄女性由于活动量较少,导致皮肤与肌肉的松弛现象比较普遍,手臂部分相对更为明显,所以很多中高龄女性不得不选择短袖或中袖服装来遮掩手臂部分曲线。然而,传统基本袖型比较呆板、不时尚,修整曲线的效果也有限。而许多新颖款式的无袖礼服、旗袍或者时装再加上不适合的袖型之后会变得造型不统一、不和谐。所以,在不同的要求下,设计出功能性、美观度、舒适性兼备的袖型也是中高龄服装设计的重点问题之一。

2.2.1 印花真丝雪纺荷叶袖旗袍

荷叶袖即360度圆形袖,视觉效果比较显眼,使用柔软的面料制作表现效果烂漫灵动,褶皱自然细腻;使用较硬的面料制作则效果膨胀夸张不失优雅,因此荷叶袖在礼服、新潮时装上使用较多见。

图3旗袍为真丝雪纺长款旗袍,袖型设计就采用了荷叶袖。

考虑此款真丝雪纺印花图案颜色大胆却不失和谐,图案大小略大却不失细腻,总体风格比较现代且异域风情浓厚,所以在旗袍的款式设计上采用延续经典款式,略增加现代元素。保留旗袍的高领、长衩和小开襟等经典部分,侧边用拉链代替大开襟的8颗扣子以增加现代感和穿着的便利性。荷叶形的袖子便是增加了此件旗袍的女性浪漫特质与海派旗袍的风情,同时,单层雪纺制作的袖子柔软飘逸,360度斜裁更是使得褶皱动感十足,完全没有古板保守的陈旧感,不但能巧妙遮挡穿着者的手臂赘肉,更不会使穿着者在手臂动作过程中感受到任何拘束或者不适,完全摆脱了带袖旗袍对穿着者手臂运动幅度限制大的问题。

图3 印花真丝雪纺荷叶袖旗袍(Silk chiffon cheongsam)

2.2.2 宝蓝色绣花绣珠长袖礼服

灯笼袖算是一种历史悠久的古典袖型,曾在欧洲时装发展的长河中风靡数次,如今也经常出现在一线服装品牌设计之中,以其高雅、神秘、梦幻等特性令人着迷。

图4为宝蓝色绣花绣珠中式礼服,袖型设计便采用了灯笼长袖。

宝蓝色是中高龄女性最能接受的高饱和度颜色之一,因为宝蓝色优雅、富贵、沉稳而靓丽,而且能很好地对比衬托出穿着者皮肤白皙,气质高贵。此件短礼服采用宝蓝色缎面真丝,配合同色

图4 宝蓝色绣花绣珠长袖礼服(Royal Blue Embroidery Dress)

的抽象刺绣与绣珠,效果华丽却不浮夸,现代感与古典内蕴兼备,这对中高龄穿着者来说恰到好处。

通常礼服设计中不会使用长袖的造型,因为长袖古板、保守、沉闷。然而,此款礼服使用真丝雪纺制作灯笼长袖,不但在隐约间遮挡了穿着者的手臂曲线,更增加了整件礼服的神秘感,为中高龄穿着者增加了不少动感和时尚感,完全不会有死板厚重的感觉。

2.3 改善领型在中高龄服装中的运用

由于中高龄人群体型和体质等问题,中高龄人群挑选服装以舒适为主,所以比较偏爱无领或低领等领型,不能接受比较拘束或暴露的领型,如旗袍领、悬荡领、大V领等。这使得中高龄人群服饰领型较单一,大多设计牺牲美感来满足舒适性。所以,通过对一些时尚领型的局部调整改良,使更多时尚化的领型也能穿着舒适且被中高龄人群接受也是一个课题。

2.3.1 豆沙色一字领绣珠礼服

图5 豆沙色一字领珠绣礼服(Purple Dress)

很多中高龄女性体型丰满,颈部较粗,就比较排斥立领、旗袍领、拉夫领等领型。特别是下颚饱满且较易出汗的人。所以,适合中高龄人群的礼服款式较为鲜有——既要庄重,不可过于暴露,且具有遮挡身体部分不完美曲线的局部设计,又要舒适,时尚。

图5为豆沙色绣蕾丝绣珠旗袍所修改成的一字领短袖礼服。

豆沙色并不被大多中高龄女性所喜爱,但却是非常适合中高龄人群的巧妙颜色。它包含丁香紫的清淡脱俗,却不如丁香紫靓丽青春;它包含暖灰色的优雅高贵,却不如暖灰肃穆寡淡,它的特性正如中高龄人群的特性,成熟而雅致。

此款礼服原本是设计成旗袍领款式,但考虑到穿着者颈部较短,下颚和后背部较饱满,且比较容易出汗,所以并不适合高领和旗袍领。同时,考虑到中高龄穿着者的保守性、内敛沉稳的特性,所以并不适合低胸造型设计。因此最终选用一字领的设计,不但大方典雅,更能通过视觉错觉来拉长穿着者的颈部曲线,有较好的视觉效果。同时,后领曲线上翘1.5厘米,巧妙遮挡穿着者后颈曲线。

2.3.2 印花真丝高领旗袍

近几年,在世界流行趋势以及国内文化氛围的鼓励与带动下,部分中高龄女性开始勇于回归关注中国传统服饰,然而在定制中式立领服装时却不得不因为自己的体型问题而放弃中式立领、旗袍领或者大大降低立领的高度来换取脖颈部位的舒适度,这恰恰减少了中式服装的肃穆、端庄程度,减弱了中装的特色。所以,如何修整版型和裁剪,使中高龄人群也能穿上好看的立领也是一个难关。

图6为印花真丝旗袍,旗袍领部分使用了修改版型。

考虑到穿着者的舒适性,领围放量增加了少许,领围的增加会在视觉效果上造成脖颈粗短的效果,所以需要在其他地方进行调节和休整,比如,把领围放大的量集中在前领围,也就是前领围下挖少许,这样不但保证了穿着者的脖颈部分舒适度,同时不过多降低旗袍领的高度,借位造成脖颈修长的视觉错觉,而达到最好的视觉效果。另一方面,增加了旗袍领领口的弧度,正好巧妙地避开穿着者下颚的曲线,从而避免了穿着高旗袍领搁到下颚造成的不舒适感。这样的设计,使脖颈不够纤长的穿着者也体验到穿高领旗袍的优雅感。

图6 印花真丝高领旗袍
（Printed silk Cheongsam）

3 流行元素对于中高龄人群的重要性

随着社会不断发展,人们对服装的审美要求不断加深,对服装的要求也产生了变化。在现代的社会生活中时尚以多变的方式展现着其魅力,人人追逐向往,可以说这是一个涉及精神与物质的社会问题。

所以,流行元素对中高龄人群尤为重要,让中高龄人群穿上适合自己,又具有潮流特征的服饰不单单只是在视觉效果上靓丽得体,更能在心理上、思想上激发中高龄人群的活力,使他们与时代、与社会保持一样的步伐,而不是感觉自身被主流群体所抛弃,被社会所隔离。

服装产品作为快速消费品,以超越现实的速度在发展,服装设计师的责任是将最新的、最前沿的流行元素通过服装产品展现给大家。好的设计师会针对不同的人群用不同的表达手法来传达时代信息和理念,引导人们追随,从而形成一种潮流,而不是只针对某些人群传达。

对时尚与美丽的追求是人类的天性,并不应该受到年龄、环境等客观因素的拘泥。而设计师就是这样一个为大众圆梦的职业。让不同年龄段的人群都能穿上令人满意的服装是服装设计师的职责。

参考文献:

[1] 刘瑜,张祖芳.基于中老年体型特征的服装号型研究[J].东华大学学报,2004,04:20-23.

[2] 岑昱玮.上海人口老龄化现状分析——以两次人口普查结果为视角[J].全国商情:经济理论研究,2012,07:p8-9.

[3] 唐亮.中老年服装的市场前景和投资分析[J].东方企业文化,2010,03.

[4] 熊玛琍.中老年妇女服饰市场的思考[M].北京邮电大学出版社,2004.

清末民初江南传统女装面料纹样的研究

◆ 张 岚[①]　（上海视觉艺术学院，上海，201620）

【摘　要】　本文以清末民初女装实物及女装用面料小样实物为主要研究对象。通过对现有资料的整理，并结合当时织造、印染工艺对面料纹样的影响，系统规纳出当时女装面料纹样题材的风貌和变化规律，进而透彻解析当时的流行时尚和审美文化的变迁。女装面料纹样题材由较单一的写实、装饰风格，发展为多元风格并存的局面。纹样的内涵由伦理的、吉祥的意念逐渐转化为以审美、装饰为唯一目的。

【关键词】　清末民初；江南传统女装；面料纹样；题材

The research of traditional women's dress fabric and pattern of southern Yangtze River in the late Qing dynasty and early Republic of China

◆ Zhang Lan　*(Shanghai Institute of Visual Art, Shanghai，201620)*

Abstract: *This article uses real women's dresses and samples in the late Qing Dynasty and early Republic of China as main research objects. Through study of current documents as well as combining the influence on fabric and pattern of weave and printing and dyeing technology at that time, this article systematically induces the style and rule of change of women's dress fabric and pattern theme at that time, it further thoroughly analyzes the change of fashion and aesthetic culture at that time. The theme of women's dress fabric and pattern develops from a relatively single realistic and decorative style to a situation of co-existing multi-style. The implication of pattern gradually transforms from ethical and auspicious thoughts into a unique purpose with aesthetics and decoration.*

[①] 张岚，女，上海视觉艺术学院讲师，硕士。E-mail:zhanglan_666@163.com

Keywords: *late Qing Dynasty and early Republic of China, traditional women's dress of southern Yangtze River, fabric and pattern, theme*

清末民初的中国,西洋文化东渐,服饰变革巨大。随着封建制度的瓦解,等级森严的服饰礼制也退出历史舞台。到了民国时期,人们可以按照自己的意愿选择喜爱的服饰,款式、颜色、纹样等方面均不受限制。

本文以清末民初约 200 件女装实物及约 100 件女装用面料小样实物为主要研究对象(苏州李品德先生提供),以保留的大量图片和文字记载为补充研究资料。通过对现有资料的整理,并结合当时织造、印染工艺对面料纹样的影响,可以系统规纳出当时女装面料纹样题材的风貌和变化规律,进而透彻解析当时的流行时尚和审美文化的变迁。

晚清到民国,女装面料纹样题材由较单一的写实、装饰风格,发展为多元风格并存的局面。纹样的内涵由伦理的、吉祥的意念逐渐转化为以审美、装饰为唯一目的。

1 晚清传统女装纹样的题材

1.1 晚清女装纹样题材的基本内容

1.1.1 植物题材

植物题材是晚清女装中最为常见的题材,搜集的 56 件晚清女装实例中,有 49 件涉及植物纹样。其内容又包括如下几方面:花卉纹样、叶形纹样、蔬果纹样。其中花卉纹样是最为常见的题材,49 件植物纹样题材的女装中,有 33 件涉及花卉纹样,其中会辅以叶形纹样,及动物纹样。以叶形纹样为主要题材的女装共有 13 件。此外 3 件以蔬果纹样为主。

晚清女装的纹样具有程式化的特点,因此通过对搜集资料的分析,可以较为明确的归纳出纹样的题材分类。花卉纹样通常包括以下几种:1、牡丹、芙蓉类团形花;2、梅花;3、菊花;4、牵牛花。其中第一类花型占绝大多数。33 件花卉纹样女装中,30 件都以牡丹、芙蓉类的团花为题材。另有两件为菊花纹样。

1.1.2 动物题材

除龙、凤外,动物纹样单纯作为纹样题材的情况较少,通常会辅以植物及其他纹样。从搜集的资料看,一共出现如下几种动物纹样:正统服装中的龙、凤,最为常见的蝙蝠、蝴蝶、鹤,以及只出现一例的松鼠。

1.1.3 文字题材

所搜集资料中,只出现两种文字纹样,一种为寿字纹,一种为万字纹。寿字纹有单独出现的情况,也有辅助菊花、蝙蝠等纹样的情况。万字纹则常作为一种底纹或辅助纹样出现。

1.1.4 云水纹

指云气纹及水纹。前者常作为一种底纹或辅助纹样；后者除作为底纹或辅助纹样外，还出现在晚清典型纹样——海水江崖纹及落花流水纹之中。

1.1.5 其他题材纹样

以上所说的纹样题材占据所研究纹样的绝大多数。除此之外我们还在传统女装中发现了如下一些纹样题材。如杂宝纹（灯笼团凤如意杂宝纹）、暗八仙、十二生肖纹、博古纹、人物风景纹（红楼梦人物纹挽袖）。这些纹样虽数量较少，但可以传统出晚清女装纹样的波及范围。但是诸如十二生肖纹、博古纹、人物风景纹这类较为独特的纹样，在挽袖上出现较多。大身及滚边仍为上述常见纹样。

1.2 晚清女装纹样不同题材的选择及应用

1.2.1 外形柔美、华丽、多变的题材运用较为广泛

由于是女性服饰，受到时代对女性审美的要求，所选用纹样的形式特点，多倾向于华丽、柔美。例如花卉纹样在晚清女装中运用最为广泛。而且考虑到效果变化生动的要求，百花蝴蝶的组合是晚清女装纹样中最为多见的一种题材搭配方式。搜集的56件女装实物中，有21件女装运用了花与蝶组合的搭配方式。百花蝴蝶纹的每一单位纹样中有粉蓝色枝梗的白色梅花，粉红色的桃花，层次分明的牡丹和蓝黄绿各色的花叶，还有月季、海棠、芙蓉等百花丛集灿烂夺目。再穿插十种以上大小不同，姿态各异的蝴蝶，及随风飘舞的花瓣，充分展示了多变的华丽与柔美。[1] 外形方正、变化较少、质感坚硬的纹样题材则较少在女装面料中采用。例如在晚清其他场合会出现的器皿纹样、人物纹样、风景纹样等。

1.2.2 对纹样所反映的吉祥意义有所选择

清代纹样号称"图必有意，意必吉祥"[2]，在服装中也不例外。但晚清女装中并不是任何吉祥纹样都可以出现，而是针对着衣者的地位、年龄及着衣场合有所变化。通过对所搜集资料的分析，发现诸如祈求学而优的、做官的、富裕的此类主要针对封建时代男性的美好祝愿通常不会出现在女性服装中。而女性服装中的纹样又以祝愿幸福、多子、长寿三种意义的题材最为多见。

1.2.3 挽袖的纹样题材比较宽范

前面所提到的女装纹样的特点，主要针对衣身、滚边等部分，而挽袖的纹样题材在选择上要相对其他部分自由得多。例如杂宝纹、博古纹、人物风景纹、十二生肖纹，[3] 这些在衣身、滚边等部分不可能出现的纹样，在挽袖中都有所体现。挽袖纹样题材在风格上的变化，增添了

女装纹样的趣味性和文化性。

总体看来,晚清女装纹样风格豪放,与早期典雅、中期豪华、外来风格明显的特点有所区别,有繁琐、浮躁之嫌,预示着新风格纹样的出现。

2 20世纪10—20年代传统女装纹样的题材

搜集到的10—20年代的女装实物资料相对较少,因此在研究上存在局限。通过对有限资料的分析,可以发现如下一些特点。

2.1 延续晚清纹样题材

这一阶段的女装面料处于一个过度时期。一定程度延续前代风格。因此前面提到的花蝶纹样、四君子纹样、福寿纹样仍占据一定地位。

2.1.1 花卉的题材更宽范

晚清时期花卉纹样题材具有明显的程式化的特点。几乎可以明确的概况几种花形。但这一时期女装的花卉纹样,出现了不同以往的题材。

2.1.2 纹样搭配方式有所突破

晚清时期花与蝶的搭配方式极为流行,但这一阶段这样的搭配方式逐渐减少。取而代之的是非吉祥含义的搭配,对视觉效果的追求比以往强烈,对纹样内涵的要求降低。但是,"既注重纹样的社会功利性,又注重纹样的审美愉悦性;既注重纹样形式美的创造,又注重情感意念的传达。……是中国丝绸纹样的审美特征"。[4]从这个角度看,中国传统女装纹样的审美特征受到影响。

3 20世纪20—30年代传统女装纹样的题材

20到30年代的资料较多,搜集并选择100件作为研究对象。如果说10—20年代纹样题材还在谨慎的追求突破,这个时代的纹样题材终于可以大胆创新,产生巨大变化。

3.1 花卉纹样

100件女装中36件女装以花卉纹样为题材,占总数36%。而花卉纹样又分为以下几种情况。

3.1.1 装饰花卉纹样

这类花卉纹样以某一种花卉为原型,经过概况处理,形成装饰效果强烈的花卉纹样。这种花卉纹样,其形式感强,视觉效果好,同时在实际染织过程中容易实现,因此是为数较多的一类花卉纹样。36件女装中,有18件女装以此类花卉纹样为题材。而所绘具体内容,有如牡丹、小菊花等以及一些看不出原型,但形式感突出的花型,及一部分手绘广告。其中以牡丹团花的装饰形为题材的女装面料为数最多,可见这在当时是一种比较流行的纹样。一方面数量上看,牡丹花卉纹样最多;另一方面,这一时期大多数手绘广告中的女性着装上装饰的花卉为团形的牡丹花变形图案。这说明,人们对当时女装面料上装饰形牡丹花的接受程度比较高,也可以看作对传统女装纹样题材的变化延用。

3.1.2 写实花卉纹样

这类花卉纹样以西方写实手法绘制,通过明暗光景的描绘,表现出具有真实质感的花卉纹样,与现代面料中出现的花卉纹样非常接近,通常出现在印花面料上。因此也可以说是印花技术的出现,促使了这类花卉题材的诞生。所绘制内容也与现代手绘花卉的题材相近。如牡丹、百合、小菊花、茶花等。36件花卉纹样中,有10件以写实花卉为题材。

3.1.3 传统花卉纹样

这类纹样指对晚清时流行的花卉题材改造利用,再结合色彩、布局的不同方式,使其具有一定的时代特色。但是这类纹样更多的是对以往花卉纹样形式感的继续,甚至加入了晚清花卉纹样中不会出现的题材,如花篮等。而对晚清纹样的吉祥意义、或特定的不同题材的搭配则不如以往要求严格。

这样的花卉纹样实物共有8件。而且此类花卉纹样题材的来源集中于女装实物资料,而前两种花卉纹样主要存在的《申报》《良友画报》等资料中。因此我们可以推断,当时民间依然比较广泛的存在对晚清特点纹样的延续改造的使用,但时尚媒体中比较多的宣传前两种花卉纹样。反映出当时的流行与大众实际操作之间的差距。

3.2 抽象几何纹样

女装面料中,对没有含义的简洁抽象几何纹样的利用,是这一时代非常重要的突破。100件女装实物中,有31件女装运用了几何纹样题材。几乎与花卉纹样平分秋色,可见其发展之迅猛。预示着之后几何纹样大行其道的流行趋势。

几何纹样可以概况为以下几类来分析。

3.2.1 条格纹

所谓条格纹,指以横条或竖条纹样,或格子纹样满地装饰面料的纹样类型。这类纹样在几何纹样中最为常见,31件以几何纹样为题材的女装中,有18件为条格纹样,约占60%。

这类条格纹样中,以竖条纹为装饰元素的面料,通过采用宽窄不同的条格纹样,形成或强烈或柔和的视觉效果。同样格形纹样,也会因单位纹样大小的不同,创造出丰富多变的效果。同时条格纹样也会与简洁风格的滚边结合,形成统一、简洁的视觉效果。在一些真人图片资料或手绘广告作品中,也存在将条格纹样与装饰花绘纹样结合的情况,效果简洁中有变化,比较成功。

3.2.2 其他几何纹样

除条格纹样外。其他几何纹样如心形、圆形、多边形、三角形、或多种几何形穿插排列形成装饰效果。这些几何纹样中有些比条格纹样丰富而活泼,但又有因规则秩序感而产生的美感,有花卉纹样不可比拟的优势。

中国纹样自唐代至近代,装饰以花草纹为主,动物纹样为辅,是为花草纹时期。花草纹的应用,标志着人的觉醒,摆脱了天或神的精神束缚。与此之前,以动物纹样为主的格局相对照,可见人们开始认识自我,并追求自我,把欣赏提到了重要的地位。[5]而从近代开始,几何纹样的盛行,说明人们审美方式的又一次进步。

女装面料中的几何纹样有两大特点:1) 简洁性。几何纹样多数是客观事物的规律的抽象化结果,所以它舍弃了复杂的现实形象,只保留其突出形态特征的部分。2) 明晰性,几何纹因经过艺术取舍达到简洁效果,故剔除了更多的理性思维因素,而归总在较单一的意义中,具有明晰的效果。[6]因此,对几何纹样的欣赏,标志着对抽象形象欣赏能力的提升。

3.3 传统风格纹样

虽然这一阶段的女装面料纹样的题材得到了前所未有的发展,但是传统纹样依然以一种改头换面的姿态,存在于这一阶段的女装面料纹样中,并非突然消失。100 件研究对象中,有 19 件体现了这种特点。可见这一类纹样也在当时占据一席之地。19 件女装中,除 8 件是对传统花卉纹样的延续使用外,其他 11 件有对传统龙凤纹样的沿用、有对如意纹、云纹、水纹等传统纹样的改造使用。但是这类纹样脱离了原本的存在环境,同时其内含在这个时代不再被重视,虽然做了迎合时代的改造,但视觉效果并不理想,尤其是如意纹、云纹、水纹等。可以看出其必将被历史淘汰的趋势。

3.4 素面无纹样面料

100 件研究对象中,有 10 件运用了素面无纹样的面料。但是这类素面无纹样面料,往往搭配立体装饰或较为复杂的滚边。繁简对比,效果鲜明。

这一阶段女装纹样对形式美感的追求成为主要目的,机械化的生产方式也对纹样特点产生新的影响。纹样涵义不如以往丰富,情感意念的传达限于对以往的继承部分,符合当时时代特征的新涵义几乎没有体现,纹样的这一审美特征受到影响。但纹样变化快、种类繁多,装饰意义逐渐占据主导地位,且生产效率提高,产量扩大。这应该是面料商品化、平民化过程中一个必然的趋势。

"服饰图案一般可分为平面和立体两种形式。平面装饰指在平面物体上所表现的各种装饰,效果是平面形,如服装与配件所用的面料,包括匹料和件料的装饰均属于此类。……立体的是指服饰上的装饰具有立体效果,如利用面料制成的立体花、蝴蝶结等。有的则有浮雕效果,如盘花纽等。此外,鞋、帽、戒指……甚至有些运动装上的拉绒图案装饰,都具有立体效果。"[7]

4 20 世纪 30—40 年代传统女装纹样的题材

4.1 花卉纹样

30—40 年搜集的研究资料最为丰富,共有 419 件。其中花卉纹样仍然是最常用的题材。总共有 193 件,占总数约 46%,大大超过之前已经很高的 36% 的比例。而之前曾出现的蔬菜瓜果类及一部分动物纹样,在这一时段很少甚至没有出现,说明了花卉纹样在女装面料中顽强的生命力,经过时间的选择终于沉淀下来。

花卉纹样按照不同的绘制、设计方法及风格,分为写实花卉、装饰花卉。写实花卉指按照西方绘画方法,以花卉实物光影明暗变化为基础进行造型的花卉纹样。装饰花卉指按照构成规律,以视觉效果为追求,经过变形加工而成花卉纹样。装饰花卉纹样中一部分延续中国传统风格,称其为传统装饰花卉纹样,另一部分采用新装饰手法,称其为新装饰花卉纹样。

4.1.1 新装饰花卉纹样

193 件以花卉纹样为题材的女装中,164 件女装为新装饰风格纹样。约占所有花卉纹样的 85%,而占所有研究对象的 39%。说明新装饰花卉纹样虽然年轻,但以其多变的形态,鲜明的视觉效果,广泛的应用环境,迎得了当时市场的普遍认可。

这一时段的装饰花卉纹样种类繁多,风格多变,已经相当成熟。其中一些纹样或纹样风格至今仍在延用。

数量最多的是外形规整、形象概况的花形。这类花形或简或繁,或大或小,视觉效果各异。而这其中的大部分纹样,脱胎于中国传统纹样,同时结合来外来纹样的设计方法,效果即保留了中国传统纹样繁密为主的特点,又不落入俗套。

谈到外国风格纹样的影响,不能不提当时日本及欧美对中国女装纹样的影响。新装饰花卉纹样中,部分效果具有明显的日本风格。也有一部分花形突破规整形态的束缚,随意勾画,以自然而成的笔触为美,有写意画的感觉,突破传统方式,给人以耳目一新的感觉。这一类花形以其独特的风格占据一席之地,虽然在当时唯数仍然较少,但是一种非常具时尚感和生命力的装饰手法,在今天的面料设计中广泛存在。

花卉纹样有时也呈现出几何效果。如有一部分装饰花卉纹样,以小花为题材,整齐排列。整体观之,会呈现如条格纹样般规整效果。也有一些装饰花纹样,花形极为概况,规纳为近乎几何形的简单形态,也使花卉纹样呈现出几何效果。

4.1.2 传统装饰花卉纹样

传统装饰花卉纹样,经过迎合时代的改造,在基本保留传统特色的情况下,存在于这个时代的女装纹样中。成为这个时代特有的传统装饰花卉纹样。

另外,传统面料以及装饰手法,如缎、刺绣,在这一时代仍然延用。在这类传统工艺的影响下,所织造的花卉纹样也不可避免的呈现出传统的特点。

4.2 抽象几何纹样

419件研究对象中,各类几何纹共172件,约占41%,略低于花卉纹样的46%,与之前情况相近。(之前花卉纹样36%,几何纹样31%)但是要大大高于之前几何纹样31%的比例。也就是说,花卉纹样、几何纹样在较长一段时间内一直是较重要的纹样题材,经过发展,一部分纹样,如蔬果、动物等比重下降,但花卉与几何纹样比重均有较大幅度增加,但仍保持花卉纹样略高于几何纹样的情况。

4.2.1 条格纹

条格纹可以说自诞生之日起,就因为其效果大方,生产便利,受到了广泛的传播,并为民间所接受。30—40年间,419件女装中,有100件女装为条格纹样,约占总数24%。较之前20—30年间占总数18%的比例,有所提高。有趣的是,20—30年间条格纹样占几何纹样数的58.10%,而30—40年间,条格纹样占几何纹样总数的58.13%,几乎相等。说明,条格纹样是一类重要的几何纹样题材,而伴随条格纹样的发展,其他几何纹样也在同步开发。

条格纹样的变化不仅仅在数量上,装饰手法相对前代也有明显进步。其变化之一,对条格纹样的使用更巧妙多变,创造出了丰富的效果。例如在服装裁剪过程中,运用疏密不同,或方向不同的条格纹样,来制造多变的视觉效果。而面料本身的设计,也不再拘泥于横平竖直的条格纹,而是设计出有角度变化的条格纹,效果流畅而多变。

条格纹样与花卉纹样结合的方法,在之前已有出现。但主要集中于画报等媒体所见明星等人的着装,条格纹与花卉纹样结合的实物并没有见到,而且结合方式也较为局限,通常是将装饰花绘设计为长方形适合纹样间隔置于条格内,整体效果仍为条格布局,但细处观察会发现条格内的花卉纹样。30—40年间,条格纹样与花卉纹样结合的女装面料主要体现见到了很多面料实物中存在条格与花卉纹样结合的情况以及结合方式多种多样这两个变化,不拘泥于前代。所以条格与花卉纹样结合的女装面料,经历了由明星到普通民众的进化过程,装饰手法也不断推陈出新。

4.2.2 其他几何纹样

除条格纹样外,圆点形纹样、折线纹样等几何形,或规则排列,或大小不一、叠压穿插,均以几何纹样简洁、多变、秩序感强、形式感突出的特点,取得良好视觉效果,不一一赘述。

4.3 传统风格纹样

搜集资料中,30—40年传统风格纹样有40件,占总数约9.5%,其中存在与装饰花绘的交叉部分。除去19件传统装饰花卉纹样，传统风格纹样只占总数约5%。与之前19%的比例相比，明显下降，可见接纳新风格之初，传统风格受到了比较大的冲击。

与之前相同，传统风格纹样在这个时代，继续以新的方式保留。但形式美感较前代有明显的进步。之前20—30年间，可以看出有些女装面料在设计过程中希望保留传统元素，但又想不被潮流抛弃，因此对传统纹样进行了一些改造。但可能是因为当时的设计手法不够成熟，对新式纹样的理解不够准确，因而出现了一部分女装纹样不伦不类，传统味道不浓，现代特点不够。而到了30—40年间，这种问题基本得到解决，可以看出明显的进步。

将传统纹样模糊化，再运用，是这一时段延续采用传统纹样较为常见，也较为成功的方式。而所谓模糊化的方式，最常见的是将传统纹样单位纹样的面积缩小，满地排列。也有将传统纹样的形象更加概况，忽略细节。或者将传统纹样打散，节取其中部他，使其形象不完整而降低视觉冲击。

还有一类女装纹样比较完整的延用了传统纹样，纹样本身没有进行太多改变，传统风格浓厚，在新潮流中，也具有相当的存在价值。

4.4 素面无纹样

所搜集的30—40年间的实物中，素面无纹女装35件，约占总数8.4%。与之前10%的比例没有太大变化。素面无纹女装的装饰手段还是有微妙的改变。在之前，采用素面无纹样面料的女装，往往会装饰较复杂的立体花或滚边。30—40年间，这种情况依然存在，但无花、无滚边素面无纹女装也成为一种风格。可见民众对单纯统一的风格也开始欣赏。表明女装纹样由繁到简的审美趋势之一。

这一时段内的女装面料纹样，在色彩、风格、形态、绘制方式等诸多方面，均随着时间的推移，呈现出阶段性的特色，具有强烈的时代特征和风貌。它蕴含着丰富的信息，映射出这个时代的工艺水平和审美观点。作为历史的一面镜子，甚至可以从面料的特征和变化中看出时代变迁的历程。

5 结束语

本文通过对清末民初大量女装实物和面料小样的研究，发现女装面料纹样题材由较单一的写实、装饰风格，发展为多元风格并存的局面。纹样的内涵由伦理的、吉祥的意念逐渐转化为以审美、装饰为唯一目的。有助于人们更好的了解当时女装面料纹样题材的风貌和变化规律，进而透彻解析当时的流行时尚和审美文化的变迁。

参考文献：

[1] 吴淑生，田自秉.中国文化史丛书·中国染织史[M].上海：上海人民出版社，1986.6.
[2] 陈娟娟.中国织绣服饰论集[M]，北京：紫禁城出版社，2005.30.

［3］黄能馥,陈娟娟.中国丝绸科技艺术七千年［M］.北京:中国纺织出版社,2002.
［4］回顾.中国丝绸纹样史［M］.黑龙江:黑龙江美术出版社,1990.9.
［5］吴淑生,田自秉.中国文化史丛书·中国纹样史［M］.上海:上海人民出版社,1986.5.
［6］吴淑生,田自秉.中国文化史丛书·中国纹样史［M］.上海:上海人民出版社,1986.5.
［7］上海服装行业协会.中国服饰大典［M］.上海:文汇出版社出版,1999.9.16.

中高龄服饰功能性研究

老年防摔功能服装设计分析研究

◆ 朱达辉[①] 宇 锋[②] （1，2.东华大学，上海市延安西路1882号，200051）

【摘 要】 随着老龄化社会的到来，独居老人或暂时无人陪伴的老人发生跌伤、坠落等占意外伤害的首要位置，目前老年人防摔功能服装存在功效差、设计感不足、服用性不强等诸多问题；了解老年防摔功能服装设计要求，更要对易发摔跌伤害部位加以科学原理分析，通过色彩的把握和创新防护服用材料使用，有针对地确保功能，兼顾造型的美观；同时根据老年人的安全需求辅助其他功能模块，设计中结合电子、通讯等定位与预警等多功能高新技术，以附加更多服用功能。通过科技的运用，结合创新设计理念，来服务老年人群。

【关键词】 部位；防摔结构；功能服装；科技；设计

The Design and Analysis of the Aging Clothing with Fall Prevention Function

◆ Zhu Dahui[1] Yu Feng[2] *(Donghua University, Shanghai, P.R.China, 200051) (E-mail: zhudahui2000@163.com, yufengemily@163.com)*

Abstract: *As the arrival of the aging society, the old man who live alone for long-term or just temporarily are easy to get injured by the accidental fall, which happens most frequently among all other injuries. Currently, the aging clothing with fall prevention function has arisen many problems, such as poor efficacy, bad design, inconvenience, and so on. Firstly, the most important issue is certainly to meet the basic function of prevention about falling. Then, it is necessary to analyze scientifically, more specifically, suitable color choice and creative protective materials, which all provide pertinent measures both for function and appearance. Meanwhile, design should combine with electronics, communication and positioning, warning*

[①] 朱达辉，男，副教授。E-mail: zhudahui2000@163.com
[②] 宇锋，女，硕士研究生。E-mail: yufengemily@163.com

techniques, etc., in other words, multi-function design can afford better service for the elderly people with the applying of hi-tech and creative theories.

Keywords: *Body parts; Anti drop structure; Functional clothing; Science and technology; Design.*

随着老龄化社会的到来，传统家庭养老功能渐弱，独居老人或暂时无人陪伴的老人发生跌伤坠落等占意外伤害的首要位置，容易造成严重的后果。现有的专业研究主要集中在骨折术后老人防护方面，本文以设计方法入手，采用分析法来探讨研究老年人防摔结构功能服装设计。

1 老年人防摔功能服装需求

1.1 老年人意外伤害的主要原因

老年人因为生理机能的退化，容易引起诸多意外伤害，其原因主要有跌伤坠落、交通伤害、动物伤害、钝器伤害等等。浙江省疾控中心的数据显示，2009—2010年浙江省60岁以上老人意外伤害造成主因为跌伤、坠落，占比率为41.63%（图1），排在首位，可见做好防摔保护措施对老年人健康安全尤为重要。

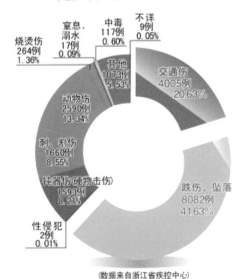

图1　2009—2010年浙江60岁以上老人意外伤害原因构成

1.2 老年人防摔服饰的种类用途

如何预防和保护老年人摔伤成为老年人自己和每个家庭关心的问题。从老年人生活息息相关的衣食住行考虑，老年人除了生活中加强注意外，更需要一件既能保障日常生活又不妨碍坐卧起居的防护衣。随着科技的飞速发展，运用先进技术手段针对老年人不同需求的防摔服

装与护具等各类产品不断推陈出新,服务老年人群,主要有下面几类:

1.2.1 防摔结构服装

防摔结构服装对象是普通易摔老人,用来加强保护或者缓解摔倒伤害强度。按照季节分,防摔服可以分为夏季轻便款和秋冬重型款。按照穿着分类,可以分为内衣、内搭和外套。

1.2.2 专业防摔护具

专业防摔护具对象是摔伤术后老人,用来加强保护术后部位与防止复摔伤害。其分类根据摔伤部位来划分。

1.2.3 防摔预警系统服饰

目前跌倒识别和预警主要是基于可穿戴式传感技术,其主要是通过对加速度计量工具、陀螺仪等对人体动作进行捕捉。目前市场上的可穿戴式传感技术主要是基于 MEMS 惯性传感器来实现的。此类是针对摔倒后家人或医护机构及时得到信息施救,防摔服装设计可以附加此项功能。

1.2.4 安全气囊防摔服装

目前防摔服装中有一种能防止老年人摔倒后受伤的安全气袋,一旦使用者正在朝地面加速倾倒的话,气囊就会在十分之一秒的时间内充气。用以保护关键部位,十分适用于癫痫症等突发失控的病患老人。

1.3 老年人防摔服饰存在的问题

防摔结构广泛应用于登山、滑雪、赛车等着装,但在日常生活中十分需要其保护的国内中老年人群中却接受度极低。究其主要原因如下:(1)设计感差,笨重丑陋突兀;(2)不能完全满足穿着者实际需求,服用性差;(3)价格不合理;(4)功能有效性差。综合以上,防摔功能服装的设计需要从设计依据、功能要求、造型要素方面综合考虑,重新审视。

2 老年防摔功能服装设计依据

2.1 防摔功能服装穿着环境

虽然对于骨质疏松症高发的老年群体,防摔是时时刻刻要警惕的事,但是受日照、温度、饮食、地表硬度等一系列环境因素的影响,冬季成为老年人摔伤频率最高的季节。而且考虑到专业护具的厚度和老年人的接受度,防摔服装更适合温度不高的春秋冬季。当然,老年人夏季防摔也是不可忽视的,夏季护具应考虑服装轻薄透气又具备功能。

2.2 防摔功能服装穿着对象

穿着对象主要为易摔倒和易摔伤的老年人。老年人随着年龄的增长,随之而来的视力下降、听力减弱、关节退化、身体协调性差、反应迟钝和癫痫、白内障、心脑血管疾病等一些老年人常见病,以及安眠药、抗高血压药、降糖药等一些药物的影响,导致老年人更易跌伤、坠落。作为骨质疏松症高发的老年人群体,一旦摔倒极易发生骨折或导致更严重的后果。另外,术后老人发生跌倒,更容易发生危险,日常生活也更需要防摔功能服装保护。

2.3 防摔功能服装穿着要求

作为功能服装的防摔服其第一要义是功能性,防摔作为首要功能应该凸显出来。目前市场上的防摔服不被中老年消费者看好的重要原因就是其丑陋突兀、服用性差,老年人感觉自身被捆绑束缚、着装异于常人。因此美观和舒适性的考量在防摔服的设计中也是极其重要的。

3 老年防摔功能服装设计要求

3.1 防摔服装功能

防摔服是保护人体在各种体位的摔倒中不受伤害的服装。老年人摔倒主要伤害为骨折,北京积水潭医院曾对 11 000 例 60 岁以上老年骨折患者进行统计,其中最常见的三种骨折分别是髋部骨折(占 19.8%)、胸腰椎骨折(占 18.8%)、桡骨远端骨折(占 17.6%)[1]。尤其是髋部骨折和胸腰椎骨折,后果严重且治愈率低,因此这两个部位的保护尤其重要。此外,也可根据老年人的安全需求,防摔服还辅助其他功能模块,例如全球定位系统(Global Positioning System, GPS)、全球移动通信系统(Global System For Mobile, GSM)等电路模块功能。用以防摔预警与摔后报警施救等功能。

3.2 防摔保护部位

防摔分不同部位的防护,重点保护的部位根据不同部位的骨折频率和伤害大小而定。例如癫痫患者就要重点保护其后倾摔倒的动作,头部、胸腰椎、髋部的防护尤其重要。有些偏瘫的患者可能造成侧摔的,防护的重点又有不同。依据防摔部位的不同主要分为髋部防护结构、胸腰椎防护结构、胸骨肋骨防护结构,肩肘防护结构,膝盖防摔结构(图 2)。

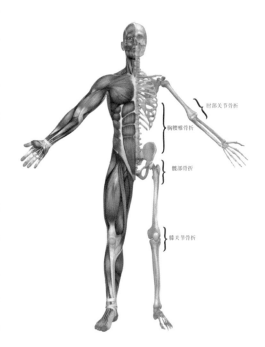

图2 防摔服装保护的主要部位

3.2.1 髋部防护结构

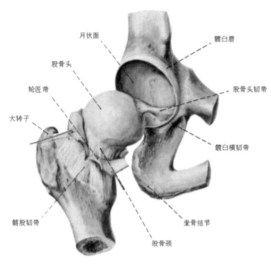

图3 髋关节分解图

髋部(图3)骨折是老年人最常见骨折类型,主要包括股骨颈骨折和股骨转子间骨折两种类型[2]。由于后摔倒和侧摔倒对髋部的损伤最大,所以侧腰后臀位置是髋骨保护的重点。当在直立位站立或行走中发生侧方跌倒时,跌倒者有较大发生股骨近端骨折的风险;当侧方跌倒发生时,触地侧的股骨近端最先发生骨折的部位可能为股骨颈位于股骨头与大转子之间且靠近大转子一侧的部位。业内公认效果最好的髋关节保护器是马蹄形髋关节保护器,其正中央空心位置对准股骨大转子部位贴合大腿进行防护,经临床测试能够有效对抗侧跌倒对髋部带来的伤害[3]。

3.2.2 胸腰椎防护结构

胸腰段位于固定的胸椎和活动的腰椎之间,由 T11-L2 共 4 个节段组成,外力作用时椎体的刚度急剧增加[4]。脊柱承受上肢及躯干垂直载荷后迅速传导至胸腰段生理弯曲汇聚后经由骨盆放射至双侧下肢,形成一个 X 形应力分布特点,而胸腰段恰好位于应力极度高度集中的 X 形中点上。由于 X 中心集中过度的由上肢及躯干传递下来的超负荷有害应力,而无法迅速及时地向骨盆及双下肢分散,故在胸腰段容易造成骨折破坏[5]。胸腰椎保护器的设计应该尽量贴合人体,这样能够将碰撞产生的冲击力通过更大面积软组织和骨骼分流和吸收,减小胸腰椎的负荷。例如贴合腰腹区域的设计,这样腹部作为一个人体自带的密闭水囊具有缓冲作用,吸收了部分应力并向周围肌肉分流,从而减少对胸腰椎的伤害。也可以通过肩带将冲击力分散至肩胛骨处的肌肉和骨骼。此外,胸腰椎保护器需要尽可能控制肢体前屈,协同人体后部肌肉和韧带,产生对抗力,使摔倒时的人体重心后移,从而维持平衡[6](图4)。

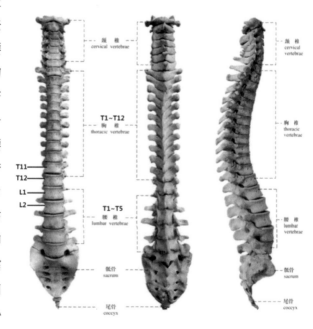

图4 胸腰椎防护部位

3.2.3 桡骨远端骨折

摔倒时,人会下意识地用手掌支撑地面阻止跌倒,因此,应力迅速由手掌软组织传递至前臂尺桡骨远端,并沿着尺桡骨向近端传递。应力在尺桡骨远端集中,故易于产生桡骨远端 Colles 骨折及尺骨茎突骨折[7]。行之有效的腕保护器须贴合手掌、手背、手腕和前臂轮廓,根据暴力分流和暴力吸收原理,外部采用较硬材质,内部采用柔软内衬。在吸收暴力的同时,向身体其他软组织分流,因此掌骨和前臂重点区域也产生较高应力[8](图5)。

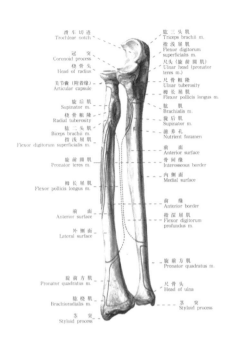

图5 桡尺骨(前面)

4 老年防摔服装设计电子科技运用

根据老年人的安全需求辅助其他功能模块,老年防摔服装设计结合高科技电子、通讯等定位与预警等多功能高新技术,以附加更多服用功能。另外还可以结合服饰产品来设计,如腕表、腰带、鞋子、帽子等,整体系列化设计,功能各不相同。

4.1 防摔预警系统方法分类

目前跌倒识别和预警的方法主要包括以下三类:基于视频监测的技术,其主要通过摄像器材获取人体运动影像来判断;基于环境感知的技术,其主要通过声波、压力、振动、红外等来实现对跌倒的判断;基于可穿戴式式的传感技术,其主要是通过对加速度计量工具、陀螺仪等对人体动作进行捕捉。其中,可穿戴式传感技术因其成本低、实用性强和性能高等技术优势成为首选方案[9]。目前市场上的可穿戴式传感技术主要是基于 MEMS 惯性传感器来实现的。

4.2 防摔预警系统服饰产品

国外对跌倒预警系统研究较早,目前国外市场上已经有很多相关产品,主要分为腕表式、腰带式、鞋子、服装等。主要功能是为了让跌倒伤患能够被及时发现从而得到及时救护,避免因救治不及而带来的严重后果。

4.3 安全气囊防摔服装

日本公司 Prop 发明一种能防止老年人摔倒后受伤的安全气袋。使用者一旦倾倒,气囊就会在十分之一秒的时间内充气。此外该产品还置入了 GPS[10]。虽然这种绑在腰间的安全气袋不能保护向前的倾倒,但是能够有效避免摔倒时伤害巨大的髋骨、胸腰椎、颈椎骨折和头部伤害,十分适用于突发失控的癫痫症患者。

4.4 防摔预警服装的不足

目前搭载防跌倒预警系统的服装和配件在中国没有形成大的购买需求的原因主要有以下几点：（1）产品性能稳定性还没有达到预期，容易产生误报警或报警失灵；（2）产品设计不合理，不符合人体工学，笨重累赘；（3）价格过高；（4）老年人自身对防摔问题重视度不够以及此类科技产品关注甚少。

防摔系统未来的研究重点将是灵敏度、准确度的提高和体积重量的减小，以及可穿戴性的设计和更多模块功能的构想。

5 老年防摔功能服装设计造型要素

5.1 防摔功能服装配色

柔和不突兀的色彩，让防摔服装或者护具看起来更像是一件日常着装。也可以依据结构做一些拼色，使产品看起来功能结构明确，更具科技感和时尚感。

5.2 防摔功能服装材料

防摔功能服装材料分为外层的面料和内部的护具材料两部分。近年来，智能纺织面料受到广泛关注。智能调温材料、抗菌防臭材料、防紫外线材料、防水透湿材料、保健纤维等都是此类功能服装外层面料的首选。专业护具原理主要分为暴力吸收、暴力分流和混合型三种，暴力吸收的材质较软，主要为高弹性的聚乙烯材料；暴力分流材质主要为聚乙烯泡沫材料；混合型是前两者兼而有之。这两种材料质地轻盈、廉价且防护效果较好，分别安装在肩、肘、臂、背部、膝、臀等衣体关键部位，在摔倒时能够有效地避免或减缓对上述部位的伤害。值得一提的是被誉为十大未来材料之一的 D3O 材料，D3O 材料是一种由"智能分子"组成的抗冲击材料。在常态下，这种材料柔软而具有弹性，一旦遇到高速的冲撞或挤压，分子间立刻相互锁定，材料变得坚硬从而消化外力。当外力消失后，材料会回复到它最初的柔性状态。它有着极强的吸收冲击的能力。这种材料同传统的护具相比轻巧许多而且与防护部位贴合感很好。它是能将自由活动与碰撞打击保护结合在一起的一种很理想的材料，此材料还有人称之为"软铠甲"。

5.3 防摔功能服装款式

有研究表明，防摔护具佩戴舒适性欠佳，负担感很重，患者不愿意在日常生活中佩戴，护具临床依从性差，从而使护具防摔效果得不到证实。在国外的研究统计中，在开始阶段 85% 老年人愿意佩戴髋关节护具，但只有 29% 愿意长期佩戴；经验使用率为 14.3%～80%[11]。如此一来，在防摔护具的效果得不到证实之前，任何轻微的不舒适和丑陋，都会成为老年人拒绝防摔护具的原因。

考虑到护具重量、体积量以及价格的问题，防摔服宜设计成分拆组合功能式，这样针对不同保护区域、不同温度、不同防摔级别,以及不同科技功能的需求都可以兼顾考虑。使用者可以随意组合功能,同时也方便拆卸清洗护具。

防摔服装款式设计中,主要针对易发伤害需要保护重点部位入手,最大限度地使躯干活动自如的同时,又要确保功能,兼顾造型的美观是款式设计的重点。在款式设计中,运用形式美法则中的一些原理,使设计功能结构部位"变成"装饰部位,提高消费者接受度。防摔服装上下装兼具不同保护功能。上衣结构设计中,胸腰椎、肘关节是保护重点,裤子结构设计中,髋关节、膝关节是保护重点(图6),盆骨的保护在上下装中都有体现。

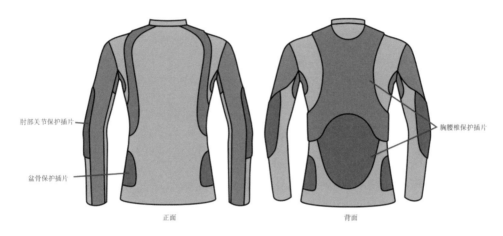

图5 防摔服装上衣设计重点部位

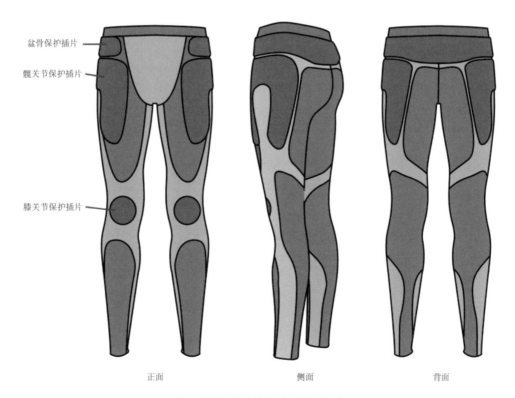

图6 防摔服装裤子设计重点部位

6 结论

老人防摔功能服装的设计充分考虑老年人人体工学原理，了解老年防摔功能服装设计要求，更要对易发摔跌伤害部位加以科学原理分析，通过色彩的把握和创新防护服用材料使用，有针对的确保功能，兼顾造型的美观是款式设计。未来老年防摔功能服装设计发展趋势，一是加强功能研发，防摔功能性是老人防摔服装主要目标，将来可以多功能、实效性加强研发；二是注重审美设计，丑陋突兀与异于常类服装是当下产品不足；三是运用科技手段，结合电子、通讯等定位与预警等多功能高新技术，以附加更多功能模块；四是提高服用效果，舒适性、灵活性与轻便性是未来防摔服装服用功能指标。通过科技的运用，结合创新设计理念，来服务老年人。

参考文献：

［1］董克芳.老年人最常见的三种骨折［J］.老年人，2012（05）：54.

［2］李玉彬，谢利民，徐颖鹏.髋部骨折住院患者266例流行病学调查分析［J］.山西医药杂志，2009-05，38（5）：430-432.

［3］孙培栋.侧方跌倒高度及髋保护器对髋部冲击影响的实验及有限元分析［D］.广州：南方医科大学博士论文，2012.

［4］王向阳，戴力扬.胸腰椎爆裂性骨折的生物力学研究进展［J］.中华骨科杂志，2006，26（7）：487-489.

［5］何剑颖，董谢平.脊柱生物力学的有限元法研究进展［J］.中国组织工程研究，2011，15（26）：4936-4940.

［6］蓝霞，王冬梅，周立义，等.动力性胸腰保护器对中上胸椎保护的有限元研究［J］.中国矫形外科杂志，2011，19（16）：1361-1364.

［7］何剑颖，吴小辉，邓亮，李晨，舒勇，董谢平.腕保护器预防桡骨远端骨折的有限元分析［J］.中华手外科杂志，2015-04，31（2）：142-144.

［8］梁丁.基于MEMS惯性传感器的跌到检测与预警研究［D］.大连：大连理工大学硕士论文，2012.

［9］尹芹.潮科技：B-Shoe智能鞋让老人抛弃拐杖［EB/OL］.中关村在线.
http://smartwear.zol.com.cn/422/4224801.html，2013-12-23.

［10］石一鸣.日本发明老人防摔气袋售价1400美元［EB/OL］.中国新闻网.
http://news.sohu.com/20080925/n259738027.shtml，2008-09-25.

［11］孙培栋，欧阳钧.髋关节保护器的研究现况及展望［J］.中国修复重建外科杂志，2012-01，26（1）.

老年人服装功能性实现方法分析

◆ 张梦莹[1]　李　俊[1,2][①]　（1.东华大学功能防护服装研究中心，上海，200051；2.东华大学现代服装设计与技术教育部重点实验室，上海，200051）

【摘　要】　随着年龄增长，老年人的生理、心理各方面均发生变化，并衍生出对服装的特殊需求，对服装功能性的要求更高。本文首先从生理学、生物力学、人体工学和心理学4个方面分析老年人对服装的特殊需求，然后提出老年服装功能性设计的方法，包括服装材料的选择、服装结构的优化设计方法以及服装的智能化实现等多个角度。针对中老年服装的功能性设计有利于老年人的着装健康和舒适，并能满足其心理需求，提高其生活品质。

【关键词】　中老年服装；功能设计；功能面料；服装结构；智能服装

Analysis on Implementation Methods of Functional Design of the Clothing for Old-aged Group

◆ Zhang Mengying[1]　Li Jun[1,2]　*(1. Protective Clothing Research Center, Donghua University, Shanghai 200051　2. Key Laboratory of Clothing Design & Technology, Ministry of Education, Shanghai 200051)* (E-mail:mirandazmy@hotmail.com, lijun@dhu.edu.cn)

Abstract: *Elderly people have special requirements for clothing as they grow older because their physiological and psychological features changed, and they have higher requirements on the functionality of clothing. In this paper, the old people's special requirements for clothing function, including physiological, biomechanical, ergonomic and psychological requirements are analyzed. And then the functional design methods are put forward, including the selection of materials, the optimal design methods of garment structure and the function implementation of smart garments. The functional design of aged population is conducive to health and comfort, as*

[①] 李俊，男，东华大学，博士，教授，博士生导师。E-mail: lijun@dhu.edu.cn，mirandazmy@hotmail.com

well as the psychology requirements, so as to improve the quality of life.

Keywords: *old people's clothing; functional design; functional textile; garment structure; smart garment*

根据世界卫生组织定义，老年人是年龄在60周岁以上的人群[1]。1999年，我国提前进入老龄化社会，目前我国的老年人口占全球的1/5左右，是世界老年人口最多的国家。2015年以后我国将进入人口老龄化迅速发展时期，到2050年，老年人口总量将超过4亿，高龄老人达9500亿[2]。老年人口日益增多，由于老年人在60岁以后出现生理机能下降、抵抗力下降、体型变化等问题[3]，日常生活中会遇到各种困难，生理和心理等方面的变化对其着装的功能性和舒适性的要求也引起关注，相应的个性化的设计方法也成为独立的研究课题。本文根据老年人群在生理、心理等方面发生的变化，老年人对服装的需求分为生理需求、生物力学需求、人体工学需求以及心理需求四个方面进行分析，根据其特殊的需求，分析服装功能性设计要点，从面料的选择、服装结构设计以及服装的智能化3个方面探讨老年服装功能性的实现方法。

1 老年服装需求分析

由于生活模式变化、生理机能下降、心理状态改变，老年人的生理和心理状况与其他年龄段的群体相比发生显著改变。老年人对服装的穿脱方便性、舒适合体性等要求更高，并且随着生活水平及健康意识提高，中老年人在服装的功能性和防护性方面也有了更高的要求。

1.1 生理需求

老年人体温调节、新陈代谢等生理机能下降，体温容易随外界环境温度的变化而变化，外界环境温度较高时，老年人体温较高；外界环境温度较低时，其体温较低。老年人皮肤老化，水分含量降低，易与服装产生静电，引发心血管疾病[4]。身体反应较为迟钝，触觉功能下降，对疼痛感觉不灵敏，常常被烫伤、擦破、灼伤等。另外，老年人由于心脑血管、消化系统、内分泌系统等生理功能下降，对细菌、疾病的抵抗能力减弱，生理疾病增加，一些老年人常患有高血压、颈/腰椎病和心脏病、冠心病等慢性疾病[5]。

因此老年服装应具有补助人体生理机能的不足、保持身体处于舒适状态、保护人体不被外界环境所伤害以及一定的保健功能。

1.2 生物力学需求

生物力学涉及人体力学特性、运动学和人体运动行为分析[6]。老年人的协调能力和平衡能力降低，导致其容易摔倒，且由于老年人的肌肉退化、柔韧性变差，摔倒后容易发生骨折，严重时更会影响其生命安全。

因此，老年服装应具有摔倒防护功能，在摔倒前进行预判，在摔倒后进行保护，防止老年人摔倒,最大限度地降低伤害。

1.3 人体工学需求

服装人体工学通过服装优化设计,使生活、工作更加舒适,改善或提高人体能力[7]。对于着装人体,其运动灵活性、运动幅度、速度与未着装人体相比,均有不同程度的降低。尤其对于60岁以后的老年人,其肩关节活动范围减小,四肢体力下降,灵敏性显著降低,服装的阻碍作用更加明显,对穿脱方便性要求较高。另外,老年人的体型开始变化,如脊椎弯曲、腹部隆起、腰部变粗等,由于没有通用的号型标准,市场上的服装往往不能满足老年人的特殊体型需求。

因此,老年服装应通过在不同部位配置不同面料、进行合理的结构设计,以改善老年服装的舒适性、灵活性以及方便性。

1.4 心理需求

温勇等学者[1]对5002名老年人进行调查,研究发现有23.4%的人未检查过身体,不知道是否有健康问题。一些老年人不重视自身的健康问题,导致不能及时发现疾病,当发现时可能为时已晚。另外,老年人随着年龄的增长,容易出现一些负面情绪,如焦虑、孤独感等。

服装作为人每天穿着的装备,可以通过微电子技术实现智能化,使其能够实时检测人体血压、体温、情绪等生理指标,安抚老年人情绪,具有"陪伴"功能,达到满足老年人的特殊心理需求的目的。

2 功能性优化方法

针对以上老年人在生理、生物力学、人体工学以及心理方面的需求,老年服装的功能性优化设计方法可以从以下三个方面入手:服装材料选择、服装结构设计以及服装智能化设计。

2.1 服装材料选择

织物的性能受多种因素影响,如纤维或纱线的种类、织物结构及后整理方法等。老年服装材料要求柔软舒适、吸湿透气、有一定的弹性及伸缩性能、质轻、安全性好、难燃、对皮肤的刺激小、不产生静电等。随着新型纤维的不断出现与应用,多功能织物越来越多,在老年服装材料的选择方面自由度更大。

(1)调温织物

由于老年人对外界环境温度变化比较敏感,因此服装应具有一定的调温功能。冬季服装应注意保暖,选择质轻保暖的材料,保证中老年人体温恒定且不因服装材料过于厚重而感到压迫;夏季服装应具有散热、透气的功能,防止体温上升。天然纤维中,动物的绒毛保温效果较好,如羊毛、鸭绒等;传统的合成纤维中,丙纶、腈纶的保暖性较好;在新型服装材料方面,还可选择新型发热保暖面料,通过将电能、太阳能使纤维发热,达到保暖的效果[8];另外,在智能服装材料方面,可选择温度变化范围接近中老年人体皮肤温度变化范围的相变材料,制作自动调温服装,为人体提供舒适的衣下微气候环境。

(2)保健类织物

保健类织物对于预防及调养老年人慢性疾病有一定的效果。天然纤维中,蚕丝、罗布麻、竹纤维中的有效成分作用于人体皮肤,对皮肤病、慢性病均有一定的缓解作用;一些新型的功能性纤维,如海丝纤维、远红外纤维、碳纤维、中草药纤维等也具有一定的保健、抗菌、护肤的作用。尤其对于远红外纤维,它能够促进人体血液循环和新陈代谢,增强免疫力,同时具有抗菌、防臭、保暖等作用,十分适用于制作中老年保健内衣[9]。

(3)防护型织物

在抗静电方面,化纤的电阻较大,易产生静电,因此选用天然纤维制成的织物较为合适,如棉、麻、丝等;一些新型服装材料也具有防静电功能,如纺织品中含有少量的镀银纤维(1%左右),即可兼具防静电与抗菌的功效[10]。另外,镀银纤维还具有除臭的功效,能够消除人体汗液、皮脂分解所产生的臭味,减少皮肤病;能够反射太阳和人体发射的远红外线,调节体温,达到冬暖夏凉的效果,并且无毒无害,因此镀银纤维用于老年人尤其是高龄人群的服装材料效果良好[11]。

在摔倒、碰撞防护方面,应用最广的材料是 D30 凝胶,它由英国研究人员研发制得,并成功应用于运动防护领域。D30 凝胶是一种胶状物质,在受到外力撞击后迅速凝固,有效吸收动能,外力消失后,D30 凝胶恢复柔软状态[12]。因此,将 D30 凝胶应用于老年服装的肘部、膝部、臀部等摔倒后易受伤的部位,能够起到一定的防护作用,并且由于其柔软性良好,制作的服装不会阻碍人体活动,运动舒适性良好。

2.2 服装结构设计

(1)服装开口设计

在服装温度调节方面,可通过服装的结构设计,尤其是服装开口的设计,优化服装的热湿舒适性。张向辉等[13]的研究表明,服装腋下部位的开口能够促进衣下微环境与外界环境之间的热交换,从而促进人体热湿生理调节,着装舒适性较好。因此,对于老年人夏季服装,可以在服装的背部、腋下等部位增加开口设计,增加衣下与外部环境的通风,从而提高服装的热湿舒适性。对于冬季服装,可采用多层配置的方法,保证各层服装之间存在一定的空气层,最大限度地增加服装的保暖性。韩笑[14]对冬季保暖服的开口进行优化设计,证明无运动时,服装腋下开口闭合状态的舒适性更好;运动时,服装腋下开口打开状态的舒适性更好。因此,中老年冬季服装在保证原有保暖性不受影响的基础上,可以适当增加开口(如腋下开口),使运动后多余的热量及汗液及时排出。

(2)服装紧固件设计

在服装穿脱方便性方面,根据华南理工大学陈庚笙[15]的调查研究,老年人对服装的穿脱方便性要求较高,在服装设计方面,可采用操作容易的大扣、拉链、尼龙扣等,避免套头设计、后拉链和侧拉链设计。由于体型随年龄的变化,老年服装在领部、袖子、臀部的设计也与其他年龄阶段人群不同,这些细节部位的设计对于服装穿脱方便性起到至关重要的作用。

（3）服装规格设计

在服装合体性方面,根据齐齐哈尔大学李晖的调研结果,老年人最喜欢穿宽松的服装,其次是合体的服装[16]。确定服装各部位的放松量时,可使用三维人体扫描仪,得到立体的人体形态,通过计算机精确测量人体尺寸,进行服装结构设计,制成成衣后试穿,之后将着装人体再次进行三维扫描,研究服装的合体度,对活动受限部位进行调整,最终制成合体度、灵活性更好的服装[7]。

2.3 服装智能化设计

智能服装是在服装中嵌入微型传感器、芯片、电源等装置,能够检测人体或外部环境的信息,由服装中的中央处理器对检测到的信息进行处理,并将其输出至服装中的执行装置或通过一定的通讯技术将信息传递至终端接收器。智能服装具有感知、反馈和反应的功能[17]。本文将从老年人服装的健康监护功能、防走失智能、防摔倒智能三个方面进行介绍。

（1）健康监护功能

一些学者研发出能够实时检测并反馈人体健康信息的服装。Gi-Soo 等[18]研制出一种数字智能纱线,将其应用于智能服装可以检测人体心率,甚至能够取代传感器。Hakyung 等[19]研究出具有心电监测功能的服装,服装包括两层材料,内层嵌入传感器及其他装置,使用弹性较好的面料,使传感器与服装充分贴合,更加准确地测量人体的生理信号;外层与普通服装相同,保证美观性;主板位于后领口处,温度传感器位于袖窿,使用导电织物进行信号传输。

健康监护服装一般是将医学传感器嵌入服装中,实时监测老年人的心率、血压、体温、呼吸等重要生理指标,监测到的生理指标通过一定的通讯手段发送给用户及医疗保健站,根据监测到的生理信号判断用户身体是否处于健康状态。

（2）防走失智能

北京服装学院的刘力源等人[20]将 GPS 定位技术与服装相结合,研制出防丢失服装,该服装能够实时监测着装者所处位置,实现防止老年人走失的功能。朱建新等[21]基于 Zigbee 技术研发出老年人防走失装置,能够对老年人的相对位置进行动态监测,当老年人脱离一定的区域即进行报警提示,主动预防走失,并且通过红外传感器感知该装置是否处于正常佩戴状态,避免装置脱离人体后失去作用。

防走失智能服装通过在服装中配置定位传感器、红外传感器及其他装置,对老年人进行实时定位,在着装者走失时,将位置信息发送给亲属或警察局,以保证其人身安全。

（3）防摔倒智能

韩国首尔大学的研究人员研究了一款佩戴于人体胸部的摔倒探测系统,它由加速度传感器、陀螺仪、倾角传感器组成,如果人体的胸部角度变化超过 70°,并且加速度变化非常剧烈则认为人体摔倒[22]。美国的相关人员研制出一款多位置佩戴的监测系统,将三轴加速度传感器和陀螺仪分别佩戴于胸部和腿部,通过阈值算法对摔倒的检测成功率达到了92%[23]。哈尔滨工业大学的闫俊泽[24]利用三轴加速度传感器研发了一套老年人摔倒监测系统,该系统可以准确检测出老人摔倒事件,并及时发出蜂鸣警报。

防摔倒服装一般是将加速度传感器、角速度传感器、倾角传感器嵌入服装的胸部、腰部以及大腿部位,根据摔倒前的动作进行跌到预判,提醒用户远离危险环境。一些防摔倒服装中还配有保护装置(如保护气囊、D30凝胶)和求救装置,在摔倒后加以保护并通过嵌入服装中的报警装置及时通知医院和家人进行救援。

此外,由于在服装智能化设计时通常会采用较多的电子元件,这些部件在刚柔性方面会区别于传统的服装材料,因此,对于腹部突出、胸部下垂等体型特征,在满足功能性的基础上,应尽量避免在这些部位添加过多的部件,可以将设计转向领部。如在不影响智能服装功效的前提下,可将电池、中央处理器等电子元件置于领部,避免过多的元件堆积在腹部造成活动不便及不舒适。

3　总结

随着我国人口老龄化日益严重,中老年服装市场具有广阔的发展空间,由于中老年人特殊的生理和心理特征,其对服装的舒适性和功能性需求较大。本文首先根据中老年生理、心理等方面的变化分析其功能性着装需求,然后从服装材料的选择、服装结构优化设计方法以及智能服装的设计三个方面对中老年服装的功能性实现方法进行分析和探讨。目前对于老年人服装消费态度及市场的研究较多,并日臻成熟,而有关服装功能性方面的研究较少,中老年功能服装的整体及细节部位设计尚需进行大量研究。合理的功能性设计不仅能提高中老年人的着装舒适性,而且具有一定的防护和保健作用,满足老年人的身心需求,提高其生命品质。另外,目前中老年功能服装在款式和外观方面的研究较为缺乏,中老年功能服装的美观度不够,在实现各种服装功能性的同时,也应提高其美观性,使得功能性、实用性、美观性综合发展,最终实现中老年服装的"安全、健康、时尚"。

参考文献:

[1] 温勇,宗占红,舒星宇,周建芳,孙晓明,汝小美.中老年人的健康状况、健康服务的需求与提供——依据中西部5省12县调查数据的分析[J].人口研究,2014,05:72-86.

[2] 穆光宗,张团.我国人口老龄化的发展趋势及其战略应对[J].华中师范大学学报(人文社会科学版),2011,05:29-36.

[3] Civitci S S. An Ergonomic Garment Design for Elderly Turkish Men.[J]. Applied Ergonomics, 2004, 35.

[4] 王洁.老年人谨防静电伤身[J].解放军健康,2005,(2):27-27.

[5] 吕雅男.城市老年人健康状况及其影响因素研究[D].中南大学硕士学位论文,2012.

［6］ 汪世奎,谢红,曹蕊超,李露,叶航.基于运动生物力学防护服装的研究［J］.纺织导报,2014,01:91-93.

［7］ 刘可.服装人体工学的应用与发展［J］.纺织科技进展,2013,01:81-85.

［8］ 王敏,李俊.发热保暖服装材料的开发现状及发展趋势［J］.产业用纺织品,2009,04:6-9.

［9］ 白亚琴,孟家光.新型医疗保健织物的开发［J］.针织工业,2006,06:49-54.

［10］ 段守江.防静电纺织品的现状与发展［J］.非织造布,2013,06:72-75.

［11］ 叶卉,李东平,夏芝林.镀银纤维的研发进展及应用［J］.纺织导报,2006,06:54-56.

［12］ D30凝胶［J］.现代纺织技术,2010,03:60.

［13］ 张向辉,李俊,王云仪.服装开口部位对着装热舒适性的影响［J］.东华大学学报(自然科学版),2012,02:190-195.

［14］ 韩笑.冬季防寒服装的开口结构设计研究［D］.北京服装学院硕士学位论文,2010.

［15］ 陈庚笙.老年人功能性服装的应用分析［D］.华南理工大学硕士学位论文,2014.

［16］ 李晖.老年服装的人性化设计研究［D］.齐齐哈尔大学硕士学位论文,2012.

［17］ 田苗,李俊.智能服装的设计模式与发展趋势［J］.纺织学报,2014,02:109-115.

［18］ Chung G S, Kim H C. Smart Clothes Are New Interactive Devices［J］. Communications in Computer & Information Science, 2011:18-21.

［19］ Cho H, Lee J. A Development of Design Prototype of Smart Healthcare Clothing for Silver Generation Based on Bio-medical Sensor Technology［M］// Human-Computer Interaction. Interaction Platforms and Techniques. Springer Berlin Heidelberg, 2007.

［20］ 刘力源,范秀娟.防走失服装中GPS定位数据的采集与处理［J］.北京服装学院学报(自然科学版),2015,01:20-25.

［21］ 朱建新,高蕾娜,田杰,张新访.基于Zigbee技术的老年人防走失装置［J］.计算机工程与科学,2009,05:144-146.

［22］ Purwar A, Jeong D U, Chung W Y. Activity Monitoring from Real-time Triaxial Accelerometer Data Using Sensor Network［C］// Control, Automation and Systems, 2007. ICCAS '07. International Conference on. IEEE, 2007:2402-2406.

［23］ Qiang Li, Stankovic, J.A, Hanson, M.A, et al. Accurate, Fast Fall Detection Using Gyroscopes and Accelerometer-Derived Posture Information［C］// Wearable and Implantable Body Sensor Networks, 2009. BSN 2009. Sixth International Workshop on. IEEE, 2009:138-143.

［24］ 闫俊泽.基于三轴加速度传感器的老年人跌倒监测系统的开发［D］.哈尔滨工业大学硕士学位论文,2012.

Kinetic Garment Construction for variable, aging bodies

◆ Rickard Lindqvist[①]　(Swedish School of Textiles, University of Borås, Swedish)

Abstract: Fashion designers are presented with a range of methods and concepts for pattern cutting, however, the main body of these methods, both traditional and contemporary, is predominately based on a theoretical approximation of the body that is derived from horizontal and vertical measurements of a an ideal static body in an upright position: 'the tailoring matrix'. As a consequence, there is a lack of interactive and dynamic qualities in methods connected to this paradigm of garment construction, from both expressional and functional perspectives. In order to meet challenges of comfort and functions while constructing garments for various aging, variable body types and postures with different demands of shape and functionality a need for an alternative theory for understanding the body arises.

'Kinetic Garment Construction' (Lindqvist 2015) proposes an alternative paradigm for pattern cutting that includes a new theoretical approximation of the body as well as a more dynamic method for garment construction that, unlike the prevalent theory and its related methods, takes as its point of origin the interaction between the anisotropic fabric and the biomechanical structure of the body.

This paper investigates possibilities of congruence between the shear forces in the human skin and the anisotropic qualities of woven fabric through visual analysis of two draping experiments carried out on live models. Thus connecting the balance directions and key biomechanical points of the 'Kinetic Garment Construction' theory to the notion of Langer's lines. Langer's lines, utilised in surgery, denote the skins anisotropic qualities and is hence also a notation of the movements of the body. Through a visual comparison of the grain directions of the fabric in draping experiments and notations of Langer's lines it is argued that 'Kinetic Garment Construction' may be a suitable framework for designing garment for mature and aging bodies.

---

① Rickard Lindqvist, Swedish School of Textiles, University of Borås, PHD.　E-mail: rickard. lindqvist @ hb.se

中高龄人体的动力学服装结构设计研究

◆ Rickard Lindqvist　（瑞典布罗斯大学，瑞典）

【摘　要】　时装设计师采用不同的裁剪方法呈现他们的设计理念。然而,这些传统和现代的裁剪方法,主要是基于人体的理论近似值。这些数据来自于理想的静态人体在直立的状态下的水平和垂直测量这样得到的结果,从表达和功能角度看,与动态的人体缺乏链接。为了满足舒适性和功能性的挑战,构建可以适应老龄化身体,多变的身体类型和姿势对于外形和功能的需求,一种可替代的理论产生了。

"动力学服装结构"（*Lindqvist 2015*）提出了一个可选的纸样裁剪范例。它包括一个新的理念近似人体以及服装的的动态构成方法。该方法不同于当前普遍运用的理论,它依据织物的各种异性和人体生物力学结构之间的相互作用的相互作用确立关键点。

本文在两个模型上进行立体剪裁实验,运用视觉分析,研究人体皮肤与织物间剪切力和织物各向异性之间一致性的可能性。因此连接方向平衡和"动力学服装结构"理论的生物学关键点形成朗格氏线。朗格氏线在外科手术中表示皮肤的各向异性,因此也可作身体的运动符号。通过立体裁剪实验和朗格氏线记号法所形成的织物纹理,进行视觉比较,讨论认为"动力学服装结构"可以用于为老龄化人体进行服装的结构设计。

（中文翻译：范振毅）

1 Introduction

The mathematical systems developed during the 19th century (Waugh, 1964) in attempts to turn pattern cutting into a scientific practice have had a huge impact on the development of the field of garment construction. Theories of construction based on horizontal and vertical measurements of the body, the tailoring matrix (see fig.1) and the methods of drafting patterns based on mathematical instructions has proven to be successful and operational ways of representing the static body. The practice of drafting from a tailoring matrix of straight lines in combination with given measurements makes it relatively easy to communicate the procedures of drafting in literature, utilising scaled sketches and verbal explanations.

As Almond (2010:16) notes, understanding the fundamentals of cutting is an essential part of designing garments, both from a creative and commercial point of view. Working with patterns requires "*a sound knowledge of the human form*" (Hulme, 1945:23) and an ability to envision the body while shaping the pattern pieces. However, Fischer (2008:25) notes that, in educational settings, "*pattern cutting can at first seem difficult and intimidating but with a basic understanding of the rules to be followed – and broken! – the aspiring designer will soon learn interesting, challenging and creative approaches to pattern cutting.*" May this difficulty of understanding and applying pattern cutting rules relate to the foundations of the prevalent theoretical framework? Does the tailoring matrix promote such abilities as envisioning the body while shaping pattern pieces? It can be assumed that the tailoring matrix is a valid way of representing a static body correctly, but whether or not this theoretical framework is a preferable way of describing and understanding the interaction between soft fabric and the living, variable body is questionable.

As clothing is made for living bodies, problems of inactivity and rigidity – as may occur when exhibiting fashion in museum contexts as a result of bodies being replaced by mannequins (Debo, 2003:9) – is also a problem present in the design studio. Broby–Johansen (1953:2) points out that to thoroughly know anything about clothing, we must first discover what the clothes conceal – the body – and then observe them in use. This is clear for the paradigm of ancient drapes, as only the rectangular piece constituting, for example, a sari is not enough to understand the sari: the body cannot be left out. However, the body is just as essential for understanding any kind of garment: the garment in itself is not enough, and neither are the garment and its pattern. Without the body, the garment can neither be entirely understood, nor fulfilled, or as Yamamoto notes: "*Clothing is, ultimately, made to be worn. It is complete only at the instant it is donned by a living human being*" (2010:68). Dress historian Dorothy Burnham further notes, "The body with its need for movement is a variable constant in the development of clothing" (1997:2). As valid as this point is from a historical perspective for developments in dress, it is equally essential for the creation of dress and for future developments in the field of pattern cutting.

For practical reasons, fashion designers do not typically work directly on a human body while draping or constructing a garment. Hence, theoretical approximations of the body has been developed – theories that do not represent the body exactly, yet close enough to be useful for cutting and draping. These approximated theories aid in the construction and design process and are also used to make predictions of the result easier.

The static "tailoring matrix", with an approximation of the body materialised through block patterns, and the usage of them as the foundation for pattern cutting inclines pattern cutters to work from the pattern towards to the body – from the outside inward – rather than from the inside out, i.e., starting from the core of dressmaking, which is the living, dynamic body. These dynamics of the body are easily neglected while working within the tailoring matrix paradigm, just as they are when the (block) pattern is viewed as a tool for designing. To move from this static approach towards one that focuses on the body, a need for a more dynamic theory of the body as the base for pattern cutting arises. Such a model may be based on how the moving body interacts with fabric while being dressed in a manner similar to ancient ways of dressing. This calls for a new approximation of the body that is derived from qualitative measurements of the moving body instead of quantitative measurements of a static body, in order to allow for previously unconsidered aesthetic values, both functional and expressional.

2 Materials and Methods

Instead of making alterations of block patterns assumed to fit a 'standard' body i.e. adjust methods within the tailoring matrix to measurements of elderly people (cf. Zhaohui 2014) this study concerns questions directly related to the interaction between the variable body and the fabric. How does fabric react to the body? Where does it drape? How does it fall? What happens to the fabric when the body moves?

"Kinetic Garment Construction" (Lindqvist, 2015) is system of qualitative measurements – i.e., one that studies how fabric drapes on a living body – that are created to explain and achieve what cannot be accessed through quantitative measurements, i.e., numeric measurements that can easily be compared and used for various calculations; therefore, it is a radically different approximation of the body for garment construction. Thus, it is also a theoretical foundation that emphasizes the expression and biomechanical functions of the body rather than the pattern. It proposes an alternative way of understanding the relationship between the body and the fabric in order to allow for not previously seen aesthetic developments in dress, both functional and expressional.

Considering the biomechanical functions of the body and the way that the fabric interacted with the living body in ancient ways of dressing, both full cloth wrap dressing and rectangular-cut garments, combined with reverse engineering and the design recovery of Geneviève Sevin–Doering's work (Sevin–Doering 2004, 2007)), have presented an alternative relationship between the body and dress. As Chikofsky (1990) explains, reverse engineering is a process of analysing a system to create representations of the system at a higher level of abstraction, so as to bring about new development. Thus, design recovery may be understood as a subset to reverse engineering in which recovery means: reproducing all of the information that is required for a person to fully understand what a system (or a design programme) does, how it does it, why it does it, and so forth – as explained by Biggerstaff (1989). By quoting principal parts of Sevin–Doering's work, various representative types of dress were recreated on a live body by using the same working method that she had instructed me in. In this manner, a hypothesis for a kinetic or biomechanical theory for garment construction took form. This hypothesis was then developed and refined through concrete experimentation while the human kinetic construction method was developed and defined.

2.1 Kinetic Garment Construction theory

The human kinetic construction theory builds on an alternative qualitative and biomechanical approximation of the body in contrast to the traditional quantitative theory, which is based on the horizontal and vertical measurement grid. The actual living body is the point of departure for the theory's development and its applied construction method, which is an essential difference as compared to theories that are derived from outside measurements of an upright body.

The directions suggest the places on the body where the fabric "wants" to fall and where it may be draped so as to neither fall off the body nor restrain its movements. These lines are not suggestions for where to place seams, nor are they guides for where to measure the body, but, rather, proposals for how the fabric may be draped around the body (see Fig. 2 and 3). The way in which the fabric falls and the places where it "breaks" or folds also highlights certain points towards which the cuts in a piece of fabric are supposed to be directed in order to construct garments that move along with the body and create shapes that relate to the body. As distinct from conventional draping, in this case, the 'break–lines' are not just 'beautiful' lines that exist because of how a fabric hangs when it is draped based on the traditional grid. Instead, these marginalised, "beautiful" break–lines are, in themselves, part of the fundamental structure – a grid – of a more dynamic approximation of the body (cf. Lindqvist, 2015).

2.2 Shear forces of human skin and anisotropic material qualities

The prevalent utilisation of fabric grain direction in pattern cutting is arguably strongly connected to the paradigm of the tailoring matrix. Due to the construction of weft and warp, few woven fabrics are isotropic, i.e., "the same in all directions", but instead – to a higher or lower degree – are anisotropic, i.e., "different in different directions" (Gordon 1981:251). In garment construction it is typical to "cut on the square" – i.e., to place the grain line vertically in relation to the body to minimise stretch and distortion (cf. Efrat 1982:55) – or alternatively to cut "on the bias", as introduced by Vionnet in the 1920's (cf. Kirke 1998; Bunka Fashion College 2002), in order to take advantage of the elastic qualities that appear and to the different drape or fall that emerges when cutting the fabric on the bias.

Langer's lines (Langer, 1861) utilised in surgery denote the skins anisotropic qualities, along the lines the skin is less flexible and across the lines it has more flexible qualities, thus Langer's line may also be seen as a notation of the movement scheme of the body. The directions of Langer's lines shifts across the body (see fig X.) as does the amount of flexibility in the skin. The anisotropic qualities of woven fabrics, i.e. the inelasticity on the straight grain and flexibility on the bias, inhabits an aesthetic potential if draping and constructing garments in a manner where the fabric relates to the body in congruence with Langer's lines instead of cutting on the square or on the true bias.

2.3 Visual analysis of draping experiments

The general grain directions and seam lines in contemporary fashion are connected to various archetypal garments, and large parts of the fashion design industry today refer to and elaborate on these archetypes. This fundamental difference in the positioning of seams and utilisation of grain direction within the kinetic garment construction method – beginning from the structure of the body and the material qualities of the fabric – presents an alternative both from an expressional and functional perspective. Thus, in addition to presenting an alternative theoretical approximation of the body and new methods of constructing garments, the kinetic garment construction theory questions these conventional utilisations of the grain direction of fabric. While wrapping the fabric around the body, the direction of the grain varies (instead of running straight vertically or being on a 45° bias) over the garment, creating diverse expressions and functions in different ways across the garment. Arguably, these asymmetrical distortions, addressed by Gordon as *"weird and almost certainly unwelcome"* instead have a great potential to create enhanced functions and new diverse expressions if applied in a manner congruent with the biomechanical and kinetic functions of the body. The visual analysis of draping experiments bellow explores and suggests how this application may be utilised. This notion of grain direction is here connected and related to the notion of Langer's lines (Langer 1861; Li 2006:113) and the biomechanical properties of the skin (cf. Wilkes et al. 1973) that are utilised in plastic surgery and which denote the "grain direction" of the human anisotropic skin, as well as the influence of body posture and gravitational forces on the skin (Nizet et al. 2001).

Visual analysis of two experiments, a pair of trousers and a simple chemise has been carried out.

The analysis presents the directions of the straight grain, the cross grain and the bias of the fabric on top of Langer's lines and the directions from the .Kinetic Garment Theory' (see fig 7),.

The first experiment, a pair of trousers (see fig 8), addresses the fundamental directions for the lower part of the body as well as the derived points at the back of the seat. By addressing these points, the trousers are shaped over the seat and the hips. The direction of the grain is straight only at the top part of the centre back, keeping the trousers in position. A lining with the grain straight at the front prevents the waistline from stretching. Everywhere else but at the top–centre back, the grain is on the bias to various degrees, which gives the fabric different characteristics compared to conventional trousers cut on a straight grain; for example, it adds stretch qualities at the inside of the legs and below the seat at the back.

This second experiment, a chemise (see fig. 9), connects only to the fundamental points and directions on the upper part of the body. The fabric is draped according to the shape of the shoulders and the way of constructing three-dimensionality and the scye and sleeve seams are shaped accordingly, still allowing full movement for the arms. This results in a generous drape at the front and back of the sleeves. The direction of the grain shifts gradually, from straight-horizontal at the back to a bias at the centre front. Due to the anisotropic qualities of non-stretch woven fabric – i.e., fabric that is generally non-elastic along the grain and stretchy to different degrees on the bias – it is possible on most male bodies and many female bodies to drape the fabric in this manner without creases appearing at the chest or bust.

3 Results and Discussion

The analysis shows a clear correlation between the elasticity of the skin across Langer's lines and the elasticity of the woven fabric along the bias in the experiments. As the movement scheme of the body varies across the body so does the directions of Langer's lines. In a similar manner the aging process affects different parts of the body in different ways and an alternative framework may be useful for addressing the challenges of constructing garments for aging bodies.

As the 'Kinetic Construction Theory' is derived from studies of living bodies in interaction with fabric, these analysis implies it may be a suitable framework for constructing flexible garments for aging bodies, and as an alternative to the tailoring matrix, it may fuel further creative works and unexplored paths of development and research.

The "Kinetic Construction Theory" may contribute to a change of predisposed beliefs regarding grain utilisations of woven fabrics with anisotropic qualities (see fig. 10) it may also change the notions of fit that are derived from the tailoring matrix, standards for the positioning of seams, etc. More generally, today's industrial manufacturing techniques are often closely correlated with the theoretical framework of the tailoring matrix; for example, the vertical straight seams along the grain are utilised as they are easy to stitch due to the fact that the fabric does not stretch, there are special machines for setting tailored sleeves, etc. The introduction of alternative theories may consequently also lead to new manufacturing and assembling methods where traditional ones fall short.

4 Conclusion

This study explores and defines connections between shear forces in the skin, bodily movements and anisotropic fabric qualities. It proposes how the utilisation of fabric grain may shift across the garment in congruence with the movement of the body and thus opens up for new unseen expressional and functional values in garment construction. As such the 'Kinetic Construction Theory' is a general theory and can form, just as the traditio¬nal tailoring matrix does, the foundation for any kind of dress, standardized or customized to a aging bodies. Like any other system, this framework is assumed to hold possibilities for creating any kind of shapes for garments. Further studies are suggested of how the notion Langer's lines can be used to understand the aging of the human body from a shape and movement perspective and thus also present design examples connected to such studies.

Text for figures

Fig. 1

The prevalent theoretical model, the 'Tailoring Matrix' displayed on a body, in a flat construction and on a dress–stand.

Fig. 2

Fundamental directions and points for the kinetic garment construction theory, visualised on the human body. The blue points indicate starting points, and green points indicate fundamental (structural) points.

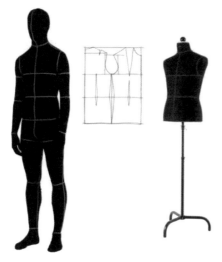

Fig. 1
The prevalent theoretical model, the 'Tailoring Matrix' displayed on a body, in a flat construction and on a dress-stand.

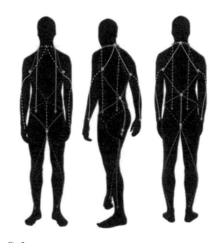

Fig. 2
Fundamental directions and points for the kinetic garment construction theory, visualised on the human body. The blue points indicate starting points, and green points indicate fundamental (structural) points.

Fig. 3

Flat visualisation of directions and fundamental points for the upper part of the body from the kinetic garment construction theory, based on experiments with living bodies in interaction with fabric.

Fig. 4 (top) and Fig. 5 (bottom)

Close up of skin at the upper part of the chest (top) and under the arm pit (bottom) of a hundred years old woman. Langer's line are clearly visible. Photo: Anastasia Pottinger.

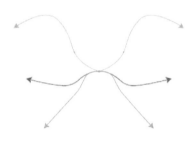

Fig. 3
Flat visualisation of directions and fundamental points for the upper part of the body from the kinetic garment construction theory, based on experiments with living bodies in interaction with fabric.

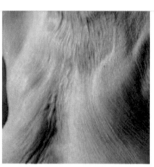

Fig. 4 (top) and Fig. 5 (bottom)
Close up of skin at the upper part of the chest (top) and under the arm pit (bottom) of a hundred years old woman. Langer's line are clearly visible. Photo: Anastasia Pottinger

Fig. 6

Langer's lines denoting the "grain direction" of the skin marked with thin red lines. Across the lines the skin is more elastic than along the lines.

Fig. 7

Langer's lines denoting the "grain direction" of the skin marked with thin red lines. Along these lines the skin is less elastic than across the lines. The fundamental directions and points of the Kinetic Garment Construction theory (Lindqvist, 2015) are marked on top of the red lines.

Fig. 8

Trousers prototype constructed after the Kinetic Garment Construction principals.

Langer's lines are marked with thin red lines on the body. On the trousers the warp and weft are marked with dotted green lines and the bias (with elastic qualities) is marked with solid black lines to highlight the congruence with the elastic qualities of the skin.

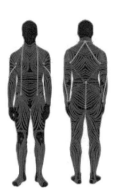
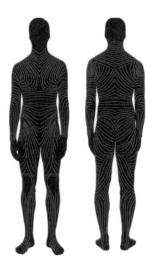
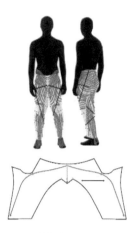

Fig. 7
Langer's lines denoting the "grain direction" of the skin marked with thin red lines. Along these lines the skin is less elastic than across the lines. The fundamental directions and points of the Kinetic Garment Construction theory (Lindqvist 2015) are marked on top of the red lines.

Fig. 6
Langer's lines denoting the "grain direction" of the skin marked with thin red lines. Across the lines the skin is more elastic than along the lines.

Fig. 8
Trousers prototype constructed after the Kinetic Garment Construction principals.
Langer's lines are marked with thin red lines on the body. On the trousers the warp and weft are marked with dotted green lines and the bias (with elastic qualities) is marked with solid black lines to highlight the congruence with the elastic qualities of the skin.

Fig. 9

Top garment prototype in three stages constructed after the Kinetic Garment Construction principals. Langer's lines are marked with thin red lines on parts of the body. On the garment the warp and weft are marked with dotted green lines and the bias (with elastic qualities) is marked with solid black lines to highlight the congruence with the elastic qualities of the skin.

Fig. 10

The blue lines suggest position of seams and matrices suggests grain directions. The green direction indicate the grain direction and the yellow the bias.

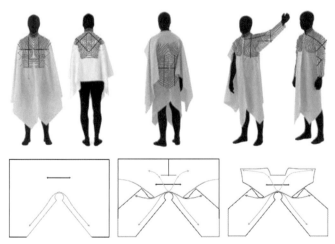
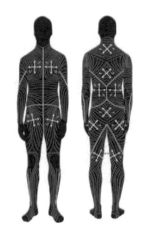

Fig. 9
Top garment prototype in three stages constructed after the Kinetic Garment Construction principals.
Langer's lines are marked with thin red lines on parts of the body. On the garment the warp and weft are marked with dotted green lines and the bias (with elastic qualities) is marked with solid black lines to highlight the congruence with the elastic qualities of the skin.

Fig. 10
The blue lines suggest position of seams and matrices suggests grain directions. The green direction indicate the grain direction and the yellow the bias.

References

[1] Almond, K. Insufficient allure: the luxurious art and cost of creative pattern cutting, *International Journal of Fashion Design, Technology and Education, 2010,* 3:1, 15–24.

[2] Biggerstaff, T. Design recovery for maintenance and reuse. *Computer*, 1989, 22(7):36–49.

[3] Broby-Johansen, R. *Kropp och kläder*. Stockholm: Rabén & Sjögren, 1953.

[4] Burnham, D. *Cut my cote*. Ontario: Royal Ontario Museum, 1973/1997.

[5] Bunka Fashion College *Vionnet*. Tokyo: Bunka Fashion College, 2002.

[6] Chikofsky, E. J; Cross, J. H. *Reverse engineering and design recovery: A taxonomy*. IEEE Software, 1990, 7:13–17.

[7] Debo, K. *Patronen. Patterns*. Gent: Ludion, 2003.

[8] Gordon, J.E. *Structures or why things don't fall down. Boston:* Da Capo Press, 1981.

[9] Hulme, W.H. *The theory of garment—pattern making – A text book for clothing designers, teachers of clothing technology, and senior students. London:* The National Trade Press, 1944.

[10] Efrat, S. The Development of a Method for Generating Patterns for Garments That Conform to the Shape of the Human Body. Ph.D. thesis. Leicester: Leicester Polytechnic, 1982.

[11] Fischer, A. *Basics fashion design: Construction*. London: Ava Publishing, 2008.

[12] Kirke. M. Madeleine Vionnet. San Franciscio: Cronicle Books, 1998.

[13] Langer, 'On the anatomy and physiology of the skin', The Imperial Academy of Science, Vienna. Reprinted in: *British Journal of Plastic Surgery,* 1978, 17(31): 93–106.

[14] Li, Y; Dai, X-Q *Biomechanical engineering of textiles and clothing. Cambridge:* Woodhead, 2006.

[15] Lindqvist, R. *Kinetic Garment Construction — Remarks on the Foundations of Pattern Cutting*. Ph.D. Thesis, Borås: University of Borås, 2015.

[16] Nizet, J L; Piérard-Franchimont, C; Piérard, G E. Influence of Body Posture and Gravitational Forces on Shear Wave Propagation in the Skin. *Dermatology* 2001;202:177–80.

[17] Sevin-Doering G. *Itineraire – Du costume de theater à la coupe en un seul morceau. Colombes:* Les editions du jongleur, 2007.

[18] Sevin-Doering G. *Un vetement autre.* (Updated 14 Aug 2004) Available at: http://sevindoering.free.fr/ Accessed November 30, 2012.

[19] Waugh, N. The Cut of Men's Clothes 1600–1900, New York: Taylor & Francis Group, 1964.

[20] Wilkes, G., Brown, I. & Wildnauer, R. 'The biomechanical properties of skin', *CRC*

Critical Reviews in Bioengineering, 1973, 1(4): 453–495.

[21] Yamamoto, Y. My dear bomb. Gent: Ludion, 2010.

[22] Zhaohui, W; Yingli, L. *Analysis on body feature and pattern construction for aged women.* Paper presented at the Research on Fashoin of mature people conference, Shanghai, 2014, 10.

基于3D试衣系统的中高龄服饰设计

◆ 贾镇瑜[1][①]　吴琰[1]　潘琦明[2]　（1. 上海工艺美术职业学院，上海，201808；　2. 上海派吉姆数码科技有限公司，上海，200051）

【摘　要】　中高龄服饰的消费群体体型较中青年人群有明显的差异化，老龄人体型随生理机能的衰落，而产生变化。在设计过程中要考虑到着装者的舒适度及健康状态、生活习惯及生理特征。中高龄人相对于一般服装消费者而言，属于特体型的人群，在服装结构设计上更需进行注意。而3D试衣系统则是借助计算机3D技术对人体进行着装效果的数码仿真部分，可以使消费者直观地看到未经直接生产的服装设计理念在生产缝制后呈现出来的穿着效果。通过对应的设计案例分析说明了在中高龄服装设计中应用3D试衣系统的可能与必要性。利用错视设计的着装修正设计，再现了中高龄特体人群着装在结构、色彩图案、面料特性方面的仿真效果。

【关键词】　中高龄服饰设计；3D试衣系统；错视设计；仿真效果

Middle and old-aged people's fashion design based garment 3D fitting system

◆ Jia Zhenyu[1]　Wu Yan[1]　Pan Qiming[2]　*(1. Shanghai Art & Design Academy　2. PGM INT'L Group LTD)*

Abstract: The body of the middle-aged and elderly population is significantly different from that of the middle and young people, and the old human body with the physiological function decline, and their body type had changes. In the design process we should taking account of the wearer comfort and health status, living habits and physiological characteristics. Respect to general clothing consumers, older people which belongs to the size of special body population, we should take more attention of the clothing structure design. 3D fitting system is with the aid of computer 3D technology on the human body for the dressing effect of digital simulation, so

---

① 贾镇瑜，女，副教授，硕士学位，上海工艺美术职业学院。E-mail：jzy_nnn@163.com

that consumers can intuitively see without the direct production of clothing design in sewing production presented the wearing effect. Through the analysis of design case it is easy to know the corresponding possibility and necessity of the application of 3D fitting system in the elderly consumers' garment design. Using visual illusion design to corrected dressing design, reproduction the fashion design by the elderly population in the structure, color pattern, fabric characteristics of the simulation results.

Keywords: *Middle–aged and old Clothing design; 3D clothing fitting system; The Visual Illusion design; Digital simulation*

联合国的数据显示,在 2025 年左右,年龄超过 60 岁的人口将达 20 亿,是现今老年人口的两倍。来自各方面的研究都表明中国老龄化速度之快前所未有,据悉 2011 年以后的 30 年里,中国人口老龄化将呈现加速发展态势,到 2030 年,中国将成为全球人口老龄化程度最高的国家。中国 65 岁以上人口占比将超过日本,成为全球人口老龄化程度最高的国家。到 2050 年,社会进入深度老龄化阶段,每 4 个中国人就有 1 个老年人,中国将成为高度老龄化的国家,和美国一样,中国的中老年人消费市场近年来也越来越大,所以开发中老年人市场大有可为。据统计,中国老龄人口将达 4 亿,上海现有 60 岁以上老人 233.57 万人,占总人数的 18%,老年人用品市场是夕阳产业中的朝阳市场,具有很大的发展潜力。

目前中国进入老龄化社会是一个被广泛认知的事实,一些经济学家视之为危机,认为它将使国家财政不堪重负;如何解决养老问题成为全世界各国政府的心头之患,例如近年来常被提及的"延迟退休"等政策信号的释放,正是对这一问题的应对体现。

但从另一方面而言,一些具有先锋意识的科研和企业人士则认为其中蕴含巨大商机,两鬓银发的老年人将带来巨大商机,并借此赚得大把的"银发财", 随着社会敬老风气的弘扬,上海老年用品市场呈现新亮点。 一批具有时代特点的老年人的吃、穿、用商品得到有效开发,并成为新的经济增长点。这些企业正试图重视开发针对老年人需求的生活所需产品和服务。对于老年人日常生活所需的服装设计来说,在老龄化的社会环境里其实也遇到了巨大的商机。

1 中高龄特体体型结构设计的 3D 应用

人体是服装款式造型的基础,人体的体型特征,是自然界赋以人类的个性符号,美感各有不同,优缺点不大,但当人体与服装结合后,美丑的差异就会扩大。根据不同形体的差异寻找一种适合的结构版型,更好的为人体服务。随着社会老龄化的出现,中高龄人中城市消费者日益增加,对着装美的要求也被提上了日程。服装结构设计的目的就是要体现人体美,弥补人体形态的不足。中高龄人服装造型设计首先服装结构上要这个年龄层人群的身体构造,突出舒适合体。其次服装要符合穿着者的气质身份。服装要宽松、合体、线形简练、不紧不松,即上下左右比例对称,以直线结构为主,不用或者少量运用附加装饰物,以充分体现老人的庄重、稳健。

1.1 中老年体型的变化

人到中年,随着生活工作所形成的一些习惯,体型就会发生变化,老年人的体型随生理机能的衰落,各部位关节软骨萎缩,两肩略显下降,胸廓外形也变得扁平,皮下脂肪增多。老年女性体型的变化影响衣身原型结构的变化,老年女装的结构设计与青年女子的服装结构设计会有很大的不同。套用年轻模特儿进行的服装设计产品,肯定不能适应于所有的中老年人,如何检验中老年人服装制版的正确性,以往我们都是通过样衣制作再修版的形式完成产品的打样,如今随着科技的发展,3D 试衣系统的出现可以直接让中高龄消费者根据自己的体型选择合适的服装了。

1.2 中高龄特体体型案例

下面以老年女性驼背体体型上衣结构设计进行分析:

中高龄驼背体特体体型较其他年龄层消费者出现的概率要多,这种体型的消费者穿上按正常结构裁剪的服装会出现的穿着特征:老年女性由于脊柱曲度增大,常出现驼背体,老年女性驼背体背部凸出且宽,头部略向前倾,前胸较平且窄,手臂呈推车型。

图1 女性驼背凸肚特体体型

人体骨骼与肌肉的构成形态和发达程度与服装造型关系极大,中高龄人体型的变化或特殊体型,就会引起结构设计中不同的处理方法。

如驼背体体型的老年女性穿上正常体型的服装,会出现如下弊病:

衣服下摆前长后短,后背绷紧,后身吊起,衣服的摆缝向后倾斜;头部前倾,脖子前探,导致后领口与颈部不相贴;由于脊柱弯曲凸出,衣服的后腰节被吊高,后衣片出现斜形涟纹;驼背体的手臂呈推车型,袖子与前手臂相碰,出现涟形。

造成此现象的原因是:驼背体型者由于遗传或其他生活或事故原因后背脊柱比正常体长,弓起于后方,势必要求加长服装的后背尺寸,服装打板中应该按特殊体型进行结构方面的特殊调整(图1)。

1.3 特体体型案例的3D表现

PGM 公司 CAD 试衣系统中用于打板的软件模块,能够自动记忆制作基础样板的打板步骤和公式的功能,并将众多的面料性能参数、人体测量数据进行数据分析,编成程序存入电脑软件自备的方案之中,为 3D 效果的形成提供样片信息与数据。OptiTex 3D 模块在 PGM 软件中可以结合二维平面纸样,在 3D 模特身上模拟逼真试衣效果。其操作目的是让使用者获得基本的 3D 生成技能,能够将 2D 平面纸样,转换创作出模仿真实服装的 3D 效果。

随着服装 CAD 三维系统的诞生,设计制版工作真正做到了从 3D 到 2D,直接在 3D 人体上进行款式的设计与样板的生成。这种模仿服装设计立体裁剪中先有成衣效果带来的制作和修正样板逆向思维模式也被运用到了 3D 试衣系统中。通过在 3D 人体上直接修正着装效果,只需要在 3D 人体模特上的部分服装造型线条即可获得适合人体的样片。

服装相对于其他产品而言与人体尺寸的贴合度较大。着装模特参数的尺寸修改,也就意味着私人定制量体裁衣的可能。在服装 3D 试衣系统中可打破固化的人体号型数据,因人而异地对试衣模特参数进行设置,满足各类型消费者的需求。这里主要介绍在 3D 模块中如何使用参数的人体模特模块,改变身体度量,及控制在 3D 试衣系统中人体模特的尺寸信息参数及姿势设置,展现出人体的高矮胖瘦。针对中高龄女性消费者中的驼背体特体体型消费者,在 3D 试衣系统中的人体尺寸参

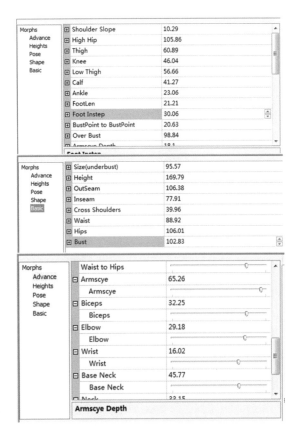

图2　3D试衣系统模特的人体尺寸参数调整

数调节及女性特体体型模拟,可为服装版型的制作提供直观可视的 3D 人体模型。在 3D 窗口点击 3D 模特儿特性按钮,分别在 Advance \Heights\shape\Basic 栏目下对胸腹部凸起和背部距离和肩部姿势给予数值增量,模拟出驼背凸肚的特体体型,而后根据基础纸样进行合理的修改,以满足中高龄人体体型的穿着要求(图2~图5)。

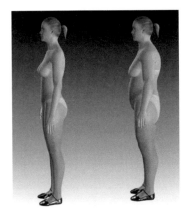 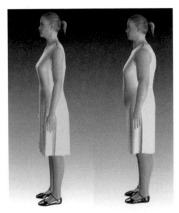 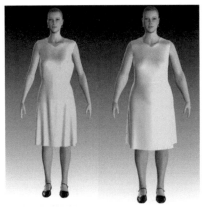

图3　模拟中高龄女性特体体型与正常体型侧面比较　　图4　模拟中高龄女性驼背体特体体型与正常体型侧面着装比较　　图5　模拟中高龄女性驼背体特体体型与正常体型正面着装比较

1.4 特体体型案例的版型结构修正

已有的服装打板步骤程序是根据二维设计效果稿款式图,分析款式造型及比例关系,按照设计制版原理进行服装衣片的二维制版。对服装制版原理知识的掌握是 2D 纸样制版工作前期所备的技能,操作人员必须具备服装或服饰产品专业制版的知识结构,所以在传统打板工作中,制版工作者必须具备专业的制版技巧及掌握服装打板公式计算及拥有丰富经验,通过对纸样的打板到后期的服装制作及修正,在侧缝及臀围处、前片的结构修正来满足对中高龄人特体对服装版型的结构需求,保证中高龄人着装的美观得体,这是从二维到 3D 修正的转换设计,在传统结构设计打板制图上的有以下调节方法用于适应穿着者的需求:

1) 通过对服装衣片(上装衣片)后中缝的修改,劈门稍微开大。在正常体型衣片后中缝加入加量,调节的具体数据视人体情况而定。加长后背长。将后衣片分别在袖窿深 1/2 和胸围线处剪开,根据测得的背长,分别拉开背长增加的量,这样不仅增加后背长,而且提高后颈点,使后领口处更贴合颈部。

2) 为了适合相对肥胖,腹部突起,背部弓起的驼背凸肚体型者穿着,制版时适当减去前衣片收省量,去除了腰省,仅保留袖窿省。放大前片下摆。后背中缝腰节以下顺势收小改进,防止下摆起空。通过袖窿和肩膀的修改整理,调节以上所述部位,达到符合驼背凸肚体型者穿着要求。

3) 改短前袖窿深。相应前领口弧线也要改低,且相应放出袖窿弧线,保持原肩缝长度不变。

4) 增加背宽。驼背体人体的厚度及背宽都有所增大,应在原型基础上适当增大背宽量,以便满足手臂和背部的活动量。

5) 前胸宽减小。在胸围不变的情况下,背宽增大,前胸宽需减小。

6) 根据背部和腹部的隆起程度来调节前后衣片的长度(图6,图7)。

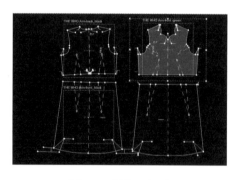 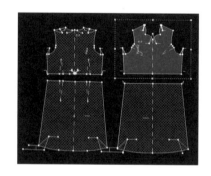

图6　正常体　版型　　　　　图7　中老年特体　版型

2　中高龄特体体型服装款式设计的 3D 应用

2.1 错视设计方法在中高龄人群服饰设计的应用

错视是设计思维的一种特殊表现形式,有背悖于正常的视觉逻辑思维但却能起到很好的美化作用。错视设计思维在平面设计中比较常用,经常会出现在矛盾空间等设计中,在现实生活中,能不能利用错视提升服装设计的款式设计美感呢? 一般来说,服装设计中深色衣服比较显瘦,比如横条纹服饰把人的视觉向横向拉伸,竖向条纹服饰就使人相对显瘦。搭配互补颜色的衣服穿,像红绿、黑白,给人以分节感觉,把人显得矮了,对个子不高的人而言,最好同种颜色或相近的颜色搭配。

2.2 3D试衣系统中错视设计表现的应用方法

在3D试衣系统的服装款式设计中,我们基于已有的服装结构造型,还可以尽可能多地进行色彩搭配、面料图案上的设计改观,以便于中高龄着装者进行商品的挑选。

PGM的3D试衣系统中的操作方法是选择所有纸样,点击【视图】—3D窗口—【阴影】选项,对其进行服装色彩图案、面料材质的调整。在编辑图案窗口Texture下空格位置进行点击,载入服装图案路径,图案图片格式可为JPG、PNG等,在【比例】位置可调整图案的大小。【种类】中的Pattern与LOGO的图案填充类别是不同的。这里在Texture中载入所示素材(图8),作为服装的面料图案,进行设计。

设计离不开色彩,设计师通过色彩表达其设计意念,色彩视错觉在设计中的应用有着不可取代的位置,如应用巧妙,不仅可以避免视错觉现象产生的消极作用,而且还可以增加作品的艺术感染力。[2] 利用同一款式采用不同的配色方法进行服装设计,同一色款式与利用有错视方法的撞色设计款式进行对比,显示出有色彩分割的撞色设计表现的着装效果将中高龄消费者衬托得更为修长与苗条(图9)。

图8　中高龄人体3D着装　物料阴影编辑

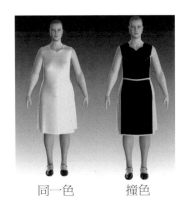

图9　中老年特体中利用色彩错视的服装设计

3　互联网时代的中高龄消费群体的网购特征

2014年我国网民数量和上网时间统计数据分析中指出50岁以上网民规模占比增加0.3个百分点,互联网继续向高龄和低龄群体渗透。[3]

随着老年社会的到来,中高龄化上网人群在未来的肯定会出现增量。通过网络购物也将成为未来老年人的生活方式。针对中高龄人的体型特点,一般尺寸的服装比较难以满足她们的需求，目前有的实体店铺已经出现了模拟试衣软件的使用，在"互联网+"的时代大背景下，网上 3D 试衣将成为可能。服装 3D，CAD 设计系统的订制发展方向应该是 3D 向二维样板的转变，真正做到以人为本，以特体体型的模型为先导，在 3D 模型的基础上再进行衣片的样式设计，而且直接可以提取和修正样板，将原来公式数据化的结构调整转变成直观的线条调整通过造型线的制作与修顺便可完成中高龄特体服装的样板制作，消费对象的细节化尺寸设计能考虑得更细致，满足特体人士的穿着，也更能直观真实地从人体出发来修正板型，提高板型制作的准确效率，提升着装者的穿着美感。

参考文献：

[1] 上海视觉艺术学院，上海中高龄时尚服饰研究中心.寻找银色光彩：2014 中高龄时尚服饰研究［M］.上海：东华大学出版社，2014.

[2] 钱洁.中老年特体服装的结构修正［J］.安徽职业技术学院学报，2008，7（2）：28-29.

[3] 张鸣艳.老年女性特殊体型服装设计研究［D］.苏州大学硕士学位论文，2013.

[4] 高峰，董兰芳.网上 3D 试衣系统技术研究［J］.计算机仿真，2006（06）：214-217.

[5] 魏立达.美国 OptiTex 服装 CAD "自动智能"操作技艺解析（上）［J］.中国制衣，2011（08）：42-45.

[6] 周飞.色彩视错觉在设计中的应用［J］.装饰 2006，（2）.

[7] 潘炳焕.关于中老年服装发展的意见［J］.上海纺织科技，1989（05）：57-58.

[8] 贾镇瑜.数码服装款式设计发展趋势探索［J］.中国科技博览，2010（27）.

[9] 李爱英.中老年服装结构设计的研究［J］.丝绸，2004（10）.

[10] 邹平.驼背体服装结构特征分析及结构修正［J］.服装科技，2000（10）.

[11] 美国：新潮老人用品"层出不穷"，企业大赚"银发美元"［OL］. http://www.cifnews.com/Article/10182. 2014-07-25.

[12] 2014 年我国网民数量和上网时间统计数据分析［OL］. http://www.chinabgao.com/stat/stats/39320.html. 2014-12-05.

[13] 3D 试衣系统［OL］. http://baike.baidu.com/link?url=PoX16cGy1jaHGL9IzHxM58F0H8pK-8d2G5wgp7SncHjNXkZxGYOpRS2kDVL6jjhNCEoRnyqigipwWCrniGsA_a. 2013-06-04.

社交型老年女性服装结构设计探讨

◆ 陈 萍[1] 李 俊[1,2]① （1.东华大学功能防护服装研究中心，上海，200051；2.东华大学现代服装设计与技术教育部重点实验室，上海，200051）

【摘 要】 女性作为我国服装消费的重要群体之一，在步入老年阶段后，其生理和心理都会产生变化，但其对服装的需求仍然存在巨大潜力。特别是对于社交型老年女性，对老年服装有更高的要求，包括合体性、舒适性、美观性和个性化等。因此，应根据老年女性不同的体型变化和社交活动，对相应的服装进行结构上的优化，以满足其生理和心理上的需求，进一步提高其生活品质。

【关键词】 老年女性服装；社交活动；体型；服装结构

Study on the Structure Design of the Social Old Women's Garments

◆ Chen Ping[1] Li Jun[1,2] *(1 Protective Clothing Research Center, Donghua University, Shanghai 200051 2 Key Laboratory of Clothing Design & Technology, Ministry of Education, Shanghai 200051)(E-mail: lijun@dhu.edu.cn)*

Abstract: *Women as one of the important groups of garment consumption in our country, when they walk into old age, their physiological and psychological will have changes, but the demands of clothing still exist great potential. Especially for the sociable old women, they have higher requirements for the old garments, including fitness, comfort, aesthetics and personality, etc. Therefore we should be according to the old women's different somatotype changes and social activities to make the appropriate garments for structure optimization, in order to meet their physiological and psychological needs, to further improve the quality of their life.*

Keywords: *Old Women's Garments; social activities; somatotype; garment structure*

---

① 李俊，男，东华大学，博士、教授，博士生导师。E-mail: lijun@dhu.edu.cn, 739581430@qq.com

20世纪90年代以来,中国的老龄化进程不断加快,2014年我国60岁以上老年人口已突破2亿,到2020年全国老年人口将达2.48亿,老龄化水平为17%[1],21世纪的中国将是一个不可逆转的老龄社会。同时出生于20世纪60年代到70年代中期第二次生育高峰的婴儿潮将进入老年阶段[2]。这些接受高等教育,并在全球经济一体化时代下进入老年阶段的人群,将不同于以往老年人旧式的生活方式,他们有更新的时尚意识、消费意识和交际意识。

现代老年人的人际交往及各种休闲娱乐活动已不仅仅局限于亲朋好友,而是变得更加丰富多彩,老年服装的选购也从单一色调以及加大加肥的传统要求向品位、时尚、功能等方向发展。其中,社交型老年人,特别是社交型老年女性对老年服装有更高的需求,例如近年来比较流行的广场舞、暴走运动等使得老年女性更注重个人服装的合体性、舒适性、个性化和美观性。而目前市场上的老年女性服装无法满足其参与不同社交活动的特殊需求。因此,有必要针对不同社交场合的老年女性服装进行结构设计上的优化,并基于老年女性的体型变化,设计出更加合体、舒适、美观的老年服装。

1 老年女性社交方式

人们在步入老年队伍之后,需要面对生理和心理上的落差,如生物调节功能下降、适应能力减弱、感觉功能减退、运动能力下降等生理变化以及心理上精神寄托对象的转移等。特别是对老年女性而言,在生理上,她们的体型、容貌发生变化;在心理上,生活中大大小小的事情会牵扯她们的精力,孤独、死亡等压力也使得她们更渴望与人交流、愿意参加社交活动,从社会生活中寻找友谊、精神寄托和生活的动力。

目前,现代老年女性的社交意识主要体现为交际范围的扩大和交际方式的多样化。主要的社交方式可以分为文化娱乐型社交和体育锻炼型社交[3-5]。

(1)文化娱乐型社交。在内容上,主要可以分为老年舞蹈、老年合唱团、老年协会、老年志愿者等集体活动;锻炼时间一般比较固定;锻炼地点主要为文化中心等室内场所。

(2)体育锻炼型社交。在锻炼内容上,主要多以散步、慢跑、广场舞、气功、舞剑、健身操等节奏较慢的体育项目和传统型体育项目为主;锻炼时间主要为清晨和傍晚;锻炼地点主要分布在不收费的场所,如公园、广场等空地;锻炼形式主要分为自发性锻炼和有组织的锻炼。

社交型老年女性经常出现在这些公共场所,形成某些新的社会群体。在这些特定的群体中,老年女性很注重自己的个人形象,在意自己的穿着打扮。她们希望通过服装来展现个人的修养、品位及曾有的社会地位,增加其自身在社会交往活动中的信心和勇气。

2 老年女性服装需求分析

社交型老年女性对服装的需求主要包括生理需求和心理需求。生理需求主要是针对老年女性老龄化后体型上的变化所引发的对服装的需求变化,包括服装的合体性和舒适性;心理需求主要是针对不同的社交场合对服装有特殊的需求,包括服装的美观性和个性化。

2.1 合体性和舒适性

一般来说,进入老年以后,人们在皮肤、体态方面必然会发生变化。其中,60岁以后的老年女性体型特征变化尤为明显,人体体型由扁平逐渐变圆,老年女性的身体围度普遍大于中、青年女性,特别是腹部和腰部,由于脂肪的堆积,围度增大,呈现明显的外凸状;胸腰差和臀腰差明显小于中、青年女性;颈部的脂肪增厚,呈现颈粗的形象;背部厚实,部分老人有驼背现象[6];同时由于老年女性乳房的萎缩,胸高点随着全身肌肉的松弛有所下降,如图1所示。

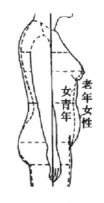

图1 老年女性体型侧面图
Fig.1 Profile of Old women's somatotype

根据服装号型国家标准,老年女性普遍属于B、C体型,占4种体型总分布的75%[7]。但由于老年女性特殊的体型,在穿上正常体型的服装后,会出现有关服装合体性和舒适性的问题。在国家标准GB1335—2008中未对老年人号型进行标准化划分,因此,需根据生理上不同的体型变化,将老年女性的体型大致分为塌肩体、驼背体、凸腹体、凸臀体、驼背凸腹体和凸腹凸臀体这6大类[8-9],如表1所示。

表1 老年女性特殊体型
Tab.1 Old women's special somatotype

序号	特殊体型名称	体型特征	序号	特殊体型名称	体型特征
1	塌肩体		4	凸臀体	
2	驼背体		5	驼背凸腹体	
3	凸腹体		6	凸腹凸臀体	

总之,需要根据不同的老年女性体型特征,对相应的服装结构进行优化,提升服装整体穿着的舒适性和合体性,包括:尺寸适度,无束缚与压迫感;易于穿脱,方便活动;穿着不易变形、走样;有优良的吸湿排汗性等,以满足老年女性生理上对服装的需求。

2.2 美观性和个性化

社交型老年女性通常会参加许多文化娱乐活动和体育锻炼活动,针对这类老年女性所设计的服装就需要适应她们参与各种活动的需要,将服装的功能性、舒适性、安全性、时尚性与老年女性的体态特征相结合。主要的服装的款式可分为[10]:老年中式服装、老年运动服装、老年职业服装及老年演出服装等；主要的面料以棉、麻、丝、毛等天然纤维为主要材质，采用纯纺或混纺织成的面料设计相应的服装；主要的色彩选择还是以稳重、含蓄、漂亮、高雅为主，结合老年女性的体态、肤色、心理及性格等来搭配艳丽颜色与中性颜色的度，以达到服装色彩个性与着装人的个性相协调，让老年人的精神面貌更加健康。

3 老年女性服装结构设计优化

3.1 特殊体型服装结构设计优化

人体的肩部、胸部、背部、腹部、臀部结构直接影响上衣和下装的结构设计，特别是对于有体型特殊的老年女性而言，针对各种体型进行服装结构优化以满足合体性和舒适性需求是需要关注的重点。

（1）塌肩体

老年女性塌肩体体型的肩斜角度超过正常体型的角度（19°~20°），一般肩斜角度大于22°则称为"塌肩"，从背面观察呈"个"字形状。该类体型的老年女性在穿上正常体型的服装后，在上衣肩外端会起斜绺，衣领显紧，衣片下沉，出现门襟止口搅盖等现象[8]。

优化方法：降低前后肩斜线，增大肩斜度；放低袖窿深线；减小前胸宽；收肩省以使肩部符合塌肩体肩部结构；同时为了确保肩部的美观，可在肩部增加垫肩，确保与正常体型肩部相仿。因此，不建议在老年女性服装结构设计时采用插肩袖。

（2）驼背体

老年女性由于脊柱曲度增大，常出现驼背体，驼背体体型背部隆起且宽，前胸较平，颈部、背部身体均向前倾，肩胛骨成弓形，后腰节比正常后腰节长，手臂呈推车型[11]。该类体型的老年女性在穿上正常体型的服装后，上衣下摆前长后短，背部绷紧吊起，后衣片易出现斜形涟纹，后领口易起空等。

优化方法：改小胸省、劈门；减小前胸宽，增加后背宽；改短前腰节线长，加长后腰节线长；前袖山头降低，后袖山头放高；并适当减小前领口。

（3）凸腹体

凸腹体体型的老年女性腹部肥满前挺，后背略后倾，常为粗腰[12]。该类体型的老年女性在穿上正常体型的服装后，上衣下摆前短后长，前领易起空，后领易触脖，上下装的腹部绷紧，腹部位置扣钮扣会比较困难，下装裤子易出现吊裆现象，裙子下摆前短后长[13]。

优化方法：对于上装，应增大劈门，放宽前胸宽、前腹宽和腹围；增加前腰节长，减小后背长；增大门襟下端起翘量。对于裤装，应加长前裆缝，增加上裆斜度；减短后裆缝；降低后翘高度。对于裙装，应增大前裙片臀围量；增加前裙片中心线长和前省量，使前裙片贴合人体。

（4）凸臀体

凸臀体体型的老年女性臀部丰满、宽大，腰显细，腰臀差大，臀部略有下垂。该类体型的老年女性在穿上正常体型的服装后，对于长度过臀部的上装，会出现上衣下摆前长后短，臀部处紧绷,后衣片下半段上缩状态,裤装的后臀部紧绷,后裆宽卡紧,后腰部有空瘪感,出现横向涟形,裙子下摆前长后短[14]。

优化方法:对于上装,应放大前、后衣片臀围,加长后片衣长;对于裤装,应放大臀围,增加后裆缝斜度、弧线长,提高后翘,重新设定后裤片腰省的位置,使后臀部更为舒适、隆起;对于裙装,应减小前片省量,增加后片省量,开落后腰中部,并加大后裙片长度,可以改善裙装后臀部较松的现象。

（5）驼背凸腹体

驼背凸腹体属于特殊复合体体型,需注意老年女性背部、腹部的凸起状态。该类体型的老年女性在穿上正常体型的服装后,上衣前胸呈凹陷状,后背部、腹部紧绷,前后衣片均吊起[8-9]。

优化方法:需基于驼背体和凸腹体体型服装的优化方法进行服装结构上的优化,应改短前袖窿深,减小前胸宽,增加后背宽,以满足手臂和背部的活动量;根据背部和腹部的隆起程度来调节前、后衣片的长度,防止上衣下摆上缩;并提高后颈点,使后领口更贴合颈部;袖子优化方案参考驼背体袖子结构上的优化。

（6）凸腹凸臀体

凸腹凸臀体体型的老年女性腹部和臀部均凸起,也属于特殊复合体体型。该类体型的老年女性在穿上正常体型的服装后,上衣腹部、臀部均有紧绷感,下摆上缩,腰部两侧出现涟形,下装腹部、臀部绷紧,裤子门襟处隆起,出现吊裆现象[8-9]。

优化方法:与凸腹体体型相似,区别在于其臀部与后腰部落差要比凸腹体体型大,需注意臀部与腹部隆起程度来确定放出胖势多少。

综上所述,特殊体型服装结构设计优化,要考虑各种特殊体型的凹凸程度,灵活运用结构上的尺寸、省道等变化,使老年女性的服装处于较为理想的状态,同时使着装既合体舒适又美观。

3.2 不同社交场合服装结构设计优化

（1）老年中式服装

中式服装一般拥有庄重、含蓄之美,主要适合在一些传统的体育锻炼活动中穿着,款式主要为前开型的斜襟和对襟式样,多用带子固定衣服,女装多为上衣下赏的式样,方便穿脱。

在设计老年女性的中式服装时,需注意其体型变化,将设计重点放在服装的细节上。由于老年女性的围度相对较大,特别是腹围、臀围,比较适合 A 型、H 型的衣身廓型;领口围度需多加 3~4cm 松量;袖子不宜过长。衣身的造型强调流畅感,自衣领部位开始自然下垂,不夸张肩部。

（2）老年运动服装

运动服装是方便从事各种活动所穿的服装,主要适合在体育锻炼活动中穿着。目前,随着老年人生活水平的提高,运动服装与休闲服装有相当大比例的重合使用部分,常常可以互换使用,运动服装已经成为与现代老年生活方式高度相关的服装款式。

在设计老年女性的休闲服装时,其结构、款式应该简单大方,易穿、易脱,适用于 A 型、H

型、X 型和 O 型，给人修长、简约、舒适的感觉。

（3）老年"职业"服装

该"职业"服装包括一些同一款式的特定社交场所服装以及老年演出服装等。通常老年女性会参与一些有组织的老年团队，如老年大学、老年舞蹈队、老年合唱团、老年棋牌队等等。这些老年团队在参加一些社会活动时，需要统一着装。

在服装结构上的设计就需根据老年女性客户不同的要求，结合团队文化、年龄结构、体态特征、穿着习惯等[10]，为老年女性顾客打造适合每个人体型的形象服装、演出服，以增强老年女性顾客展示自己的自信心。

综上所述，对于不同社交场所的老年女性服装仍需根据不同社交场合对服装的要求以及老年女性的体态特征，进行结构上的优化设计，以满足其生理和心理上的舒适感和优越感。

4 小结

随着我国老龄化社会的快速发展，应更加注重老年人的社会生活，特别是老年女性在进入老年阶段后，在社交活动中，由于生理和心理变化对老年服装的需求变化情况。本文主要通过对老年女性的社交方式进行分析，基于老年女性对老年服装的需求，对其特殊体型和社交服装进行分类，并对各个特殊体型所需的服装和社交服装进行结构上的优化，以满足老年女性对服装的合体性、舒适性、个性化和美观性的不同需求，提升其在各种社交活动中的自信心，有效减少或预防由于老年服装所引起的老年抑郁症等心理疾病。

参考文献：

[1] 阎珺.老年人功能服装的研究现状综述[J].科学论坛，2014（4）：179-181.

[2] 王露.关注老龄需求，设计我们的"未来"[J].中国纺织，2011：121-123.

[3] 杨玲.广场舞对中老年人的健身娱乐作用以及存在的问题[J].大众文艺，2012（7）：196-197.

[4] 李敬姬.中韩老年人生活方式、休闲活动及生活质量的研究——以上海与首尔的老年人为例[D].上海：复旦大学博士论文，2011.

[5] 于建新.老年服装及其市场发展研究[D].天津：天津工业大学硕士论文，2008.

[6] 陈明艳.成年女性体型特征及其服装样板设计[J].纺织学报，2005，26（3）：121-124.

[7] 李兴刚，沈卫勤.中老年女子体型变化分析和原型设计研究[J].东华大学学报（自然科学版），2001，27（5）：121-125.

[8] 张鸣艳.老年女性特殊体型服装设计研究[D].苏州：苏州大学硕士论文，2013.

[9] 邵晨霞.关于老年女性特殊体型纸样修正的研究[J].苏州大学学报（工科版），2012，32（6）：17-21.

[10] 李晖.老年服装的人性化设计研究[D].齐齐哈尔：齐齐哈尔大学硕士论文，2012.

［11］麻乐乐.从人体工学谈驼背体型服装的结构设计［J］.武汉科技学院学报,2008,21（3）:35-37.

［12］肖柳庆.不同凸腹体服装的结构设计［J］.纺织科技进展,2012,（2）:67-68,72.

［13］曲长荣,吴爱荣.凸肚体型服装纸样的设计技巧［J］.职业,2010,（20）:148-149.

［14］王秀芝.凸臀体服装结构解析［J］.国际纺织导报,2007,35（1）:74-76.

电加热技术在老年人冬季保暖服中的应用

◆ 许静娴[1,2] 李俊[1,2]① （1.东华大学功能防护服装研究中心，上海，200051；2.东华大学现代服装设计与技术教育部重点实验室，上海，200051）

【摘　要】 考虑到冬季防寒保暖对于老年人群体来说尤为重要，从服装角度出发，以使防寒保暖服更好地满足老年人需求为目的，提出了将电加热技术运用到老年人服装中。从生理及病理、心理两个层面对老年人冬季着装需求进行了分析，总结出了老年人需要重点保暖的8大部位，并据此指出了现有服装的不足。然后引入电加热技术与服装相结合，并进行服装款式、加热部位、加热系统与服装结合方式的选择。其次，进行了电加热系统智能化的简单探讨，指出通过温度传感器、信息处理器及电源驱动器来实现加热系统开关智能化，当皮肤温度低于32℃时则自动开启，高于37℃则自动关闭。可为设计更加人性化的老年人服装提供参考。

【关键词】 老年人；保暖服；电加热；智能化

Application of electrically-heated technology in warm clothing for the old

◆ Xu Jingxian[1] Li Jun[1,2] (1. Protective Clothing Research Center, Donghua University, Shanghai 200051   2. Key Laboratory of Clothing Design & Technology, Ministry of Education, Shanghai 200051) (E-mail:1034160272@qq.com, lijun@dhu.edu.cn)

Abstract: Keeping warm is of great significance for the aged in winter. This paper, from the perspective of garment, proposed to apply electrically heated technology to clothes for the aged, with the intention of making warm clothes satisfy the old people's needs better. Their needs for garments in winter were analyzed from the aspects of physiology, pathology and psychology and eight parts which needs better protection were concluded. Based on the analysis of needs, the shortcomings of present garments were also pointed out. Then, the technology of electrical

---

① 李俊，男，东华大学，博士，教授，博士生导师。E-mail：lijun@dhu.edu.cn，1034160272@qq.com

heating was put forward and the style of the garment, the part to be heated and the way of combination between heating system and clothing were analyzed either. At last, an intelligent heating system was briefly discussed and the results indicated that the application of temperature sensor, information processor and power drive can help make the heating system intelligent. When the skin temperature is lower than 32℃, the system will launch automatically and when the temperature is higher than 37℃, the system will shut down automatically. This research can provide reference for the development of more humanized garments for the old.

Keywords: *The old; Warm clothing; Electric heating; Intelligent*

全国第六次人口普查[1]数据显示,截止2010年我国65岁及以上人口已超1.1亿人次,占全国总人口的8.87%,相比2000年第五次全国人口普查,占比上升了1.91%,预计到2030年,我国65岁及以上的老年人口占比将提高至18.2%[2]。由此可见,我国正面临着人口老龄化日趋严重的现象,关注"银发群体"也将成为我国社会一大重要议题。

一旦步入老龄阶段,人体将会面临各种生理机能下降、行动缓慢、灵敏性降低等问题,尤其到了冬季,由于人体体温调节功能下降,身体产热减少,防寒保暖对老年人来说显得尤为重要[3]。然而市场上现有的老年人防寒保暖服多为厚重臃肿型,这对本来就行动不便的老年人来说无疑是增加了他们的负担。针对这一问题,近年来服装领域已有不少研究人员展开了研究,且相关研究多集中在面料上,他们[4-5]指出通过使用远红外纤维及相变储能纤维等功能纤维制作老年人保暖服,可达到既轻便又保暖的目的。然而,这一想法太过理想化,对老年人这一特殊弱势群体来说,这类功能纤维产生的热量是有限的,并不足以弥补人体热量散失。而电加热服装作为一种新型主动产热式保暖服,具备轻便、保暖效果好、可按需产热等优点[6],这对老年人来说是非常理想的防寒保暖手段。但现有的电加热服装多以普通大众作为消费对象,对于老年人这个特殊群体来说缺乏针对性。

本文从老年人冬季着装需求出发,将电加热系统应用到老年人服装中,并针对老年群体的特殊需求,对服装款式、电加热系统与服装结合的部位及结合的方式展开研究,同时提出了电加热系统智能化的概念并进行了初步探讨,以寻求使电加热保暖技术更好地为老年人服务,为设计出真正符合老年人需求的冬季防寒保暖服提供参考。

1 老年人冬季着装需求分析

考虑到老年群体生理以及心理上的特殊性,借鉴功能服装设计模式[7],从生理及病理、心理两大角度出发,对老年人冬季着装需求展开分析,并基于需求分析指出市场上现有服装的不足。

1.1 生理及病理需求

随着年龄的增长,人体内部各项功能每况愈下,其中表现最为明显的就是体温调节功能减弱。到了冬天,许多老年人难以抵抗严寒,易患上"低温症",四肢冰凉,严重影响人体健康[8]。另外,据统计[9],关节类疾病属老年群体多发疾病,50岁以上中老年人群中,患病率高达80%,尤其在寒冷天气,老年人时常要忍受关节肿胀疼痛之苦。因此,注意加强关节等部位的保暖对老年人度过健康舒适的冬季尤为重要。

通过查阅文献资料[10-11],总结出了老年人冬季需重点保暖的八大部位,包括手、脚、头、踝关节、膝关节、腰部、腹部、肩关节,其中后五个部位可以通过服装进行保暖,剩下的部位多通过手套、鞋、帽子进行保暖。

1.2 心理需求

步入老年阶段,很多老年人在心理上会变得异常敏感[12],大多数老年人会有"不服老"的心理,他们不希望别人把自己当做老人来看,因此在冬季着装上应避免外观厚重或者对于一些特殊功能性设计应采用隐藏式设计方法,避免将特殊器件暴露在外,使着装者引人侧目。

1.3 现有老年人冬季保暖服的不足

市场上现有的老年人冬季防寒保暖服,如图1所示,廓型臃肿,多是通过增加服装厚度阻止人体热量散失来达到保暖的目的,而事实上很多老人就算穿很多件都无法满足保暖需求,结合上述需求分析,总结出以下几点不足:

(1)保暖效果不好。老年人身体产热机制衰退严重,本身产热量就不足以维持人体热平衡,所以光靠增加服装厚度并不能满足老年人对保暖的需求。

(2)厚重,有压迫感,限制人体行动。现有冬装大部分有一定的重量,而老年人骨骼疏松,承重能力差[13],厚重的衣服对老年人来讲,无疑会产生一定的压迫感致使其不舒适,而且厚重臃肿的服装会限制老年人活动。对老年人这类特殊群体而言,防寒保暖服应最大限度满足质轻和保暖两大需求。

(3)功能性设计缺乏针对性。现有的老年人保暖服与年轻人服装除了款式、色彩上有所区别外,在功能上并无二致,而事实上由于老年人生理上的特殊性,需要保暖的部位以及保暖力度与年轻人是存在差别的,因此,老年人保暖服对于保暖部位的选择应更具针对性,且对其保暖效果的要求应更加严格。

图1 老年人冬季着装现状

Fig.1 Wearing status of the aged in winter

2 电加热技术的应用

通过将电加热技术运用到服装中可使服装具备主动加热的功能,且产热量高、可持续加热、加热部位可选,是开发老年人保暖服的一大优秀选择。其次,由于产热量高,电加热技术的运用可帮助老年人少穿几件衣服,从而避免出现穿着臃肿、活动不便的现象。另外,考虑到老年人需求的特殊性,电加热技术在老年服装中的应用应区别于年轻人,需重点注意加热部位和电加热系统与服装结合方式的选择。

2.1 电加热技术简介

电加热技术[6]属于主动发热保暖技术中的一种,整个技术系统由电源、温控装置、安全保护装置、发热元件、导线组成,如图2所示,通过导线将电源、发热元件、温控装置、安全保护装置连接起来,接通电路,再根据电流的热效应原理,发热元件即可将电能转化为热能,从而起到发热保暖的功能,通常在运用到服装中时发热温度可达30～80℃[14]。发热元件作为电加热系统的核心部件通常由导电发热材料制成,根据材料的种类可分为电热金属丝、电热膜、电热织物三大类,其中电热织物以其优异的可纺性、柔软性和与服装面料的相容性等优点成为服装用发热元件的最佳选择[6]。电加热系统各组成部件的功能、特点概括见表1。

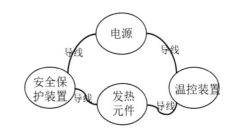

图2 电加热系统构成
Fig.2 Constitute of electric heating system

表1 电加热系统各组件的功能及特点
Tab.1 Functions and characteristics of the composites of electric heating system

组成部件	功能及特点
电源	为系统提供电能
	通常为12V以下人体安全电压[15]
发热元件	将电能转化为热能
	使用较多的为碳纤维织物
温控装置、安全保护装置	用于保护电路
	当电路过载或温度过高时自动切断电路
导线	充当电流传输载体

2.2 电加热在老年服装中的应用

为了使电加热技术更好地为老年人所用,结合市场上现有电加热服装的特点,以老年人实际需求为依据,进一步明确电加热技术与老年人服装结合时服装款式、加热部位、结合方式的选择。

2.2.1 服装款式

电加热技术在服装中的应用并不罕见,目前,市场上已有深圳八宝科技、深圳晶海科技、上海玖健康复器材等科技类和康健类公司投入研发[16],成品款式多为背心、马甲、羽绒服三大类,缺少下装,而老年人的下肢是非常需要御寒保暖的,所以应为老年人开发电加热裤。另外,应注意冬季老年人皮肤干燥松弛、易受损伤、受到刺激性作用后易瘙痒疼痛[17],所以电加热服装不可贴体穿着或者电加热系统中的加热元件不可直接贴近皮肤。本文选择背心和保暖裤两款服装为例,进一步说明电加热技术在老年服装中的应用。

2.2.2 加热部位

电加热系统运用到老年人服装中是为了起到发热保暖功能,帮助老年人抵抗冬季严寒,减少相关疾病引发的疼痛,因此加热部位的选择应是有所依据的,根据上文的分析,老年人冬季应注意8大部位的防寒保暖,其中能通过服装加以保护的部位包括肩关节、腰部、腹部、膝关节、踝关节。因此,加热位置的选择应以这些部位为主。电加热系统与服装结合的结构示意图如图3所示,其中,黄色部件表示发热元件,放置在对应于人体需要保暖的部位,电源及其他辅助部件则放置在口袋里,两者通过导线相连接。

2.2.3 结合方式

电加热系统与服装的结合包括三大部分,一是加热元件与服装的结合,二是电源与服装的结合,三是连接电源和加热元件的导线与服装的结合。在选择具体结合方式时应明确三点要求:可拆卸、舒适性、安全性。关于加热元件与服装的结合,通常选择在服装内侧需要加热的部位缝制小口袋,然后将加热元件放置在小口袋内;电源通常是直接放置在服装自带的口袋里;导线与服装结合的方式应特别注意隐藏式设计且应保证着装舒适性,建议在服装内侧的侧缝或接缝处缝制导线专用通道,可将导线穿过通道连接加热元件和电源。在有加热保暖需求时则将电加热系统三大部件与服装结合,不需要保暖或服装需要清洗时则可将加热系统拆卸下来。

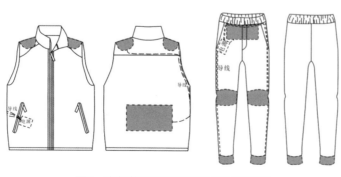

图3 电加热服装加热部位结构示意图
Fig.3 Heating parts of the electrical heating garment

3 老年人电加热保暖服智能化探讨

在将电加热技术应用到老年人保暖服的基础上,考虑再增加智能化设计,使电加热系统不仅能够为人体提供热量,而且可以根据人体温度的变化实现自动加热和自动停止加热,可减少了老年人手动控制加热开关的繁杂过程,而且考虑到老年人皮肤反应迟钝[17],如果单纯依靠老年人主观感受来决定何时加热或者何时停止加热,可能会对人体产生一定的危害,所以电加热系统的智能化设计还有助于实现对老年人的安全监护。电加热系统智能化需要借助温度传感器、信息处理器及电源驱动器三个器械[18],具体工作原理如图4所示。

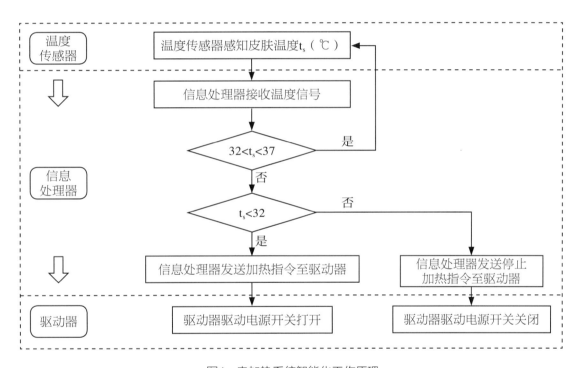

图4 电加热系统智能化工作原理
Fig.4 Operating principle of the intelligent heating system

(1)温度传感器。用于获取人体皮肤温度,皮肤温度作为加热系统工作与否的衡量指标。相关研究表明[19],人体皮肤温度的较舒适温度范围为32~37℃,当皮肤温度低于32℃时,人体会产生不舒适的冷感,当皮肤温度高于37℃时,人体会产生不舒适的热感,因此加热系统应在皮肤温度低于32℃时开启加热功能,高于37℃时停止加热。

(2)信息处理器。用于接收温度传感器传达的温度信号,并比较人体皮肤温度与临界温度,进而生成加热或停止加热的指令,并将指令发送给驱动器。

(3)驱动器。驱动器用于接收信息处理器发送的指令,来控制电源的开关。

4 总结

鉴于老年群体生理及心理上的特殊性,以老年人为目标对象的服装设计在款式、色彩尤其是功能上应更具针对性。本文以老年人冬季防寒保暖服为研究对象,基于对老年人生理、病理及心理需求的分析,总结出了市面上现有服装的不足,包括保暖效果不好、厚重、功能性设计缺乏针对性,然后提出了将电加热技术运用到老年人冬季防寒保暖服中,开发出专属于老年人的防寒保暖服,最后为使电加热保暖服更好地为老年人服务,提出了电加热系统智能化设计方案,通过温度传感器、信息处理器及驱动器来实现加热功能随人体皮肤温度变化自动开启和关闭。

本文提出的基于电加热系统的老年人保暖服设计方案可为开发专属于老年群体的服装提供参考,另外,本文最后对于老年人电加热系统智能化的设计只进行了简单探讨,并未有过多深入的研究,希望今后在智能化可穿戴设备盛行的大背景下,能在此方面有进一步突破,设计出更加以人为本的服装。

参考文献:

[1] 国家统计局.2010年第六次全国人口普查主要数据公报(第1号)[J].北京报周报,2011,54(22).

[2] 联合国开发计划署,中国社会科学院城市发展与环境研究所.2013中国人类发展报告[M].北京:中国对外翻译出版有限公司,2013.

[3] 陈庚笙.老年人功能性服装的应用分析[D].硕士论文.广州:华南理工大学,2014.

[4] 王文娟.老年人服装的功能设计研究[J].艺术教育,2015(2):257-258.

[5] 吴亚红.老年人服装市场对新型纤维的需求分析[J].纺织导报,2013(12):40-41.

[6] 任萍,刘静.可加热服装技术的研究[J].纺织科学研究,2008(3):12-18.

[7] 苏云,何佳臻,李俊.中老年服装功能设计模式研究[J].2014年中高龄时尚服饰研究,2014:174-184.

[8] 王荣华.老年人度寒冬需防低温症[J].山西老年,2013(12):62.

[9] 邓廉夫,任步方.软骨决定您的关节寿命[J].家庭用药,2007增刊.

[10] 王金水.冬季老年人易患关节炎注意保暖预防"老寒腿"[J].今日科苑:146-148.

[11] 郭旭光.老年人冬季保暖要得法[J].以诗会友,2013(17):41.

[12] Xiaoping Hu, Xia Feng, Delai Men et al. Demands and needs of elderly Chinese people for garment [J]. User and Context Diversity,2013:88-95.

[13] 孟祥令.中高龄服装的生理舒适性设计初探[J].2014年中高龄时尚服饰研究,2014:225-232.

[14] 李峻,刘晓刚等.碳纤维发热服装设计的研究[J].江苏纺织,2007:48-51.

[15] 唐世君,郭诗珧.电加热服装的研制[J].防护装备技术研究,2013(6):5-8.

[16] 深圳八宝科技有限公司 http://babao898.1688.com/

［17］王宛华,沈健. 老年人皮肤特点与护理［J］. 中国中医药现代远程教育,2012（21）.

［18］Snjezana First Rogale, Dubravk Rpgale, Zvonko Dagcevic et al. Technical systems in intelligenr clothing with active thermal protection［J］. International Journal of Clothing Science and Technology, 2007, 19（3）: 222-233.

［19］周浩. 人体皮肤温度影响因素试验研究［D］. 西安：西安建筑科技大学硕士论文, 2013.

老年人着装舒适性的影响因素分析

◆ 王雅芝[①]　（苏州大学纺织与服装工程学院，苏州，215006）

【摘　要】　老年人着装不舒适是服装行业亟待解决的问题。基于人体工效学,分析影响老年人着装舒适性的三个因素：人、服装、环境,并就每个因素的研究现状分别进行综述。分析得到老年人对服装的舒适性较为敏感,其服装材料应以天然纤维为主；老年人为典型的腹凸体,适合较宽松的服装廓型,且其服装号型应重新设定；环境因素要求老年服装保温性好,具有个性化。因此,老年服装需从舒适性、安全性、美观性等方面去改善。

【关键词】　老年人；舒适性；服装；环境

Discussion on the influential factors in the old people's clothing comfort

◆ Wang Yazhi　*(College of Textile and Clothing Engineering, Soochow University, Suzhou, 215006) (E-mail: 531148326@qq.com)*

Abstract: *The old people's clothing discomfort has been a problem needed to be solved. Based on ergonomics, this article analyzed the 3 factors affecting the old people's clothing comfort, which are human, clothing, environment, then summarized the current research status of each factor respectively. The old is sensitive to clothing comfort, which explains natural fibers are better; the representative somatotype is abdomen bulged, so, resetting garment size and slightly loose clothing profile are needed; besides, guarantee high efficiency in heat preservation with personalized style. So, improve the old people's clothing from comfort, safety, decency.*

Keywords: *The old; Comfort; Clothing; Environment*

　　2015年的人口统计数据显示,2014年60周岁及以上的老年人口数相比2013年,增长近

[①] 王雅芝,女,硕士研究生,苏州大学。E-mail: 531148326@qq.com

1000万人,老年人口数占全国人口数的比重也高达15.5%,表明我国人口老龄化现象日趋严重。这对我国包括老年服装行业在内的健康养老产业提出了高要求。

中国老年服装的消费市场充满潜力,但老年服装行业的发展却不尽如人意。于建新[1]调查了河北地区老年人的服装消费态度,76.6%的老年人对目前市场上的老年服装不满意,其中相当不满意的占11.2%,主要集中在服装的舒适性差、品种少、款式陈旧方面。由此可知,目前老年服装市场并不成熟,不能满足老年人对服装舒适性、安全性及美观性要求。

1 老年人着装舒适性

Slater[2]将舒适性定义为人与环境间生理、心理及物理协调的一种愉悦状态。并定义了三种名称:生理舒适性与人体维持生命的能力有关;心理舒适性指人脑在外部帮助下满意地保持其自身功能的能力;物理舒适性则是外界环境对人体的作用。综上可知,人体的着装舒适性是人与服装、环境相互作用及合理结合的结果,要求服装和环境系统适合人体的心理、生理等特点,使服装达到穿着安全、健康、舒适的目的。

我们通常所研究的舒适型都是狭义上的生理舒适性,即通过人体—服装—环境之间的能量交换,促使服装微气候内的热湿达到平衡。但由于老年人的生理和心理对外环境较普通人更加敏感,老年人对着装的舒适性、安全性也因此有着更高要求。

2 影响老年人着装舒适性的因素

影响老年人着装舒适性的因素主要包括老年人本身、老年人的服装、老年人所处的环境,如图1所示,三者相互作用,相互影响。

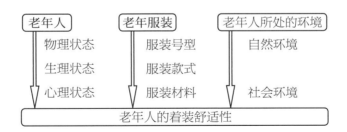

图1 影响老年人着装舒适性的因素
Fig.1 Factors affecting old people's clothing comfort

2.1 人的因素

在人—服装—环境这个系统当中,人是舒适感觉的主体,其生理因素与心理因素对服装的舒适性都具有重要的影响。

2.1.1 老年人的体型

随着年龄的增长,老年人的体型逐渐发生变化。王宝环[3]对辽东地区228名中老年女子进行体型测量,分析数据可知,中老年女性的胸围、腰围、臀围分别比国家标准体型大5.0cm、6.2cm和2.8cm,说明有较大比例的中老年女性由于身体发福,脂肪在腰腹部堆积,造成腰围、臀围显著增大。吴巧英等人[4]使用了围度差、扁平率、周长比对中老年女性与青年女性的体型进行了对比。结果显示,对比青年女性体型,中老年女性肩、颈部曲势明显,胸腰曲线、腰臀曲线起伏不明显,腹部明显突出,前上档弧线曲势大,整体上显得较为圆厚。

由此可知,女性步入老年后,腰、腹围增加,腹部明显突出,胸高点下移,胸腰围差、臀腰围差减小,肩、颈部曲度增加,腰椎前弯减小。如图2所示,虚线为青年女性体型,实线为老年女性体型。

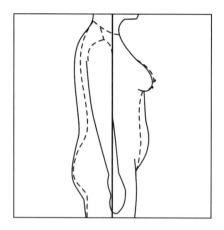

图2　青年女性与老年女性体型侧面图
Fig.2　Bodily side view of young women and old women

男性老年人随着年龄的增大,曲背情况加重,容易出现驼背、凸肚等体型,其胸围的增量相对腰围、臀围的增量小,导致胸腰落差减小,腰臀落差的绝对值也在减小,胸臀差则由负值走向正值[5]。

2.1.2 老年人的生理状态

人体衰老是涉及全身性各种细胞、组织和器官的退行性改变[6]。随着年龄的增长,老年人身体的各系统、各器官会发生程度不一的器质性或功能性改变,包括关节的退化、细胞代谢机能减退、适应力下降等。

(1)老年人的骨骼、肌肉、关节等老化和退化,导致肌力减弱、动作缓慢、手脚哆嗦等现象,容易发生骨折,行动不便,因此老年服装设计应注意穿脱容易度。

(2)老年人细胞代谢机能减退,体内色素不能及时排除,导致不少老年人肤色变深、出现棕褐色老年斑。另外,老年人黑色素无法形成,与黑色素合成相关的酶活性降低,导致新陈代谢变缓,致使老年人头发变白。所以在设计老年服装时,应注意服装的颜色能美化老年人的肤色及发色。

(3)身体机能的退化,导致老年人皮下脂肪减少、汗腺萎缩,油脂及汗液分泌减少,因此老年人皮肤干燥,衣下空气层湿度低,皮肤与服装接触摩擦易起静电,导致不适感。

(4)由于免疫力低下,老年人对自然气候变化(尤其是冷环境)的适应能力减弱,因此,在设计老年服装时,要注意服装保暖、透气及透湿性。

2.1.3 老年人的着装心理

老年人的着装心理受社会文化因素、生理因素和生活环境因素的影响[7]。

（1）中国素有"礼仪之邦"的美称，讲究体面、严谨。受此影响，老年人的服装消费观、审美观较为保守，但随着社会的发展，他们的心理需求趋向于服装的方便舒适性、管理方便性和着装讲究性，对于时尚性和价格低廉等不太在意[8]。

（2）随着年龄的增长，生理机能逐渐退化，老年人在一定程度上感到自卑抑郁。他们希望通过着装带来"年轻感"，来弥补体型的缺陷，给人以朝气又不失稳重。

（3）着装心理受周围环境的影响。有调查表明，城市中 1/3 的健康离退休老人有自己的兴趣爱好活动，兼职兼差，或从事某些与公共利益有关的活动，因此需要不同风格的服装，以适应不同场合、角色。如参加广场舞活动，老年人的服装新颖靓丽；练太极的服装严谨端庄；外出旅游的服装休闲合体等。

2.2 服装的因素

服装作为皮肤覆盖物，联系并改善着皮肤的循环机能，同时为外环境所影响。老年人的着装应与其身体各部位相适应，使人体与服装界面达到整体上的内外和谐。同时，服装还必须与外界环境相适应，具有良好的气候调节性与保护性。

2.2.1 服装号型

服装的号型标志着尺寸合体性与运动自由度等舒适程度。国家标准号型规格系列不适用于老年人的腹凸体型，而针对老年服装号型的研究也比较少，导致了大量老年服装的不合体性。例如，国标体型分类中的胸腰差量范围，男性为 2~22cm，女性为 4~24cm，但王慧娟[9]、钱露露等人[10]在研究老年人体型时，均发现了胸腰差量为负数的现象。

在大范围准确测量的基础上，科学提取中间体各部位尺寸、聚类指标、分类档差等，并对老年人的中间体体型进行重点研究，在满足老年人适体性着装的基础上，制定便于工业化生产的老年服装号型。

2.2.2 服装款式和颜色

服装的款式和颜色可修饰着装者的体型、肤色，达到适体舒适性和视觉美观舒适性的效果。老年人在着装要求上，必须首先适应生理机能衰退后的机体情况，力求宽松、舒适、简洁、方便和实用，体现出老年人的气质和成熟美。

（1）服装造型

妥善、合理的造型结构设计是解决老年人着装舒适、美观的关键。老年服装多要求宽松的版型[11]，如长方形、A 型、H 型和 O 型。A 型轮廓，肩部收窄，放宽底边线，具有安全感；长方形和 H 型轮廓，修长、款式不贴身、松腰设计，可掩盖体形变化。O 型肩部、腰部以及下摆处均没有明显棱角,并采用宽大的腰身设计,令服装整体呈现 O 型的廓型,再以简洁明了的线条和流畅的剪裁进行设计制作。

(2)服装细节

老年人的腹凸体型,容易导致服装前后衣片腰节线、下摆长的不平衡性。吴巧英、袁观洛[12]对中老年妇女上衣纸样进行了设计分析,考虑到中老年女性后背部曲度增加、前胸部乳房下垂的特点,后衣片腰节长应增加 0.5~1.5cm,前衣摆线比后衣摆线低落 1~2cm。另外,通过采用在前衣片收腹省的方法,改善了中老年人着装时前中心线下摆处向外倾斜的情况。

针对老年女性的臂围增加且胸高点下落的问题,可增加袖窿深量以及袖窿门宽度,提高衣袖的穿着舒适性。针对老年女性背部微驼并前倾的问题,可使袖子的前袖窿长小于后袖窿长,并当袖窿开度一定时,有意识地将袖山顶点前移[13]。

周萍[14]针对老年男性的腹凸体型进行裤装原型的调整。由于腹部围度的增加,适当加大裤片的前腰口宽,将前裆弧线向外放出 0.5~2cm,前门襟长度增加、直线相应变弧;前腰口线稍外弧,并减小或取消前腰口的褶裥量;减少后翘高量,并增加后裆宽 1cm 左右。

基于老年人特殊的生理特点,其服装款式应以穿脱方便为前提。内部结构装饰不能过多,外衣尽可能采用对襟式,如使用拉链、尼龙扣等,且钮扣直径宜大、数量不宜多。

(3)服装颜色

色彩可以影响人的情绪和行动。李琳琳[15]随机调查了老年人对色彩的偏爱程度。结果显示 68.7% 的老年人在意服装色彩给自身带来的影响,49.2% 的老年人喜欢的服装颜色与年轻时不同,80.6% 的老年人认为服装颜色会影响自己的情绪,59.2% 的老年人喜爱红色、黄色、橙色、棕色等暖色调的颜色,79% 的老年人喜爱蓝色、绿色、紫色等冷色调的颜色。数据说明,老年人对服装色彩的要求呈现多元化,不局限于过去的黑白灰,并重视服装颜色的选择。

2.2.3 服装面料

(1)面料的舒适性

对于老年服装消费人群而言,服装面料必须首先满足舒适性要求。老年人着装舒适性主要体现在保温、透湿透气、触觉柔和、抗静电、无过敏源等方面。

天然纤维具有良好的生物相容性,能满足老年人对服装的触感舒适性及热湿舒适性的需求,如舒适贴体的棉织物、吸汗舒爽的麻织物、柔软透气的丝织物以及质感精良的全毛织物。另外,也可以通过后整理来改善织物的舒适性,如柔软整理,能改善织物手感僵硬和粗糙的弊病,同时增加其悬垂性、平滑性。

近年来,市场上出现了用氯纶制作的毛衫裤。由于老年人皮肤干燥,其衣下空气层相对湿度通常较低,导致氯纶与皮肤摩擦后易产生静电,静电能给皮肤带来温暖,会对关节起按摩作用。但另一方面,静电也会加重老年人因皮肤干燥导致的瘙痒、脱屑等不适感。因此,老年人仍偏爱天然纤维,尤其是棉、麻织物[16]。同时也可使用一些混纺纤维和织物,在满足舒适性和安全性的同时,体现出服装的外观风格与效果。

（2）面料的安全性

服装面料经过印染整理后如含有过敏物质,或 pH 值不合适,会造成皮肤瘙痒或过敏性皮炎。由于老年人皮肤的敏感性,老年用织物应注意染料以及印染过程的健康无污染。植物染料上染率高,可生物降解,具有较强的抗菌、消炎的功能,并且许多天然植物染料具有保健功效,因此可作为老年用织物的染料[17]。另一方面,选择天然的彩色棉、彩色兔毛、彩色绵羊毛等,可回避染色加工。

老年服装的辅料应保证卫生安全性,注意避免重金属饰物及锐利、粗糙饰物的使用,注意里料、衬料、填料的质量优劣。

2.3 环境的因素

2.3.1 自然环境

自然环境因素包括环境温度、湿度、气流速度、平均辐射温度、空气品质、照明、噪声及安全卫生等,与人体的舒适感觉有着直接或间接的关系。人体通过服装与周围环境保持热湿平衡,因此,对于老年服装的选择和功能要求,环境气候起着决定性作用。

2.3.2 社会环境

随着经济的发展、社会体制的优化,老年人的社会活动向多样化发展,相应的服装行为也呈现多元化趋势。由于老年人的活动主要为群体活动,如老年舞蹈团、演唱团、老年协会等,他(她)们很注重在群体中的个人形象,希望通过着装来展现个人的修养、品位及曾有的社会地位,因此,对服装的质量、美观、功能也有更高要求。

3 结语

随着社会人口老龄化的日趋明显,老年服装产业急需从服装号型、廓型、结构以及纺织材料等方面来改善,以满足老年服装消费市场的需求,达到老年人着装的舒适性、安全性、美观性的要求。

老年人腰围增加、腹部明显突出、胸腰围差及臀腰围差减小、肩及颈部曲度增加等体型的变化,导致相应的服装号型也必须进一步细化;老年人对服装廓型也因此偏爱如长方形、A 型、H 型和 O 型等端庄稳重的廓型;对于衣片原型,后衣片腰节长比正常体服装要增加、前衣摆线比后衣摆线要低落,袖窿深量、袖窿门宽度都要增加;对于裤原型,裤片的前腰口宽、前门襟长度、后裆宽应增加,前裆弧线向外放出,前腰口的褶裥量及后翘高量应减少。老年人生理状态的改变,要求服装必须穿脱容易,具有保温、抗静电等特性。老年人心理状态的改变,要求服装色彩多元化、着装效果年轻化。

参考文献:

[1] 于建新.老年服装及其市场发展研究[D].天津:天津工业大学硕士学位论文,2008.

[2] Slater K.The assessment of comfort [J].Journal of Textile Institute, 1986, 77 (3): 157.

[3][10] 王宝环.辽东地区中老年女性体型研究[J].针织工业,2008,(12):33-35.

[4][11] 吴巧英,袁观洛.中老年与青年女性上衣纸样比较研究[J].纺织学报,2004,25(2):93-95.

[5] 李兴刚,李元虹.上海地区中老年男子体型变化和服装号型划分研究[J].东华大学学报,2003,29(6):18~22.

[6] 李广智.老年期心理特征和心理保健[J].老年医学与保健,2007,13(3):190-192.

[7] 田伟.从着装心理看中老年服装市场的发展潜力[J].西北纺织工学院学报,2000,14(1):39-42.

[8] 刘国联.东北地区老年人服装态度研究[J].纺织学报,2003,24(2):175-177.

[9] 王慧娟.陕北地区老年人体型特征及号型细分研究[D].无锡:江南大学硕士学位论文,2008.

[12] 钱露露,王宏付,胡潮江.基于三维测量的江浙地区[J].山东纺织科技,2014,(2):40-44.

[13] 孟凡瑜.中老年女装袖型问题的探讨[D].大连:大连工业大学硕士学位论文,2010.

[14] 周萍,中老年男性服装的结构造型研究[J].河南职技师院学报,1999,27(4):88-89.

[15] 李琳琳.基于老年人肤色因素的服装色彩搭配机理研究[D].上海:上海工程技术大学硕士学位论文,2010.

[16] 师华.男性老年人服装消费行为分析[J].浙江纺织服装职业技术学院学报,2009,(3):36-41.

[17] 徐腾.植物染料及其应用[J].技术进步,2004,25(11):46.

老年群体功能服装产品的设计现状及进展

◆刘慧娟[1]　王云仪[1,2]①　（1.东华大学功能防护服装研究中心，上海，200051；2.东华大学现代服装设计与技术教育部重点实验室，上海，200051）

【摘　要】　我国当前老龄化现象的加剧使得与此群体相关的研究课题得到关注，针对老年人群的功能服装产品设计和开发同样得到重视。本文从防护型功能服装、卫生保健型功能服装、气候适应型功能服装三个层面，回顾分析了该类服装产品的设计现状和进展。并对老年人群的生理和心理特征进行归纳总结，以期为针对老年人的服装多功能设计提供参考，使服装成为提高老年群体的生活品质和安全健康的有效手段。

【关键词】　老年人；生理心理；身体机能；功能服装

The Research and Analysis of Function Garments for Elders

◆ Liu Huijuan[1]　Wang Yunyi[1,2]　*(1. Protective Clothing Research Center, Donghua University, Shanghai, 200051; 2. Key Laboratory of Clothing Design & Technology, Ministry of Education, Shanghai, 200051)*

Abstract: *The increasing phenomenon of ageing in our country makes this group to be concerned by the research subject, and the design and development of functional clothing product for the aged also get attention. In this paper, the design and development of this kind of clothing products are reviewed from three aspects, which are protective function clothing, healthcare function clothing and climate adaptation clothing. And the physiological and psychological characteristics of the aged population were summarized, in order to provide reference for the multifunctional clothing design for the aged, make clothing become effective means to improve quality of life and health and safety of the aged.*

---

① 王云仪，女，东华大学，教授、博士。E-mail: wangyunyi@dhu.edu.cn

Keywords: *the elderly; physiological and psychological; body function; Functional clothing*

20世纪90年代以来,中国的老龄化进程加快。中国人口学会发布消息称,60岁及以上老年人口将从"十一五"年均净增480万,提高到"十二五"的800万左右,在2015年总量将突破2亿。80岁及以上老人正以每年5%的速度增加,到2040年将达400多万人[1]。人进入老年时期,基本的生理机能在逐渐退化,日常生活中会遇到各种障碍。作为日常必需品,服装是生活中与人体最紧密相关的事物。服装能够凭借其所能提供的功能性,诸如防跌倒、发热、卫生保健等,帮助其穿着者更好地应对生活面临一些困难和问题,不但提高其生活自理能力,更能促进其心理健康,提升生活质量。

1 研究现状

从最初的兽皮树叶遮羞,到当今的呼吸型雨衣、防跌倒服等,各种功能服饰层出不穷。在欧洲,功能服装的思想最早出现于古希腊的皮肤呼吸理论[2]。此后,人们逐渐开始对服装功能性的探究。二战期间的军事竞赛和战争需求,加速了功能服装的发展[3]。20世纪60年代开始,欧美国家开展了对老年人人体尺寸方面的研究,英国也于1969年颁布了第一部定义老年人人体尺寸的标准[4]。国外学者和服装设计师率先展开了对中老年人功能服装的研究与设计开发。

新西兰的设计师Deborah Leath创立了专为老年人及残疾人设计服装的品牌Clothing Solutions,他所设计的服装外形上与普通人的衣服并无差别,但在细节上帮助老年人及残疾人解决了由于自身行动困难引起的穿脱难的问题。美国的Adaptive Clothing专为生活不能自理、丧失自理能力的老人及残疾人设计服装,其品类包括从睡衣到防雨衣等各种功能服装。加拿大的Silvert's也是为相同群体设计制作服装的品牌,做到最大限度地增强服装的功能性和安全性。

国内学者对老年人功能服装的研究起步较晚。清华大学的胡海涛对老年人的人体尺寸、力量和关节活动的内容进行了测试和分析,并讨论了老年人失禁护理产品的号型划分方法[5]。张向辉、徐继红、张文斌等人将老年女性与青年女性的身体数据进行对比分析,得出了老年女性身体部位的变化规律并得出实用性老年服装原型[6]。王海毅、冯伟等人探讨了智能纺织品在老年服装产品开发上的应用前景,提出以环境适应和安全为中心,从舒适、技能、便利的角度把握纺织品设计开发的空间和发展方向[7]。

2 产品设计开发进展

面向老年人群的功能服装设计包含多种需求满足,本文将其中主要的类型归纳为防护性功能服装、卫生保健型功能服装和环境适应型功能服装。

2.1 防护型功能服装

（1）防走失服

2014年的一项调查显示我国65岁以上老年人老年性痴呆患病率为5.9%，在2020年将达到1020万。对于患有老年痴呆的老人，在独自外出时经常没办法自己回家，因此采用"防走失服"可有效应对这种情况。"防走失服"上装有"GPS定位紧急呼叫器"，如若老人出门在外走丢，便可按下小方块上面的红色按钮，接到报警的一端便会明确老人具体的位置，从而迅速找到老人。

（2）防摔服

随着年龄的增大，老年人体内的钙会流失，从而导致骨质疏松症状的发生，这是一种全身代谢性的疾病。据报道，我国已成为世界上骨质疏松症患者最多的国家，我国老年人的跌倒发生率为14%~34%[8]，对于老年人而言，摔倒很容易造成骨折。因此，"防摔服"的作用不容轻视。

麻省理工学院的研究生伯曼利开发出了一种监测老年人平衡状态的防跌智能鞋，利用其内置的传感器来监测穿鞋者的平衡状态并收集数据。美国弗吉尼亚理工大学的工程团队设计出了防摔电子裤，通过监测关节运动的状态分析个体步态的稳定性，当显示个体步态不稳定后，系统会自动报知穿着者并提醒其尽量避开不安全路面[9]，从而有效减少老年人摔倒的概率及数量。此外，部分防摔服选择在服装的肘、胯、膝和关节等处装置D3O凝胶功能性材料，其在常态时保持松弛状态并呈现柔软的泡沫状，可被随意挤压成各种形状，一旦受到外界的高速强烈撞击或瞬间强力时，分子会迅速互相交错并锁在一起从而使得材料变得坚固，外力消失凝胶即可自动恢复柔软，从而保护老年人摔倒后肘、胯、膝和关节等容易受伤的部位[10]。

（3）特殊功能服装

根据因身体某部位出现运动功能障碍或者有特殊需要的老人而设计开发的服装，以帮助老人在非正常的身体状况下提高生活的便利性和生活自理能力。如轮椅使用者所穿着的功能服装，在保证穿着者舒适性的情况下，还需要考虑使穿着者穿脱方便。东华大学的吴黛唯[11]通过调查发现现有轮式使用者所穿着服装存在的问题，并对其进行在裤装采用双链门襟、在袖肘部增加特殊功能贴片等功能性的设计和开发，保证穿着者在各种情况下穿着舒适和方便。

2.2 卫生保健型功能服装

卫生保健型服装从纺织品加工形式上可以分为三类：第一类为通过纤维改性获得抗菌效果；第二类为通过织物后整理起到治疗或预防作用；第三类为在特制的服装中附着一定量的中草药，通过中草药的发挥起到治病目的。

我国先后研制开发了暖胃背心、护肩服、平喘背心等保健功能服，这些服装的胃、膝、背、腰等处放置不同成分的中草药袋，药袋中的中草药会发出有效成分，借人体体温使得皮肤和穴位吸收，从而对一些因受寒所引起的疾病起到预防和辅助治疗作用。此外，护肩、护膝等服装中的中草药还有祛风燥湿、活血通络的作用，因而对风寒所致的肩周炎等有显著的预防和治疗效果[12]。

另一种保健服装主要面料成分为罗布麻纤维,经专业人士鉴定,罗布麻纤维对高血压、高血脂、气管炎患者均有良好的保健效果。罗布麻根中的酚类和甙类物质,茎中的鞣质等酚类,叶中的蒽醌、黄酮类等能起到抑制细菌的生长和杀菌作用,从而促进细胞活力,抵抗紫外线对皮肤的伤害,阻止血管硬化等[13],因此具有良好的保健效果。

2.3 气候适应型功能服装

（1）调温服

中老年人身体机能的下降导其不能很好地调节自身温度以适应环境气候的变化。智能调温服装此时能在外部环境与着装人体之间发挥积极的温度调节作用,保持人体的热平衡。其所用智能调温纤维存在具有能量转换功能的相变材料(常用的为碳氢化腊),吸收储存和释放热量,在温度变化中以固态、液态互相转化而达到吸放热效果,进而维持人体温恒定[14]。

在炎热的夏季,中老年人在户外的高温下很容易中暑甚至热中风,在室内空调或风扇也会带来身体的不适,调温功能服内含有大量无纺布絮片,由高吸水率的纤维制成[3],在夏季可以起到很好的避暑降温效果。

（2）发热服

在冬季容易气候变化时中老年人易感染疾病,甚至高血压患者一旦受到寒冷刺激会使血压升高,因此,同样需要服装来调节自身与环境之间的关系。发热服是冬季很好的选择,常用的有电热服、太阳能热服、化学能热服等[15]。

电热服的原理是将带有发热功能的发热布安置在衣服的内衬里,用电池为它提供电能,使其慢慢发热,并可以根据服装内外的温度进行自动调节,保持服装内温度的正常和舒适,有效避免中老年人因温差而引起的不适。同时,安置在服装上的装置可以方便的拆卸,便于清洗和折叠。

近年来,利用太阳能的发热服装正在兴起,包括太阳能发热服和蓄热服[16],目前日本企业在蓄热保温材料领域的开发处于世界领先地位[17]。但这类服装只是针对人体局部加热,容易造成人体皮肤温度分布不均匀,血液分布异常,所以还没有得到广泛的推广。

3 需求分析及设计原则

老年人由于生理及心理的变化,与其他年龄段对产品需求有较大差异。在智能时代,老年人对服装产品不仅仅在款式、色彩、面料等基本层面有需求,更需要开发具有功能性的服装产品来满足其需要[18]。

3.1 老年人特征分析

（1）生理特征的变化

人类的衰老首先从生理方面开始。在体型特征方面,40岁以后很多人的体型会出现腹部脂肪较厚且下垂、胸部下垂从而使得腰部和胸部尺寸明显增大,胸腰差减小的现象。同时由于老年人骨骼成分的流失使得其身高逐渐变矮。随着年龄的增大,还会出现弯腰驼背、身体机体衰老,肌肉松弛、皮肤有皱纹等现象。总的来说,中老年人体型特征的改变主要表现在身高变矮,胸腰围尺寸明显增大,臀部阔而扁平,有些老年人出现驼背[19]。

体质机能方面由组织器官与生理功能退化而带来的机体储备能力减少;由于多种生理功能的减退,导致体内环境稳定性失调,出现各种功能障碍;免疫功能的衰退与紊乱造成抵抗力下降;由于心脑血管、呼吸、消化、内分泌等系统的生理功能的全面衰退,对环境的适应能力逐渐下降,易诱发疾病[20];生活自理能力减低,出现语言迟缓、耳聋眼花、手指哆嗦、运动障碍等[21]。

（2）心理特征的变化

心理学家指出,随着年龄的增大以及生理的变化,中老年人容易产生消极情绪,且其情绪体验比较强烈,出现一定程度的心理空虚、孤独、寂寞等不良的情绪[4],且相比于其他年龄层群体会更加关注自己的身体健康。

衣食住行中"衣"是排在首位的,是最贴近身体的必需品,因此中老年服装在满足其最基本的防寒保暖和审美功能之外,更应该注重其他方面的功能性。通过服装其他方面功能性的设计,延缓中老年人生理性衰老,延长其活动机能[19],为其营造良好的生理健康环境,进而促进其心理健康。

3.2 产品设计原则

功能服装最主要的作用是满足人的某种需求,只有最大限度与人的需求和特点相结合,才能充分发挥其功能。因此,老年人群功能服装的设计应遵循以下三项原则:

（1）与生理特点相结合。老年人身体机能的退化使其更看重服装的舒适性、安全性。需要对老年群体的体型进行系统的研究,建立健全个性化的号型体系,以使服装能够较好地符合人体特点;

（2）与心理需求相结合。老年人多倾向大方得体的服装,在显示其精神活力的同时又不失庄重优雅。需要对老年人的审美习惯进行深入研究,使服装的色彩、款式等符合其消费心理;

（3）与生活方式相结合。老年人中的很大一部分人已结束职业生活,有大量的闲暇时间,消费观念、时尚意识等都发生着不同程度的变化。需要对其生活内容进行细分,以是服装的品类及其功能满足其多方面的生活细节要求。

4 结论

老年群体应当被给予更多的关注,针对老年人的服装多功能设计将有助于提高老年群体的生活品质和安全健康,当前已开发出包括防护型功能服装、卫生保健型功能服装、气候适应型功能服装在内的多种功能服装。综合相关的医学、社会学、社会心理学、人体工程学以及服装材料学等相关领域的知识,建立并完善老年群体的体型和服装号型等数据库,系统研究老年人群在生理、心理等方面的特征和特性，才能为其提供满足需求的、多样化的功能服装产品。

参考文献：

［1］ 全国老龄工作委员会办公室.中国人口老龄化发展趋势预测研究报告［R］.2006-2-23.
［2］ 欧阳晔.服装卫生学［M］.北京：人民军医出版社，1985.
［3］ 陈庚笙.老年人功能性服装的应用分析［D］.华南理工大学：2014：2，29-34.
［4］ 李晖.老年服装的人性化设计研究［D］.齐齐哈尔：齐齐哈尔大学硕士学位论文，2012.
［5］ 胡海滔.老年人的人体测量［D］.北京：清华大学硕士学位论文，2005.
［6］ 张向辉，徐继红，张文斌.老年女性服装原型的分析研究［J］.扬州大学学报：2002（4）：29-31.
［7］ 王海毅，冯伟.智能纺织品在老年服装中的应用分析［J］.江苏纺织：2008（12）：55-57.
［8］ World Health Organization. Global report on falls prevention in older age［EB/OL］. http://www.who.int/ageing/projects.Halls prevention older age/en/index.html.
［9］ 巧云.预防老人跌倒的智能服装［J］.知识就是力量，2009（8）：70-71.
［10］ 王文娟.功能性服装设计在老年服装中的应用［J］.艺术教育，2014（12）：301-302.
［11］ 吴黛唯.轮椅使用者的功能服装研发及评价［D］.东华大学硕士学位论文，2011.
［12］ 曾斌平，具有防病治病功能的保健服装［J］.产业用纺织品，1988（10），39-40.
［13］ 胡斐娟，邱文，单小红等.罗布麻纤维结构及抑菌性能研究［J］.轻纺工业与技术，2010（4）：9-11.
［14］ 王立荣，史清丽，郭凤芝等.具有智能调温功能的桑蚕丝服装产品的开发［J］.针织工业，2008（2）：9-12.
［15］ 庄梅玲，张晓枫.电热服的热性能评价［J］.青岛大学学报：工程技术版，2004，19（2）：54-58.
［16］ 张小雪.基于太阳能利用的发热服装研究［J］.国际纺织导报，2014（3）：55-60.
［17］ 卓越的日本蓄热保温面料技术［EB/OL］. http://www.chinayarn.com/xianwei/showsj.asp?id=280. 2014-3-8.
［18］ 陶铁元.我国辽宁地区中老年女性体型变化规律研究［J］.国外丝绸，2008（5）：22-23.
［19］ 阎珺.老年人功能服装的研究现状综述［J］.科学中国人，2015（4）：179-181.
［20］ Cheng Yang, W Mark, Rosenberg. A Program for Research on Dimensions of an Aging Population［J］. Sedap Research Paper, 2009（241）：1-26.
［21］ 何晓佑，谢云峰.人性化设计［M］.南京：南京大学出版社，2003.

中老年功能性服装概述

◆ 洪正琳[①]　田亚楠　邢晓宇　尚笑梅　（苏州大学纺织与服装工程学院，江苏苏州，215021）

【摘　要】　随着老龄化社会进程的加速，中老年群体服装消费问题越来越受重视。中老年人服装设计应本着以老年人为本，由于老年人群体的特殊性，对功能性服装的需求很大。我国对老年人功能性服装的研究虽起步不久，但是效果显著。中老年人功能性服装设计主要从结构、色彩、面料等方面着手，文章介绍了中老年人功能性服装的现状和研究成果。

【关键词】　功能性；服装；中老年

Overview of the Quinquagenarian Functional Clothing

◆ Hong Zhenglin　Xing Xiaoyu　Tian Yanan　Shang Xiaomei　*(College of Textile and Clothing Engineering, Soochow University, Suzhou, P.R.China, 215021) (E-mail: 1547243685@qq.com, 1247562355@qq.com, 1576691783@qq.com, shangxiaomei@suda.edu.cn)*

Abstract: *With the acceleration of aging society, the consumption of the quinquagenarian group clothing is valued more and more. The quinquagenarian clothing design should be based on the elderly. Because of the particularity of the elderly, the elderly have a great demand for functional clothing. It has only recently to study functional clothing of the elderly in China, but we achieve a lot. The elderly functional clothing can be designed mainly from aspects of structure, color, fabrics and so on. This article introduces the status and research results of the elderly functional clothing.*

Keywords: *Function; Clothing; Quinquagenarian*

[①] 洪正琳，女，研究生。E-mail：1547243685@qq.com，shangxiaomei@suda.edu.cn

引言

中国社会已经步入老龄化社会,中老年人的服装消费越来越受服装界关注。现今,我国中老年群体中知识分子占的比例逐渐加大,物质条件有很大提升,消费观也发生了一些转变,中老年人对时尚的追求也越来越热爱。中老年人在体型上多存在腰腹部脂肪较多、略微驼背等问题,大多肤色比较暗沉,身体机能下降,患有某些疾病等。在中老年人服装设计时,需要考虑这些因素以及理解他们追求美好的心理。功能性服装可以给中老年人带来生活动便利以及愉悦他们的心情。随着科技的发展,越来越多科学产品被运用到服装中,智能设备、保健面料、可穿戴设备等,这些元素的加入使得在设计中老年服装设计时可以更好地实现所需功能。

1 中老年对功能性服装的需求

服装的功能主要有保护功能和装饰功能两种。从狭义的角度来定义,功能性服装一般指该服装产品具有一般服装所没有的、对消费者有特殊益处的各种特殊物理实用功能(如抗菌、防紫外线等)。从服装实用功能的角度看,这些功能同一般产品的已有基本功能相比,总体上都可以归纳为对穿着者身体的保护和保养[1]。对功能性服装的研究是始于20世纪40年代,在第二次世界大战中,参战国饱受严寒的侵袭,由于士兵们服装的防寒性能差而大大削弱了士兵的战斗力。自此,各国大力开展了对功能性服装的研究和开发[2]。

世界卫生组织在对人口素质和平均寿命测定后划定:45~59岁者为中年人,60~74岁者为年轻的老年人,75~89岁者为老年人,90岁以上者为长寿老人[3]。我国习惯上将60岁以上的人称为老年人,40~55岁的人称为中年人,一般说来,老年人口超过10%的比例就属于老龄化社会,截至2007年底,中国60岁及以上老年人口达1.53亿,占总人口的11.6%,预计到2020年,全国老年人将达到2.43亿,占总人口的17%,到21世纪中期,中国老年人口将超过4亿[4]。我国已经步入老龄化社会很多年,中老年服装市场也很广阔。然而,中国针对老年群体的设计服务远远滞后于社会需求。据中国老龄委提供的数据统计,目前中国老年人年消费需求为6 000亿元,而实际每年为老年人提供的产品不足1 000亿元[5]。

虽然老年服装在款式上多样化了,但就服装穿着的舒适度、健康性、便捷功能以及安全防护等问题上,对老年人的生理和心理需求存在明显不足。面对这些存在的缺陷,应该本着"以老人为本"的设计理念去全方位地改善[5]。服装是最贴近老年人生活的必需品,消除老人在穿衣、着装和日常生活过程中出现的不便和障碍,将功能性服装设计理念融入到中老年服装中是很有必要的。

2 中老年功能性服装研究现状

功能性服装的大规模研究始于第二次世界大战期间。由于作战服的防寒性差,在天气寒冷的情况下,士兵的战斗能力大打折扣,因严寒而导致的非战斗性士兵伤亡数目巨大。根据战争需要,各国都加强了对服装的功能性研究,并积累了大量的实验数据,为目前功能性服装的发展打下了基础[1]。我国对功能性服装的研究始于20世纪60年代,在七八十年代开始对服装功能性进行系统研究。进入90年代后,我国在服装功能性领域的研究取得了巨大的进展,涉及的研究领域也不断扩大,从款式剪裁到纤维面料、从人体工学到生理机能、从特定行业到日常生活等[6]。进入21世纪后,具有各种特殊性能的功能性服装产品层出不穷,如具有抗紫外线、高强度、抗菌消臭、阻燃等特殊性能的各类服装,纷纷走入市场,引起了广大消费者和产业界的关注[7]。

2004年65岁及以上老年人口占总人口的比重,美国为12%,日本为19%,德国为17%,英国为16%,法国为16%,意大利为19%[8]。很多国家都步入了老龄化社会,中老年人的服装设计越来越受重视。截至2013年底,我国50岁以上人口已超过3.6亿。如何服务中老年人,提高他们的生活质量事关重大。在2014年举行的全国中老年服装科技趋势研讨会上,不少业内专家指出,中国的中老年时尚产业比日本起码要落后20年,尤其是在功能服装领域,国内几乎是空白市场[9]。但是中老年服装对功能性设计的需求不断增长,随着经济水平的提高,中老年更加注重生活的品质,对服装的舒适性、行动便捷性、健康性、美观性等有很大的需求。

我国对中老年功能性服装的研究已经步入正轨,市面上也可见一些功能性服装,但是产品线不全,很多成果仍然停留在技术层面。英国一个老龄功能性服装设计课题遵循包容性设计的原则,以对老龄用户需求的调研和分析为出发点,以满足用户需要为原则,综合运用服装设计的审美因素与功能因素的综合性设计[10]。

3 中老年功能性服装研究成果

很多国家都在进行功能性服装的研究。美国的设计师Zach Hoeken制作了镶有电脑键盘的牛仔裤,只要穿起它就可以自由控制电脑。目前,世界上已经研制开发的功能性服装包括:抗菌保健服;健美减肥服;芳香催眠服;变色服;智能服装。甚至研发人员还开发了驱蚊服、可食服等[2]。早在21世纪初,我国功能性服装的研究成果就分为阻燃防护服、抗静电防护、防水服、防紫外线、防寒保暖服、防化服、抗菌防臭服、防弹衣几大类[6,11]。这些种类的功能性服装大多是为特定岗位人群研发的。近些年,随着生活水平的显著提升,功能性服装的市场前景才越来越好。例如我国在60年代就开始研究电磁辐射与防护服,70年代至80年代正式生产铜丝与柞蚕丝混纺布制成的屏蔽服和微波吸收防护服。此后有关不锈钢软化纤维屏蔽织物,特殊工艺镀膜屏蔽织物与服装也相继问世,但真正进入大众市场则是90年代中期至今[12]。功能性服装的研究到市场化总是需要一些时间的。面对各种功能性服装,抗皱和易于打理始终是全球消费者较为期望获得的两项特性[13]。从类型上来说,目前市场上的功能性服装主要分为防护型服装、调整型服装和舒适性服装3类[14]。

中老年功能性服装的设计主要从功能性材料、结构、色彩及辅助装置几个方面展开[15]。

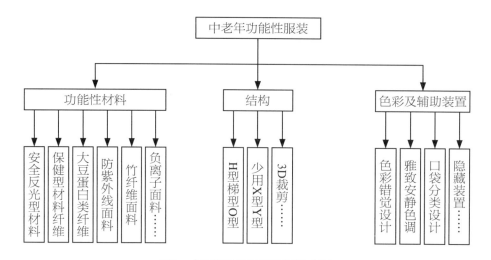

图1 中老年功能性服装设计方面
Fig.1 Functional Clothing Designed for Quinquagenarian

3.1 功能性服装材料的应用

功能性材料具有吸湿、排汗、透气、保暖、防风、防雨、防静电、可伸缩、抗皱、柔软、耐磨、阻燃、防辐射以及红外线功能等,可根据材料不同的性能应用于不同部位的功能性设计,能够更好地满足中老年人生理需求和适应环境的变化。

（1）D3O凝胶制作的老年防摔服。由英国工程师理查德·帕尔默发明的抗冲击单一材料D3O凝胶,静止或缓慢移动状态下,可被随意挤压成各种形状,性能柔软,一旦受到高速冲击,分子迅速互相交错并锁在一起变得坚固,可应用于肘、胯、膝等部位的防摔功能性设计,从而避免严重伤害[5]。用这种材料制作的老年防摔服可以有效避免老年人摔伤的情况。

图2 D3O凝胶
Fig.2 D3O Gel

（2）反光安全型材料的应用。反光材料依据高折射率玻璃珠的球体回归反射原理,将光线反射回光源处,可增强行动的人体在夜间视线不良或危急境况中的可见度,具有很强的警示效果。这种材料被广泛运用于防火服、反光背心等。用反光安全型材料制作的中老年人服饰,在夜间散步、登山徒步时穿着可以易被发现以减少危险。

（3）抗菌防臭面料的应用。抗菌防臭面料可阻止细菌和真菌的生长,可防止老人的过敏和哮喘。采用纳米铜制作的免洗内衣可以给活动不便的老人穿着,既省力又干净卫生。

图3 反光背心
Fig.3 Reflective Vests

（4）保健型面料的应用。保健型材料指纤维中含有多种具有医疗效用的物质,可在穿用过程中缓慢释放,对心脑血管病、慢性关节炎等进行辅助治疗[15]。保健型面料对老人身体可以起到健康防护作用,同时老年人穿上这种服装会产生身体在慢慢好转的心理暗示,有利于老年人的康复性治疗。

功能性材料有很多,例如大豆蛋白类纤维、防紫外线面料、竹纤维面料、负离子面料等,各种功能性材料基本满足了中老年人对服装的功能性需求。

3.2 功能性的服装结构

中老年功能性服装的款式结构上需要依据中老年的体型进行设计。中老年人大多腰腹部肌肉松弛下垂,背部略微佝偻。在服装设计时应多采用 H 型、梯型和 O 型的设计,避免采用 X 型和 Y 型的设计,且款式应以宽松为主。裁剪方面可以使用 3D 立体剪裁来增加服装的线条的流畅性,使得服装自身立体有型。

另外改变服装结构也能赋予服装特殊性能,如在 T 恤肩部和颈背部制作夹层,一方面较厚的夹层可以阻止空调冷气对皮肤的直接刺激,起到保护肩部或颈背部的效果;另一方面衣服夹层在隐蔽部位留有开口,在进行家庭理疗时,可以方便放置远红外发热材料、中药理疗包等功能性填充物[16]。这种夹层也可以在服装的肘部、膝盖部位、腰部等部位进行设计,保护中老年人的身体各部位。

3.3 色彩及辅助装置

中老年人由于岁月的沉淀,肤色大多黯淡,在进行服装设计时可采用色彩错觉设计,衬托中老年人的肤色,并采用雅致安静色调,避免活泼跳跃的颜色。

一些辅助装置的应用也能增加服装的功能性。如传感器是将各种物理化学等非电信号按照一定的规律转化成为便于处理和输出的其他物理量信号(如电压或电阻信号)的器件或装置,它在功能性服装研究中起着重要作用。针对服装的特殊需求,采用相应的传感器对功能服装的各项指标进行测试,实现对着装人体内外环境的各种信号变化进行测试、采集和显示[17]。功能性服装在满足其特有的功能外,更着重考虑人体着装舒适性。传感器在功能性服装中的应用主要体

图4 内置传感器的内衣
Fig.4 Underwear with embedded sensor

现在对服装舒适性指标的测定和信号的采集,为服装提供实时的测量、监控,实现对着装人体的动态测量与监护,例如针对老年人的生理特性及心理特性设计的老年智能化内衣。老年智能内衣带有多个传感器以及信号发射装置,利用传感器来测定老年人体温、血压、尿布内湿度变化等生理指标,当湿度超出一定的范围时,会自动开启音响系统等。智能型老年服装在设计时,更多地考虑根据老人的心律等生理指标的变化,对老人的行为作出判断,特别是对于一些突发事件和行为,如心肌梗塞、摔倒等的发生和预测,做出相应的反应,从而能够将危险降到最低[17]。

4 总结

我国步入老龄化社会已经很多年了,老龄化的趋势也在不断加深。中老年人已经成为有力的消费群体,但是在服装消费方面总是选不中心仪的产品。我国中老年功能性服装起步较晚,但是成果丰硕。设计师可以从功能性材料、结构、色彩和辅助装置三个方面着手设计中老年功能性服装。

老龄服装设计的研究与实践还面临很多问题,例如缺乏针对老龄用户生理和心理需求的全面研究;缺乏适合老年人审美习惯的款式及色彩设计的研究;缺乏对老龄人体的基础号型研究,导致服装难以符合老年人体型特点;缺乏考虑老年人的生理特点和局限性的功能性设计,如便利的穿着方式、服装人体舒适度的调节以及防护与健康方面的特殊功能的考虑[10]。每一个人都会经历人生的中老年阶段,我们设计改善功能性服装、提高中老年人的生活品质,其实是为我们的未来进行设计。中老年功能性服装发展日渐成熟,我们需要好好思考并设计自己的未来。

参考文献:

[1] 杜娟娟.功能性服装的发展趋势及建议[J].科技资讯,2010,27:246.
[2] 白玉苓.基于顾客感知价值的功能性服装开发[J].纺织导报,2012,07:146-148.
[3] 陈继红.论中老年服装设计[J].山东纺织科技,2002,06:42-45.
[4] 戴淑娇,徐蓉蓉,金慧颖.绍兴市中老年消费者服装购买心理与行为调查分析[J].绍兴文理学院学报(自然科学),2013,03:48-52.
[5] 王文娟.功能性服装设计在老年服装中的应用[J].艺术教育,2014,12:301-302.
[6] 张素英,韩月芬.功能性服装的研发现状及建议[J].中外企业家,2014,18:230-231.
[7] 苏锷.对开发功能性服装产品的思考[J].山东纺织科技,2003,01:36-38.
[8] 刘清芝.美国、日本、韩国应对人口老龄化的经验及其启示[J].西北人口,2009,04:73-75.
[9] http://www.cnfzflw.com/news/show-348197.html
[10] 王露.关注老龄需求设计我们的"未来"[J].中国纺织,2011,09:121-123.
[11] 刘丽英.功能性服装的研究现状和发展趋势[J].中国个体防护装备,2001,04:25-26,30.
[12] 文珊,刘若华.防电磁波辐射的功能性服装[J].湖南环境生物职业技术学院学报,2004,01:27-29.
[13] 晓晨.功能性服装的全球态度[J].中国纺织,2008,10:84-85.
[14] 陈国强.基于服装人体功效学的功能性服装设计[J].纺织科技进展,2015,02:77-80.
[15] 王文娟.老年服装的功能设计研究[J].艺术教育,2015,02:257-258.
[16] 李琼舟,王国书.保健型老年丝绸T恤设计思考[J].丝绸,2014,02:47-50.
[17] 闫学玲.传感器在智能老年服装设计中的应用与研发构想[J].纺织导报,2010,03:94-95.

中高龄服饰消费行为研究

上海地区针织塑身内衣市场现状及其面向中高龄消费者的思考

◆ 李 敏[1,2][①]　王 鑫[1]　(1. 东华大学服装与艺术设计学院，上海，200051；2. 东华大学现代服装设计与技术教育部重点实验室，上海，200051)

【摘　要】　针织塑身内衣是女性追求曲线美的重要途径。本文以4Ps理论为基础，通过对上海地区五个具有代表性商圈中的20个内衣品牌进行实地调研，对目前市场上内衣品牌的店铺情况、产品及价格等方面进行了分析，并针对中高龄消费者提出了产品、服务、网络化和高级定制化的建议。

【关键词】　针织塑身内衣；市场现状；中高龄消费者

Market Status of Knitted Slimming Underwear in Shanghai and Suggestion to Middle Aged Consumers Oriented

◆ Li Min[1,2]　Wang Xin[1]　(1. Fashion and Art Design Institute, Donghua University, Shanghai, P.R.China, 200051; 2. Key Laboratory of Clothing Design & Technology, Ministry of Education, Donghua University, Shanghai, P.R.China, 200051)

Abstract: *Knitted slimming underwear is the important way to keep fit for women. Based on 4Ps, marketing research of 20 underwear brands of five major shopping districts in Shanghai was conducted. Product, price, place and promotion of these brands are analyzed. Suggestion including product, service, network marketing and customization are put forward to help underwear brands to orient middle-aged consumers.*

---

① 李敏，女，教授，博士，东华大学，研究方向：服装产业经济、服装舒适性，联系方式：fidlimin@dhu.edu.cn 基金项目：上海市教育委员会科研创新项目资助（14ZS068）；上海高校知识服务平台（海派时尚设计及价值创造知识服务中心）资助项目（13S107024）

Keywords: *Knitted slimming underwear; Market status; Middle aged consumers*

女性身体特征在年龄、地心引力等因素影响下，会随着时间的推移慢慢发生改变，穿着合适、正确的内衣是保持和修饰身材的重要途径。因此，对身体曲线美的追求成为女性穿着塑身内衣的一大原因，特别是中高龄女性。目前，国内外品牌对内衣的研究重点主要放在舒适性[1,2]和工艺方面[3,4]，本文针对针织塑身内衣的市场现状，思考面向中高龄女性消费者的策略，希望对目前的针织塑身内衣品牌有一定的参考价值。

1 相关概念

1.1 内衣的概念

内衣（Under Cover 或 Under Wear）相对外衣而言，是穿在外衣里面、紧贴皮肤的服装。一般可分为基础内衣（文胸、内裤）、针织塑身内衣（塑身衣、塑裤、腹带等）、家居服（睡衣、睡裙等）三类[5]。

1.2 针织塑身内衣的概念

针织塑身内衣分为调整型针织塑身内衣和弹力型针织塑身内衣两大类[6]。

弹力型针织塑身内衣是指通过编织成型（缝迹伸长率>100%），而对人体特定部位起到牵引或约束作用，从而保持或调整人体特定部位尺寸和形态的内衣。该品种内衣可在针织圆机上缝制成型，往往在腰、腹位置上添加比例较多的起加强作用的弹性纤维，面料均具高弹性（图1）。

图1　弹力型针织塑身内衣
Fig.1　Elastic knitted slimming underwear

调整型针织塑身内衣是指通过缝制成型（可使用衬垫、骨架支撑其材料）而对人体特定部位起到牵引或约束作用，从而保持或调整人体特定部位尺寸和形态的内衣。该品种内衣大多由不同的面料拼接缝制而成，根据设计的要求，往往只是部分面料才具高弹性（图2）。

塑身内衣可以调整身材，达到丰胸、细腰、提臀等效果。塑身内衣主要是根据脂肪流动学原理，让脂肪合理、均衡地分布，从而达到塑身的效果[7]。

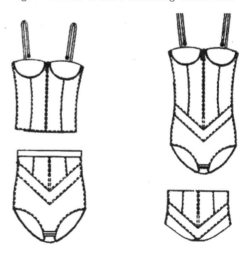

图2　调整型针织塑身内衣
Fig.2　Adjustment knitted slimming underwear

2 针织塑身内衣市场现状

调研选取了五个上海具有代表性的商圈进行考察：徐家汇商圈、中山公园商圈、南京东路商圈、人民广场商圈和陆家嘴商圈。调研对象为含有针织塑身内衣产品的内衣品牌，包括婷美、爱慕、黛安芬、曼妮芬、安莉芳、华歌尔、欧迪芬、古今等20个品牌。

调研的内容是以4Ps营销理论为基础，品牌渠道调研包括店铺位置、客流量、店铺布置等内容；品牌促销调研包括品牌在商圈内位置、POP设置、配套设施以及促销打折；产品及价格调研包括针织塑身内衣分类、件数、价格、色彩、面料、做工、设计感、产品风格等方面的内容。

2.1 店铺调研情况

（1）店铺面积、位置

内衣品牌主要位于商场三层，其次为二层和四层，一层店铺几乎没有内衣品牌。位置多位于边厅和专柜，位于专卖店和中岛的店铺较少。面积在 $12\sim25m^2$ 之间的店铺最多，较大和较小的店铺分布均不多，位于边厅的内衣品牌店面积均偏大，在 $45m^2$ 左右。

（2）店铺整体形象

同一内衣品牌位于不同商圈的店铺整体形象均很统一，可见品牌对自身形象建设十分重视，有统一的风格定位。此外，知名内衣品牌在店铺品牌名称和标志设计上清晰醒目、简洁大气，在橱窗设计上均选用了模特、海报或者灯箱展示，能够更好地表达出品牌的文化理念。并在店面外墙的色彩和装潢上注意与其他内衣品牌有所区分，不仅吸引消费者的注目，而且提升了品牌的辨识度。塑身内衣产品所占比例较大的内衣品牌，还会在其橱窗用模特出样塑身内衣，以吸引消费者注意。

（3）店铺环境

研究发现绝大部分店铺在灯光的设置上采用偏明黄色色调，给顾客以明亮舒适的购物感觉，但并没有区别于其他品牌且符合自身品牌理念的灯光设置。在这方面，伊丝艾拉品牌店铺灯光设置暖黄偏暗，性感时尚，与整个店面黑色系的神秘和品牌理念相吻合，具有明显的区分性。此外，音乐作为改善购物环境的一个重要因素，能够给顾客带来愉悦和放松的心情，调研发现各品牌店铺均没有背景音乐的设置，可以作为改善的方面。

（4）POP宣传方式

POP是英文 Point of Purchase 的缩写，即"购买点"的意思。图3为各种POP广告设置所占比例图。可见，在所有调研店铺均有POP广告设置，其中壁面POP和货架POP的广告运用最多，均超过总广告设置的20%。在调研中还发现，若POP广告中包含穿着塑身内衣前后的对比效果图，消费者极易因图片的效果对塑身内衣产生兴趣，要求导购为其试穿。这样的POP设置对提升消费者购买意愿是具有非常积极的意义的。而货架POP设置有利于吸引消费者对产品的关注，"强制"顾客接受商品信息。除此之外，在调研的内衣品牌中爱慕和黛安芬边厅店铺均设有视听POP广告，播放其品牌内衣在时装发布会和秀场的视频，从而吸引消费者注意。

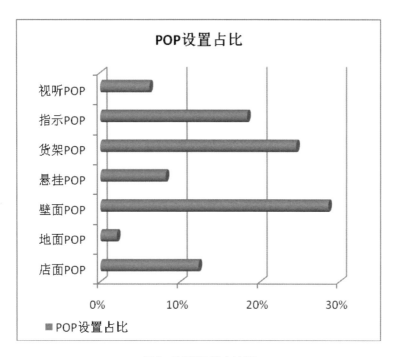

图3 POP设置占比图
Fig.3 Scale map of POP

2.2 针织塑身内衣产品调研情况

（1）针织塑身内衣所占比例

表1为各内衣品牌中针织塑身内衣所占比例，从表中可以看到，除婷美、Bradelis（日本品牌）针织塑身内衣所占比例较大（达到80%~95%）外，其他品牌的针织塑身内衣占所比均只有5%~15%。其中黛安芬、曼妮芬、达吉斯等品牌针织塑身内衣所占比例甚至低于5%。针织塑身内衣平均件数在10~20件之间。一些高端的内衣品牌通过引进日本专业针织塑身内衣品牌来吸引消费者，如欣姿芳旗下的露蒂芬（Lofan）以及其代理品牌田村驹（Yauco）。

表1 各内衣品牌中针织塑身内衣所占比例（%）
Tab.1 Result of knitted slimming underwear in underwear brands(%)

品牌	所占比例	品牌	所占比例	品牌	所占比例	品牌	所占比例
Bradelis	90~95	爱慕	10~15	古今	5~10	思薇雅	6
婷美	80~90	欧迪芬	5~15	奥黛丽	5~10	曼妮芬	3~5
欣姿芳	15~20	百利安	12	安莉芳	5~10	达吉斯	3~5
梵丝内挺	15	伊丝艾拉	10~12	奇丽尔	8	黛安芬	2~5
曼黛玛琏	12~15	华歌尔	10~12	艾黛妮	8	博尼	2~5

（2）针织塑身内衣种类

在调整型针织塑身内衣中，束腰、束裤和胸腰腹三合一内衣所占比例较多，达到总体的一半以上。调整型文胸品类也较多，且主要以聚拢效果为调整目的。一件式连体款较少，平均一家店铺只有一至两件。风格以简洁大方和高贵典雅为主，少数品牌的产品风格比较甜美性感，但所占比例不多，颜色多为黑色、白色和肉色。弹力性针织塑身内衣以长袖长裤款为主，鲜见其他款式。

（3）针织塑身内衣面料

针织塑身内衣的面料多采用棉、氨纶、锦纶、涤纶等传统内衣面料，采用新型面料和创新技术的针织塑身内衣品牌较少。塑身内衣的原料，聚酯纤维、聚酰胺纤维为其主要成分，其次为氨纶弹性纤维；铜氨、蚕丝蛋白包缠纤维等新型纤维也开始运用于塑身内衣；纤维素纤维作为内衣里料保证了穿着的舒适性。

（4）针织塑身内衣价位

针织塑身内衣产品的价位不等。束腰束裤在200~500元之间，胸腰腹三合一型内衣价位在400~700元之间，一件式连体内衣价位会更高。整体上，弹力性针织塑身内衣价位比调整型针织塑身内衣要低，一套的价格在500元左右。在针织塑身内衣中，一些品牌旗下的高端品牌产品价位偏高，平均在2000~3000千元一套。如欣姿芳的Lofan、爱慕的La cover以及伊丝艾拉的Christies等。

3 针织塑身内衣面向中高龄消费者的思考

通过调研发现，目前针织塑身内衣的消费者主要年龄在35~45岁，购买用途为日常穿着、产后塑身以及参加重大场合。从人体体型变化分析，中高龄女性对塑身的需求应该较高；从经济收入分析，中高龄女性对价格的承受能力更高。因此，内衣企业应该重视并开发这一年龄层的产品，并进行推广。

（1）针对中高龄消费者的塑身内衣产品设计

从颜色和款式的接受度看，45岁以上的大部分消费者喜爱肉色、简洁大方的款式。一半以上的消费者希望通过针织塑身内衣达到收腹、纤腰的效果。目前，一些塑身内衣还增加了养生保健功能，可以加速血液循环、增强免疫；或保护脊椎、防止腰酸背痛；甚至可以改善卵巢功能，促进女性荷尔蒙的分泌，这些都可以作为面向中高龄消费者的附加值。

（2）针对中高龄消费者的塑身内衣的销售服务

中高龄女性消费者在实体店购物时更注重消费体验，特别是塑身内衣这样的产品，需要试穿直观感受穿着效果来选择合适的尺码和产品。因此，对内衣试的设置和规划非常重要，不仅需要通风条件好、备有挂衣钩和一些女性化妆用品，而且还要有关于内衣穿着方面的"指引图"。除此之外，墙面颜色和灯光也很重要，要让女性感受到温馨和亲近。而在实际的销售中，休息区的设立和完善会增加顾客的舒适度和对品牌的好感。

（3）针对中高龄消费者的塑身内衣的网络销售

随着全民网络时代的到来，内衣消费渠道由传统店铺扩展至网络。这一内衣消费渠道的变迁，使得内衣品牌的销售从传统店铺逐渐转移到了互联网上。除一些早已为我们熟知的品牌，如婷美、爱慕、曼妮芬等已在网上开设了塑身内衣专卖店，在我国最大的C2C平台淘宝网上以"塑身内衣"作为关键词搜索时，可得到经营塑身内衣的商家12437家，共10.44万件商品，其价格从十几元至上千元不等，可见其市场巨大。另外，由于商场中塑身内衣价格偏高，很多普通收入的中高龄消费者也倾向于在网上购买，网络上以"中老年""送给妈妈的礼物"等字眼经营内衣品牌的店铺中，销售量较高的产品明显具有物美价廉的特征。

（4）针对中高龄消费者的针织塑身内衣高级定制服务

由于每个人的需求和体型都不同，商场中购买的塑身内衣基本都是已经制作完成的，消费者只能根据自己的情况来选择差不多的塑身内衣，但大多数并不能完全适合自己的要求。特别是中高龄消费者，所以根据自身情况来量身定做塑身内衣，不仅可以塑形、矫形、改善不良的身材，而且对于改善心情、增加信心大有帮助。因此，内衣品牌应该关注并尝试增加塑身内衣的定制服务，满足经济收入高、形象要求高的中高龄消费者需求。

4 结论

中国内衣市场潜力巨大，本文通过对上海地区五个具有代表性商圈中的20个内衣品牌进行实地调研，分析了上海地区针织塑身内衣市场现状，并面向中高龄消费者讨论了几点建议。为企业了解自身产品，改善存在的不足提供了帮助，并且对企业了解消费者需求起到了一定的参考作用。

市场调研结论如下：目前市场上内衣品牌中均设有针织塑身内衣这一品类，且店铺位置多在三层，面积在 $12\sim25m^2$ 的居多；针织塑身内衣产品种类齐全，其中塑腰腹的塑身内衣品类所占比例较大，但陈列不够明显，价格在 $100\sim3\,000$ 元不等；品牌的POP设置、店铺陈列均对消费者购买针织塑身内衣的意愿起着积极的影响作用。

针织塑身内衣品牌面向中高龄消费者的市场建议如下：开发肉色、简洁大方的款式，强调收腹、纤腰的效果，开发具有养身保健功能的产品；实体店应注重服务与购物体验；针对普通收入的中高龄消费者和较高收入的中高龄消费者可拓展网络销售和高级定制业务。

参考文献：

[1] 王帅.女性塑身内衣的舒适性研究[D].广州：广东工业大学硕士学位论文，2012.
[2] 姚艳菊，陈雁.塑身内衣压力舒适性的影响因素分析[J].国际纺织导报，2010（11）：76-78.
[3] 严燕连.调整形内衣的结构与工艺设计[J].针织工业，2009（3）：31-35.
[4] 廖丽霞.塑身内衣的松量设计研究[D].苏州：苏州大学硕士学位论文，2012.

［5］沈雷.针织内衣款式与装饰设计［M］.上海：东华大学出版社，2009.

［6］黄丽.塑身内衣的现状及发展趋势［J］.纺织科技进展，2009（2）：82-84.

［7］张建兴.服装设计人体工程学［M］.北京：中国轻工业出版社，2010.

Considering the Needs of Aging Baby Boomers

◆ Marilyn J. Bruin[①] Sauman Chu Lin Nelson-Mayson Juanjuan Wu Becky Yust *(Design, Housing, and Apparel, College of Design, University of Minnesota 240 McNeal, 1985 Buford, Saint Paul, MN 55108) (email: mbruin@umn.edu, schu@umn.edu)*

Abstract: As Baby Boomers become empty-nesters and imagine post-work lifestyles they change demand for housing, community amenities, and other consumer goods. Despite a predilection to deny the effects of aging, it is likely Boomers will eventually experience many of the physical and cognitive changes exhibited by their elders. Community planners need to understand Boomers' needs, expectations, and preferences as they plan the development and sustainability of resilient, vital communities that encourage healthy, active living. The purpose of the paper is to discuss public engagement research projects to describe and compare residents' expectations for housing, retail, recreation, and other amenities. We demonstrate how a public university engages with the public and business sectors provide research findings to influence community plans to mitigate the physical, psychological, and social changes that may accompany aging..

Key words: *Interactive websites, senior housing, aging,*

老龄化的婴儿潮一代需求研究

◆ Marilyn J. Bruin, Sauman Chu Lin Nelson-Mayson Juanjuan Wu, Becky Yust （明尼苏达大学，美国）

【摘 要】 婴儿潮一代成为空巢老人，退休后生活方式改变，进而改变了对住房、社区设施和其他消费品的需求。尽管偏好否认衰老的影响，但婴儿潮一代终会体验由于年龄衰老带来的许多身体和认知能力的变化。社区规划师需要了解婴儿潮一代的需求、期望和偏好来规划具有可发

---

① Marilyn J. Bruin, Design, housing and Apparel College of Design, University of Minnesota. E-mail: mbruin@umn.edu, schu@umn.edu

展性和可持续性,具有活力的社区,以鼓励健康、积极的生活方式。本文意在讨论公共参与研究性项目来描述和比较居民对住房、零售、娱乐和其他设施的期望。我们展示了一所公立大学与公共和商业部门合作的研究成果,运用于社区的规划,减缓了人们伴随年龄衰老而来的生理、心理及社会变化的影响。

【关键词】 互动网络;老年人住宅;老龄化

（中文翻译：谈金艳）

1 Introduction

Seventy-eight million Americans, 26% of the population in the United States, were born between 1946 and 1964; members of this cohort are referred to as Baby Boomers. "Boomers have left their imprint on every stage of American life they've passed through, and there's no reason to think that the senior years will be any exception" (Greenblatt, 2007). Boomers are expected to change the demand for housing, public transportation, recreation, and community-based programs as they pursue empty nest lifestyles (Greenblatt, 2007; HUD, 2005). Overall, Boomers are healthier, wealthier, better educated, expect to live and work longer, and play harder than their predecessors (HUD, 2006). Despite a predilection to deny the effects of aging, it is likely Boomers will eventually experience many of the changes in physical and cognitive abilities exhibited by their elders; although "disabilities and frailties are getting pushed back in life, they do occur" (Greenblatt, 2007, p. 883). "The likelihood that an older adult will be able to remain in their current residence, however, is quite limited, and decreases markedly with age; among individuals older than 70 years of age, only 5% can expect to age in place" (Sabia, 2008 cited in Scharlach, Lehning, & Graham, 2010, p. 1).

Countering the temptation to generalize Boomers as well-off, educated, traditionally married couples, is the realization that the large cohort is also more racially and ethnically diverse than previous generations (HUD, 2006; Hughes & O'Rand, 2004). In a group as large as the Boomers, many struggle financially and increasing health and housing costs likely make it difficult for some to maintain their level of living as they age-out of employment (Greenblatt, 2007; Joint Center of Housing Studies, 2014). Planners, policymakers, designers, and developers are pressured to address these changing housing and service needs as the proportion of aging adults increases within their communities (AARP, 2003; Ecumen, 2007; Myers & Ryu, 2008; Robison & Moen, 2000).

The Boomers are members of the first generation in the United States to actively use computers and social media (Hilt & Lipschultz, 2005; Stein, 2013). One of the goals of our interdisciplinary attempts to explore the needs, expectations, and preferences of Baby Boomers is to understand how to effectively use technology to reach Boomers and to understand how Boomers may use technology to live independently and to access consumer goods. In the following sections, we discuss three projects designed to meet provide research findings on the expectations for later life among American Baby Boomers.

2 Projects and Methods

The Smart House, Livable Community, Your Future. The purpose of *The Smart House, Livable Community, Your Future* project was to demonstrate how good design and well-planned community amenities can help Boomers to life independently, and how a wide variety of assistive technology can support safety and quality of life across the lifespan. We used a variety of strategies to share information on smart design to Baby Boomers. A 1,300 square foot exhibition designed as a home in the Goldstein Museum of Design allowed visitors to tryout assistive technology, and observe structural innovations. Targeted workshops for (a) Consumers and caregivers; (b) Developers, remodelers, contractors, and designers; and, (c) Community planners and policy makers demonstrated principles of smart design and examples of assistive technology.

We conducted usability testing of prototype websites as a foundation to design a website to showcase and accompany the exhibition (Chadwick–Dias, McNulty, and Tullis 2003). The *Smart House, Livable Community, Your Future* website provided free, downloadable activities and information to continue informing Baby Boomers and encouraging community conversations. A 30-minute video produced and distributed by the local public television stations included participants' reactions to the content as well as implications for designers and planners interested in creating intelligent environments.

Using multiple innovative formats, an interactive exhibition in the Goldstein Museum of Design, workshops, an interactive website, and a public television helped us reach members of the Baby Boomer cohort and community planners (Bruin, Chu, Riha, Smoot & Megia, 2012). Through short videos Jim and Sarah, the imaginary residents, introduced and set a context for each vignette. Through written notes exhibition visitors became privy to Jim and Sarah's challenging decision–making process as well as their motivations and emotions involved in transforming one's home. Museum visitors were encouraged to actively engage, experience, and enjoy the space, they were instructed to sit down, play with objects, and try tools. The experiences of exhibition and workshop participants were collected through a qualitative, online-survey. Survey comments suggest the participants appreciated the interactive nature of the exhibition and the workshops.

"The way it was laid out as it would be in a real home. It is better to see it like it would appear rather than in pictures."

"Visual! Great examples/exhibits and hands on experience."

"The hands on approach and the ideas generated by seeing it. Good use of technology!"

We hoped to inspire visitors to adopt technology and design solutions, including assistive technology and home modifications as well as community design that promote independent living and community integration across the lifespan. The following comments indicate that the exhibition, workshops, and website met those goals.

"I liked the way it makes the features of aging in place practical – so that those who may say 'I don't need that' can see the benefits"

"Showing how you can downsize and still do it attractively. Also, the importance of green space, parks etc. nearby."

"I liked the entry way ideas because that is the area of the home I never considered being a problem for an older person."

Overall comments suggested participants were rethinking home and community, and how to adopt more intelligent strategies in the built environment. Furthermore, the use of multiple formats, an interactive exhibition, workshops, and an interactive website, helped us reach the Baby Boomer cohort who are comfortable receiving information in multiple and innovative mediums.

Housing and Services Needs Assessment in Hennepin County

In 2010, the Hennepin County Department of Human Services partnered with the University of Minnesota to complete a needs assessment to plan the housing and human service needs of Baby Boomers. Hennepin County is the most populated county in Minnesota. Minnesota, with 5.4 million residents, is the largest state in the North Central United States; Canada touches the northern border. Hennepin County includes Minneapolis, the largest city in Minnesota, its suburbs and rural areas. Minnesota is characterized as a state populated by well-educated, highly employed, homeowners of Scandinavian descent. According to the U.S. Census 92% of Minnesota residents have a high school degree, 12% fall below the poverty line, 73% are homeowners, and 83% are white and 5% are African American (U. S. Census, 2015).

The first activity in the project was to write a background paper or literature review. Although the research literature and popular media tend to generalize Boomers as well-off, educated, traditionally married couples, who own suburban homes, however this large group is racially and ethnically diverse. Furthermore the recent recession reduced home values, home equity, and retirement savings for many Boomers. The Minnesota Department of Human Services

(DHS, 2007) estimates that up to 29% of Boomers in Minnesota have inadequate retirement savings. In another study in Minnesota (Ecumen, 2007), 72% of the respondents were concerned that they would not be able to maintain their standard of living if they left the workforce. These Boomers may lack the financial resources to self-finance home modification, in-home supportive care and services, or a move to senior housing.

The majority of Boomers believe they will be able to stay in their current homes as they age; 85% of individuals aged 50 and older would like to remain in their local community if they cannot stay in their current home (AARP, 2003). A variety of meaningful and appropriate housing options especially in areas with a high proportion of older residents would help Boomers achieve this goal.

In general, 69% of Midwesterners aged 45 and older prefer a single-story, single-family home. However, 23% are willing to move to a condominium or townhome if that means obtaining higher quality features or a better location. The majority want the most the square footage for their money, however, they do not want to sacrifice quality. Fifty-three percent say they prefer a high-quality, smaller home rather than larger, lower quality home with few amenities. Despite the pull of amenities, recreational facilities, and community features (including security) typically offered in multifamily developments, Boomers want a single-family home. However, there are Boomers who gravitate toward multifamily units in higher density, urban areas with decreased dependence on automobiles (Ecumen, 2007).

Boomers' housing preferences may prove unrealistic. It is likely Boomers will eventually experience many of the changes in physical and cognitive abilities exhibited by their elders. If so, it is likely they will need to modify their current home or move to accessible housing; only 5% can expect to age in their long term home without modification (Sabia, 2008).

Based on the literature review, we developed semi-structured interview questions for focus groups. Eight focus groups were conducted; 41 individuals participated. Focus group interviews were digitally recorded and transcribed; field notes and debriefing notes from three facilitators and note takers were consulted. The analytic strategy utilized a systematic approach, first transcripts were read several times by each researcher. Transcribed statements were organized in a matrix. Because demographic characteristics seemed similar within focus groups the matrix was organized group by interview questions. Many responses did not fit neatly into categories defined by interviewed, the matrix was reviewed to identify themes across questions. Significant statements were highlighted and theoretical categories were identified (Bloomberg & Volpe, 2012). Using the theoretical category statements we build a table help identify formulated meanings. Formulated meanings were categorized into themes, themes and statements were then compared across focus groups. The approach was organized and flexible, the matrix and tables helped identify "big ideas" without overlooking themes related to demographic subgroups (Patton, 2015). The themes were related to the concepts identified in the literature review.

Among married Baby Boomers living in suburban homes, we found they expected to remain in their current home through the end of their years. The findings mirror the literature. Some of the participants explained that their spouse resisted a change in living arrangements. No one mentioned a plan in case they lost a spouse or lost their capacity to live independently.

Because little is known about Baby Boomers who do not conform to generalizations of individuals who are married, well-educated, healthy, and wealthy, we focused our attempts to interview members of subgroups of Boomers who are likely to need income and/or housing assistance with supportive services from the public and non-profit sectors. For example, we conducted focus group interviews with individuals who struggle financially, recent immigrants to the upper Midwest, unmarried females, or the growing numbers of individuals who with the advances in medical overcame catastrophic illness and injury but deal with lingering physical issues.

We found except for those that currently reside in subsidized senior housing, the majority held negative perceptions of senior housing. They recognized that senior housing encompasses a variety of options which increased the complexity of their search. They realized that they may need it in the future, but most had not begun to explore the options. Many were not aware of the current regulations regarding senior housing and used past experiences of friends and families as their sources of information. They were confident in using the computer to conduct searches, but did not have efficient search methods or resources to find out about options available for them.

An interchange between two participants was poignant in their characterization of senior housing:

[first participant] *That's what I was talking about, the isolation, in those houses. My mother moved—she just moved out of one building in* [community removed by authors to retain confidentiality of participant] *to* [state removed by authors to retain confidentiality of participant] *because too many people were dying there. She said Death had an apartment up in there, because there was just—*

[second participant] *We've had about 10 deaths.*

[first participant] *It's too much. And it's hard to try to get to know somebody when you're older, and then all of the sudden you're talking to them today and then they're dead tomorrow. That's hard.*

Participants also expressed problems with age-segregated communities and the rights that they would lose. One woman's disabled son lives with her, and she feels constrained by options available to them in the future. Another participant helps out with her grandchildren and shared her experience:

There'd be a group of us that have grandchildren that come there, and then there was always somebody complaining who just didn't like children period, and they didn't mind letting you know. So that made it feel very uncomfortable having them there. But a lot of seniors are connected with their grandchildren, for one reason or another. Because I help my son and daughter by babysitting because they have to leave early in the morning and they have to drop my granddaughter off, and I can get her ready for school, and the school bus stops right in front. But when you have somebody complaining or threatening that you're going to be evicted if you— if they catch you with the kids all the time. But I mean, I have to do that five days a week because I'm helping them out, and they're helping me out, giving me extra change. So I'd like to be in a place where that's not a big issue. But I'd like a building that has a playground or something close by too that I wouldn't have to worry about disturbing anybody. For the most part, they're very quiet, and I always take them to the park. But there's always that fear in the back of my mind that somebody's going to say or do something that will cause me to be put out of the building. And if it's the only affordable place around—I don't like it, but I know it's affordable. I can't afford to be anywhere else.

Single female heads of households were frustrated that there were not more creative and cooperative housing options available. For instance, examples were mentioned in California and France, where young adults are paired with older adults so that the young adult can find affordable housing and the older adult has someone who can provide help with special needs (e.g., changing a light bulb in the ceiling).

The issue of accessibility was also brought up and was characterized in three ways. First, accessibility to affordable public transportation was important, i.e., all day passes to allow them to run all of their errands on one day, not limited to a two and one–half hour limit under one fare. Second, accessibility to stores, cultural amenities, the library, the hospital, etc., was important. These are the community attributes that allow them to stay engaged and receive the services they need. And, third, physical accessibility of the housing unit was important. They clearly were thinking along the lines of "aging in place" and mentioned single–level living as necessary with wider doorways and hallways. They also wanted to be sure that even if they were able–boding, that their friends who used a wheelchair could visit them. Furthermore, accessing information was a barrier to making home modifications and/or appropriate senior housing. One method available to American homeowners is a reverse mortgage. These federally–insured financial products allow seniors to convert home equity into an income stream. However, the products are complicated and misunderstood. Regarding reverse mortgages, one participant expressed, *"I think it would be good to have somebody like, or some place or a hotline or something that's real and has good intentions to really help people to understand what's available ... [not] to come into your house and to have you sign up for something that's not really appropriate."*

In summary, economically stable and/or married focus group participants tended not to forecast nor plan for aging. The more marginalized participants, including several females who had cared for aging relatives, thought about, worried, about how they would live out their years. They expressed the need for accurate, easy to understand information about senior housing, home modifications, and how to finance their later years. Most of all they requested housing in safe neighborhood with safe, convenient public transportation.

Planning and Promoting Opportunities for Lifelong Communities

The researchers have recently been funded to engage with a public health department in a suburban Minnesota county and a neighborhood association in Saint Paul to further the exploration of Boomers' needs, expectations, and preferences as communities plan the development and sustainability of resilient, vital communities that encourage healthy, active living. Lifelong Communities are carefully planned with an array of affordable safe housing options with access to health care, cultural activities, recreation, retail outlets, and transportation (Atlanta Regional Commission, n.d.; Jones, n.d., Scharlach, 2009). The purpose of the transdisciplinary, collaborative, public engagement research project is to describe and compare residents' expectations for housing, retail, recreation, and other amenities and public spaces to community plans to mitigate the physical, psychological, and social changes that may accompany aging. Research activities include surveying baby boomers in a small rural community, a small town, and an urban neighborhood to summarize and compare expectations to plans from local community leaders. The applied research findings will be translated into exhibitions, symposia, presentations, a virtual resource center on an interactive website, and reports. An important goal of the project is to encourage and facilitate conversations between community leaders, planners, developers, residents, and researchers to influence the development and sustainability of elder-friendly communities that encourage healthy, active living (Bruin, Yust, Imbertson, & Lien, 2011).

We will use a sequential mixed method research design to survey Boomers and analyze the quantitative data to develop an interview protocol for key community informants. Interview data from developers, planners, and other city staff will be analyzed for themes, coded and embedded in the quantitative data to measure differences between Boomer expectations and assumptions of the community informants. The study design allows for comparison across communities. Survey data from 400–500 Boomers will allow the UMN research to conduct rigorous statistical analysis to describe demographics, and examine relationships between behaviors and expectations and community characteristics.

We are planning formative or process evaluation and outcome evaluations. Formative evaluation will help improve and document the process for replication. We are also planning an event to introduce the outputs and collect feedback through the Goldstein Museum of Design. Invitees will include key informants, their peers, and potential funders. Attendees be asked to complete an on line survey to collect feedback on the outputs and how they might be used as data for outcome evaluation.

3 Conclusion

This paper describes the development of publically–engaged research projects within the Department of Design, Housing, and Apparel at the University of Minnesota between 2010 and 2015. The *Smart House Project* began with a broad examination of the individual housing needs for aging Baby Boomers. We learned that Baby Boomers access information in multiple ways and appear amenable to incorporating technology and smart design principles to improve daily living; aesthetic examples were important. Furthermore, Baby Boomers who accessed information through an exhibition, workshops, and public television appeared to embrace home modifications.

The Hennepin County needs assessment emphasized the importance of considering multiple points of view rather than generalizing the needs and preferences of the majority. Again, the need for clear, accessible information was an important finding. We also saw evidence that vulnerable residents think about and want to understand how to plan for their later years.

In the Lifelong Communities project, we build on previous findings and incorporate exciting new foci. First, we realize the need to understand the role of retail in communities that support quality of life for aging Baby Boomers. Second, we need to develop a strategy to present information that is accessible and understandable, but also welcome. For example, we struggle with how to present resources and information that can support aging to individuals who deny the effects of aging. Therefore, we continue our explorations on how to reach out to Baby Boomers and those who provide services to Baby Boomers. Our interests continue to expand, especially the intersections between the fields of aging, communication, community development, housing, retail, and other services and we are considering the issues across urban, suburban, and rural settings.

References

[1] AARP. These four walls...Americans 45+ talk about home and community. Retrieved from http://assets.aarp.org/rgcenter/il/four_walls.pdf, 2003.

[2] Atlanta Regional Commission, Area Agency on Aging (n.d.) Retrieved from www.atlantaregional.com/llc

[3] Bloomberg, L.D., & Volpe, M. *Completing your qualitative dissertation (2nd ed.)*. Thousand Oaks, CA: Sage Publications, 2012.

[4] Bruin, M., Riha, J., Chu, S., Smoot A., & Mejia, G.M. Smart housing: An intelligent environment for aging independently. (pp.31–39) First International Smart Design Conference, Nottingham Trent University, 2012.

[5] Bruin, M.J., Yust, B.L., Imbretson, K.M.,& Lein, L.L. Aging in Hennepin County Project: Literature Review and Focus Group Summaries. Report to the Hennepin County Department of Health and Human Services, 2011.

[6] Chadwick-Dias, A., McNulty, M., & Tullis, T. Web Usability and Age: How Design Changes can Improve Performance. *Conference on Universal usability* (pp. 30–37). Vancouver: ACM, 2003.

[7] Greenblatt, A. Aging Baby Boomers. CQ Researcher 17(3), 867–885 Retrieved from www.cqresearcher.com, 2007.

[8] Ecumen. "Age-Wave" Study. Retrieved from http://www.ecumen.org/app/webroot/files/file/White%20Papers/Age-Wave-Study.pdf, 2007.

[9] HUD Elderly housing consumption: Historical patterns and projected trends. Washington, DC: Housing and Urban Development, 2005.

[10] HUD U.S. Housing market conditions 3rd Quarter 2006. Washington, DC: Housing and Urban Development, 2006.

[11] M.E. Hughes, & A.M. O'Rand. *The Lives and Times of the Baby Boomers.* New York and Washington D.C.: Russell Sage Foundation and The Population Reference Bureau, 2004.

[12] Joint Center for Housing Studies. *Housing America's Older Adults – Meeting the needs of an Aging Populations.* Cambridge, MA: Harvard School of Design. Retrieved from http://www.jchs.harvard.edu/sites/jchs.harvard.edu/files/jchs-housing_americas_older_adults_2014_0.pdf, 2014.

[13] Jones, M. (n.d.). 5 Essential Facts From "Housing America's Older Adults": Harvard study highlights housing challenges facing older adults. Retrieved from http://www.aarp.org/aarp-foundation/our-work/housing/info-2014/5-facts-from-housing-report.html

[14] Myers, D., & Ryu, S. H. Aging baby boomers and the generational housing bubble:

Foresight and mitigation of an epic transition. Journal of the American Planning Association, 2008, 74(1), 17–33.

[15] Patton, M.Q. *Qualitative research and evaluation methods.* (4th Ed.). Thousand Oaks, CA: Sage Publications, 2015.

[16] Robinson, J.T., & Moen, P. A life-course perspective on housing expectations and shifts in late midlife. *Research on Aging,* 2000, 22(5), 499–532.

[17] Sabia, J.J. There's no place like home: A hazard model analysis of aging in place among older homeowners in the PSID. Research on Aging, 2008, 30, 3–35.

[18] Scharlach, A.E. Creating age-friendly communities: Why America's cities and towns must become better places to grow old. *Generations,* 2009, 33(2), 5 – 11.

[19] Scharlach, A., Lehring, A., & Graham, C. (June 2010). A demographic profile of village members. [PDF document]. Retrieved from http://millcitycommons.org/sites/millcitycommons.org/files/u181/Village_Demographics_Report_Final.pdf

[20] Stein, J. (1/29/2013). *It's stupid and insulting to pitch Baby Boomers as tech novices* Retrieved from http://www.forbes.com/sites/jonstein/2013/01/29/2013-the-year-your-grandpa-becomes-more-tech-savvy-than-you/ Accessed 3 Aug 2011

[21] United States Census Bureau *Quick Facts.* Retrieved from http://quickfacts.census.gov/qfd/states/27/2758000.html, 2015.

A Study on Fashion Brand by New Mature Women's Characteristics

◆ Hyesoo Yeom[①] *(Design College, Sangmyung University, S.Korea)*

Abstract: The expansion of an aging society has transformed people's lifestyles. In particular, mature women's uniform and standardized lifestyles have become more diverse and inclusive. As a result, there have been a lot of changes in the brand positioning of mature women's wear for a short period time. In fact, a brand image for mature women has shifted from elegant and refined brand to trendy and stylish brand. Mature women are from a group with high purchasing power. Focusing on their independent value and focus, new markets have been formed in various fields. Since new mature women in their 50s pursue trendy fashion which keeps changing based on their new lifestyles, it is needed to derive opinions on this fashion market and their preferred design factors.

Key words: *New senior; New generation; 50s pursue trendy fashion*

基于中高龄女性生活特征的时尚品牌研究

◆ Hyesoo Yeom （祥明大学，韩国）

【摘　要】　老龄化社会的扩展改变了人们的生活方式。特别是中高龄女性的统一、标准的生活方式变得更具多样性和包容性。因此，在短期内，中高龄女性的着装品牌定位有了很大的变化。事实上，中高龄女性的品牌形象已经从优雅、精致转向新潮、现代。中高龄女性正成为高购买力的群体。她们聚焦自生的独立价值和关注点，在各个领域都形成了新的市场。目前新型中高龄女性，50多岁，她们有新的生活方式，追求不断改变的时尚因而基于这一时尚市场与她们首选的设计要素来进行品牌定位是必要的。

【关键词】　新中年；新一代；50多岁追求新潮的时尚

（中文翻译：范振毅）

[①] Hyesoo Yeom, Professor, Design College, Sangmyung University. E-mail:yhs812@smu.ac.kr

I. Introduction

With the considerable extension of human life, the expansion of an aging society has transformed people's lifestyles. In fact, a new life pattern which cannot be explained with the conventional standpoint has been formed. In particular, there have been a lot of changes in middle-aged people's lifestyles. At the same time, mature women's dull and uniform life patterns have become more diverse as well. Escaping from the conventional simple and monotonous lifestyles, they have pursued a carefree and relaxing lifestyle through which they can accomplish self-realization. After all, they have created a brand-new active and dynamic lifestyle. With the emergence of new terms such as 'New Middle-aged,' 'Adult Lady' and 'Flower Middle-aged,' in addition, more segmented generations have been formed. Focusing on their independent value and focus, new markets have been formed in various fields. To analyze the characteristics of new mature women's wear, therefore, this study attempted to derive fashion trend in new mature women's wear by analyzing fashion brand design targeting conventional fashion brand design for middle-aged women and current fashion brand targeting new middle-aged women based on analysis on current fashion brand for middle-aged women in their 30~50s.

II. Experimental Methods and Results

For analysis, a scope was narrowed down to a total of 52 brands targeting middle-aged women in their 40–50s. For the scope setting, "2014–2015 Korea Fashion Brand Yearbook" which provides information on current brands in Korea was referred to. Based on the Korea Fashion Brand Yearbook and brands in department store, brands were classified into "designer boutique," "formal wear for middle-aged women" and "middle-aged women's casual wear." In addition, price was divided into "low price," "middle price" and "high price."

As stated in Figure 1, in terms of the number of middle-aged women's fashion brands, "casual wear" was the highest with 21, followed by "formal wear (17)" and "designer boutique (14)." According to price analysis by the brand group in Figure 2, "mid price" was the highest (32), followed by "high price" and "low price" (10 each) in terms of the number of brands. When mature women's wear by brand was investigated in 2008 Korea Fashion Brand Yearbook, "designer boutique" and "formal wear" were high with 28 each, followed by casual wear (2) in terms of the number of brands among a total of 58 brands. There were a lot of changes in the brand positioning of mature women's wear for a short period of time, which means that a brand image for mature women has shifted from elegant and refined brand to trendy and stylish brand.

II-1. Analysis on new mature women's fashion brand

Figure. 1　Brand Positioning of Mature Women's Wear

	Price	Brand Name
Desiner boutique	high price	KANGHEESOOK, KIMYEONJU BOUTIQUE, GGARBENJUNG, LUCIANO CHIO, MADAM POLLA, CHIOYUNWOK
	mid price	DOHO, SULYUNHYOUNG, SONJUNGWAN, SHINJANGKYOUNG TRANS MODE, JUNGHOJIN KNIT, JOHANUK COLLECTION, JIJU
	low price	PARKYOUNSOO
Formal Wear	high price	LEBEIGE, REVEDO, FRANSOISE
	mid price'	CARTE KNIT, RENOMA LADIES, MARIANI, BELISSIANG, PRENDANG, CHERE DAME SPECCHIO, SYSMAX, ASTUCES, ANSICHE, ESCALIER, JOSUNHEE COLLECTION, FLAMINGO, CHATELAINE
	low price	
Casual Wear	high price	RALPH LAUREN Black Label
	mid price'	DAKS LADIES, DEMIAN, DECIDER, MORADO, VELARDI TOUCH, SHWAEREUCHEE, CHIOBIKHO FASHION, CARRIES NOTE, FREE BLANCE, WHO'S NEXT, OLIVIA HASSLER
	low price	DI CLASSE, RAGELLO, MONTINI, MONTE MILANO, SPORTIVA, ALLFORYOU, LEEDONGWOO COLLECTION, ESAE, PAT

Figure. 2　Average Base Price for Mature Women's Wear

price	Item	Av. standard price(won)
high price	JACKET	698,000(won) ~
	BLOUSE	338,000(won) ~
	PANTS	398,000(won) ~
mid price	JACKET	168,000(won) ~ 698,000(won)
	BLOUSE	185,000(won) ~ 338,000(won)
	PANTS	115.000(won) ~ 398,000(won)
low price	JACKET	~168,000(won)
	BLOUSE	~185,000(won)
	PANTS	~115.000(won)

According to the Korea Economic Daily (September 2012), as more young mothers enjoy putting on stylish clothes just like their daughter, sales in "designer boutique" have considerably declined. In fact, many middle-aged women in their 40~50s have turned their eyes toward SPA and import brands. As a result, some brands have given up business in a department store. In particular, Fashion Group HYUNGJI signed the 30-year-old actress Ko Jun-hee for a model of the brand "Chatelaine." The company also chose Han Ji-hye (31) for a model of the mature women's wear "Oliviahassler." It's been common to have an actress in her 30s for a model of the middle-aged (40~50s) women's wear because of changes in target consumers' mind age. Therefore, this change has been promoted to target this new consumer group. The young generation's fashion trend of "absorb quick trend changes, prefer simple design" has spread to middle-aged women.

II-2. Comparison and analysis of changes in new mature women's fashion brand

As new mature women's preference rapidly shifts to low and mid-price trendy brands, there have been a lot of changes in design concept to meet these new consumer needs and demand. "Designer boutique" had been the mainstream in women's fashion brand until the early 2000. However, it's been forced out from department stores due to poor sales. The middle-aged women in their 40-50s have refused to be "ordinary mature ladies." Instead, they have been born into the so-called 'Ruby Group' pursuing stylish fashion by investing time and money for themselves. The conventional "designer boutique" brands have failed to keep pace with these trends and consequently disappeared. Recently, Shinsegae Department Store announced that the design boutique brands such as "Ans Mode", "JR", "Clio" and "Madame Polla" were out, introducing

Figure. 3 MADAM POLLA　　　　Figure. 4 CARTE KNIT　　　　Figure. 5 JUNGHOJIN KNIT

Figure. 6　LEBEIGE　　　　　Figure. 7　CHATELAINE　　　　Figure. 8　OLIVIA HASSLER

the select shop "New Adult Zone" targeting the "Ruby Group." In contrast, the trendy and stylish mature women's wear brands such as LEBEIGE and Oliviahassler have dramatically grown, showing 10%~15% increase in sales. In other words, middle-aged women's preference has shifted from the luxurious and elegant brands to light and trendy brands with the emergence of new mature women.

As shown in Figures 4, 5 and 6, traditional middle-aged women's wear focuses on the creation of elegance and refinement, using high-quality materials and simple patterns (or no patterns) based on voluminous silhouettes such as H-line and O-line. In the Figures 7, 8 and 9, on the contrary, the brands for new mature women use more diverse colors and patterns with the trendy characteristics of young women's wear.

III. Conclusion

The said results can be summarized as follows: First, fashion trend for new mature women has shifted from the refined and elegant designer boutique to trendy and stylish casual brand, reflecting changes in new mature women's lifestyles. Second, in terms of price positioning, price policy has shifted from the high price to mid and low-price strategy. With the emergence of new mature women pursuing a new lifestyle, in other words, reasonable price brands have become more popular. Third, as confirmed in analysis on brand preference, new mature women prefer modern and young styles (originally designed for those in their 20-30s) and casual wear, showing off their beauty and personality. In fact, they invest a lot of time and money and make themselves look younger. They have no hesitation in buying young casual wear, instead of traditional mature women's clothes. With the growth and development of medicine and science technology, the proportion of the aged population has gradually increased. At the same time, middle adulthood now lasts considerably longer than childhood and teenage years. Therefore, we cannot explain middle-aged generations with a traditional concept on them. In addition, it appears that the changes in middle-aged generations would continue. Mature women are from a group with high purchasing power. Focusing on their independent value and focus, new markets have been formed in various fields. Since new mature women in their 50s pursue trendy fashion which keeps changing based on their new lifestyles, it is needed to derive opinions on this fashion market and their preferred design factors. Therefore, there should be studies on the characteristics of diverse generations depending on the lifestyles which keep changing over time. This analysis may enhance product values by narrowing the gap fashion brands and consumers' view on them. In addition, there should be a development of brands for new mature women in their 50s, focusing on design which reflects each generation's characteristics.

References

[1] Korea Fashion Brand Annual. Apparel news. p107–203, 2014/2015.

[2] The Korea Economic Daily. (Sep. 2012).

[3] A Study on Image Preferences of Fashion Product According to Life-Style Groups – Focused on Middle-Aged Women between 35 and 59 Years Old. Sim Junhee. p81, 2007.

[4] Fashion Image Types and Design Factors for Middle-aged Korean Women. Jung Suin p11–12, 2014.

[5] Study of desired image and behavior of appearance management by their perceived age differential of middle-aged women. Pyo Minkyung. P31, 2015.

[6] http://blog.naver.com/dokdae99/110095410920.

[7] http://kstarfashion.com/archives/73700.

关于老年人的服饰消费行为——针织商品的多样性

◆ 岩佐正树[①]　（株式会社MODELLISTA，日本国东京都，136-0082）

【摘　要】 随着老龄化社会的不断深入，如何使老年人在步入人生的最后阶段充分释放自己，活出自己的精彩是值得我们深思的问题。尤其是老年人的消费，特别是对于服饰的消费有着和其他年龄段人的不同理解与认知，因此，怎样为老年人提供他们所需要的服饰商品是我们奋斗的目标。

【关键词】 服饰消费；变化；释放阶段；旅游；针织商品

The Diversity of Knitting Fashion of Aging People

◆ Iwasa Masaki*(MODELLISTA Co.,LTD.,Tokyo,)*

Abstract:

How to help the elderly fully release oneself, live out their own wonderful is worth our deep searching question in the aging society. Aging people have different understanding and cognition with other ages for consumption, especially for clothing consumption. Therefore, providing appropriate apparel for the aged is our goal.

Keywords:

Clothing Consumption, Change, Free, Travel, Knitting Fashion

引言

老年人的服饰消费行为不是由"年龄"而是由"变化"决定的，随着自身生活状态的改变，"做自己想做的事"的愿望愈发强烈。充分释放自我感受万物最好的方法就是去旅行，而对于老年人旅行时的着装选择针织商品是很适宜的。针织商品的各种特性能最大化的满足老年人的心理和生理的需求，从而激发老年人的消费力。

[①] 岩佐正树，男，株式会社MODELLISTA 取缔役社长。E-mail:masaki@modellista.work

1 老年人的生活状态中的各种变化及需求

1.1 年龄增长后的生理变化

人的体能在 20 岁时达到峰值后开始减退。虽然这种减退是无法避免的自然规律,但是速度是可以减缓得。随着退休后离开了自己的工作岗位,并且处于一种几乎不出门的生活环境下,就会出现如注意力的下降、肌肤的老化、体型的变化、体力的减退等诸多生理上的改变。

1.2 生活心理的变化

在职期间由于需要顾及到领导的指示和下属的意见及其他诸多因素,因此长期处于一种紧张的工作状态。退休以后远离职场,从这种紧张得生活状态中解脱出来,使自己的生活状态发生了巨大的改变。

1.3 社会关系的变化

长期养成的用餐时间的制约荡然无存,随之而来的是符合自己生活节奏的用餐时间及饮食习惯的改变。

1.4 兴趣爱好的变化

把自己在少年时期感受过的如文学、音乐、电影、时尚等重新定义,使之成为自己余生的兴趣爱好和一生的事业。

1.5 工作方式的变化

退休后每周花费在工作上的时间不超过 3 天,其余的时间可以用在自己的兴趣爱好和旅游等方面,充分享受自己的退休生活

1.6 除工作以外的外出目的(如图1)

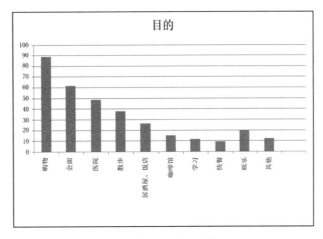

图1 老年人外出目的　　　　　　　　　　（资料来源：作者自制）

老年人日常外出的目的无非就是购物、会会朋友相互交流一下,交流的内容与工作无关,而是在日常生活中各自的收获。

1.7 关于老年人的服饰消费心理和商品企划

促使老年人消费的重要因素是能充分体现出对于消除老年人在面对疾病、看护以及年老后"对于以后生活的茫然与不安"的心理状态有引导性的商品价值,以及无法用金钱能买到的(健康、时间、乐趣、喜悦)却很想得到的主观能动性的商品。

2 年龄增长与大脑的潜在能力

大脑的外侧有叫作灰质的神经细胞,内侧有叫作白质的神经纤维。换句话说,外侧感知后通过与内侧神经纤维的衔接进行判断。

神经细胞的体积从20岁后开始逐年递减,与此同时神经纤维的体积随着年龄同步增长,峰值在60岁~70岁。

神经纤维的增长指的是逐渐有了直觉和洞察力,老年人随着"岁月的沉淀"与"智慧的积累"大脑的潜在能力逐渐发达。

2.1 心理的发展(如表1)

美国乔治华盛顿大学的心理学家Gene David Cohen提出45岁以后心理发展分成4个阶段,如表1。

表1 美国乔治·华盛顿大学的心理学家Gene David Cohen提出45岁以后心理发展分成4个阶段

年龄段	阶段	意识	行为
40~50	再评估阶段	自己也有死的一天	探索、转移
50~70	释放阶段	只有现在去做	实验、革新
60~80	总结阶段	分享自己的智慧	总结、决断、贡献
70~	安享阶段	想阐述、想主张	继续、回忆

2.2 "只有现在去做"是老年人消费行为的原动力。

这个年龄段的人迎来了退休或是早退,想要换一种活法,依附着"只有现在去做"等想法去参加料理课程后自己开店,学习畜牧业后选择在牧场工作等等这种引发一种蜕变的状态正是释放阶段的特征。戴维·科恩阐述到,这个时期是面临退休及把儿女养育成人后,自身的生活状态发生了改变,和家人在一起的生命周期容易起变化"做自己想做的事"以及促进自我释放的能量也容易产生的时期,尤其是当看到释放阶段的行为是冲动、欲望、憧憬的时候表现显著。

3 促进释放阶段消费的三要素

3.1 愉快、健康、有内涵的日常生活

只有把愉快、健康、内涵这三要素融入到商品开发和服务中,才能促进老年人的消费力。

3.2 释放阶段的生活方式的提案(如图2)

促使释放阶段消费的三要素和符合消费内容的理想生活方式就是退休后会表现出想花钱的行为,其中最具代表性的就是旅游,旅行的内容涵盖了老年人全部的欲求。

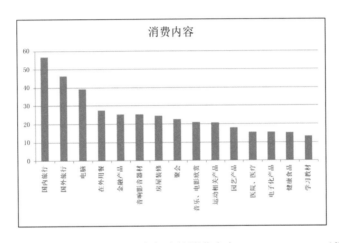

图2 老年人的消费内容　　　　(资料来源:作者自制)

(1)兴奋、蠢蠢欲动
(2)成为参与者
(3)有勇气、有活力

计划着旅行的内容,设想着场景的画面不由得使自己兴奋,为自己是正直的参与者而感到高兴。

4 愉快、健康、有内涵的日常生活

旅行计划的安排一般都通过网络来查阅,网购对于身体行动不便的老年人更加便利,同时也能将商品通过网络推荐给同样有兴趣的朋友,并能使其销售的更好。

4.1 释放阶段的网络利用率(如图3)

灵活运用网络来收集资料并有积极消费意识超前的老年人。

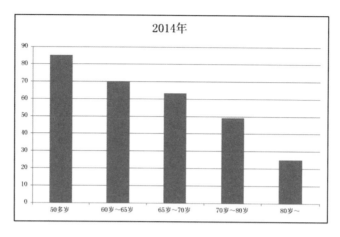

图3　老年人的网络利用率　　　　　　　　　　（资料来源：作者自制）

4.2 面向释放阶段的服饰商品的开发

老年人在过去的60年里被各种服饰包围、并作出选择。其敏感度要比任何年代的人都出色，对于冲动购物及现在拥有的东西没有兴趣。因此，对于在自己家里或是养老院里充分利用IT设备来搜索，足不出户就能得到自己想要的衣服等商品的开发很重要。

4.3 用IT设备搜索关键词

旅行需要安排的内容包括：地方、季节、人数、移动手段、住宿设施、计划等。决定了旅行的内容后，在考虑场景时选择着装的关键在于不易起皱、能洗涤、速干性好、高品质，这种范畴的商品就是针织商品。

旅行用品的用途多种多样，出发时的着装，不选择移动手段（火车、巴士、飞机）时的轻便装，在当地购物餐饮时的着装，短时间里能收纳进行李箱的服装，根据这些用途最适宜的商品就是针织商品，每当想起这些场景时，就会勾起购买新衣服的欲望。

4.4 面向老年人的商品企划

4.4.1 从安排内容到符合场景的着装（如图4）

3天2夜的国内旅行（秋季）

（1）出发

符合鞋子的休闲服

（2）移动（火车）

衣服不易起皱，肩膀和手臂无负担，不管什么姿势都能承受的衣服

（3）休息

放小物的轻便小包

（4）住宿设施

替换新买的衣服

（5）购物

转换心情的披肩、帽子

（6）晚餐

远离平常的日子,穿着符合自我释放意识的高品质衣服

（7）舒适

放松心情的衣服

（8）就寝

容易洗涤,速干性好的衣服

图4　面向老年人的针织商品　　　　　　（资料来源：作者自制）

4.4.2　针织商品的特点

在任何内容或场景里针织商品对于适宜度的把控都比较容易,丰富的色彩变化使得心情大好,作为改变自己着装的手段是最合适的。

4.4.3　电脑横机的问世

1975年以前,说到针织的主流就是认为手织的毛衣或是依靠手摇横机来生产,但是1975年以后,由于电脑横机的问世,加上丰富的原料、丰富的颜色、多彩的针数,丰富多彩的针织商品深深地渗透到了日常生活中,这种变化与进化和日常生活的场景是息息相关的。面向老年人的商品企划不是单一的设计企划,而是符合常态并能满足功能要求的商品企划。

4.5　面向老年人的商品企划

把旅行作为场景的无缝制针织商品的功能性特点

利用新型编织技术WHOLE GARMENT（全成型无缝横编机）电脑编织机进行从头至尾

的无间断,无缝合的立体编织服装的概念。这种技术的运用能充分体现针织商品的各种特性,使穿着更加舒适。如图4。

(1)出发

无需选择移动手段能放松下来并具有伸缩性的服装。

(2)购物

充分利用针织本身具有的悬垂性使之不易起皱,购物和用餐时也无须介意的高品质服装。

(3)舒适

旅途中可以洗涤,速干性优良,不走样的服装。

(4)就寝

没有一处缝头,使用抗过敏的原料,任何方向的伸缩性均良好的服装。

5 结论

老年人中的大部分都会在退休后迎来生活中心理的变化,并伴随着生理上的变化自己的社会关系也发生着巨大的改变。因为希望自己在有生之年里不留遗憾,所以想做的事"只有现在去做"的思绪很强烈,也就趋向了释放型消费的消费行为。面向老年人的商品企划需要有内容,也需要附加内容的场景来依托。丰富多彩的内容和场景组合在一起突出"旅行"概念,针织商品的多样性在不可或缺的旅行服饰商品中成为重要的商品企划。

Travel Jams: IntersectingStrategies for the Mature Chinese Female Traveler

◆ Kathryn.Hagen[①] Eulanda A. Sanders *(The wood bury University, USA)*

Abstract: This paper will examine the history and growth of domestic and overseas travel of Chinese citizens, focusing on the motivations, challenges, and successful travel strategies of and for mature Chinese women. The findings are examined in the context of intersectionality, a theoretical construct, and Chinese society; practical and sociological implications are discussed, and a plan is presented as a possible business model that could also serve the needs of these mature travelers.

Key words: *Motivation; Challenges; Identity; Self-Determination; Constraints*

旅游拥堵：中国中高龄女性的旅行策略研究

◆ Kathryn Hagen Eulanda A. Sanders （伍德伯里大学，美国）

【摘　要】　本文考察中国公民国外旅游的历史和发展。重点研究针对中国中高龄女性群体的动机、挑战和成功的旅游策略。统筹规划、理论模型和中国社会的背景下进行了检验，并探讨了实用和社会学含义。最后提出了一个可以服务于中国中高龄女性旅游者的商业模式。

【关键词】　动机；挑战；身份；自我决定；约束

（中文翻译：范振毅）

Intersectionality:A "complex of reciprocal attachments …that confront both individuals and movements as they seek to 'navigate' among the *raced, gendered, and class-based* dimensions of social and political life."

In doing research for this paper, an intriguing theoretical term, *intersectionality* emerged and resonated as a structure to inform the ideas underlying this treatise.[1] The word stems from *intersection*, which is a point where roads- or ideas- intersect or cross. Race, gender, and class, all of which may impact our intrepid female traveler, cannot be seen as separate issues, but

---

[①] Kathryn Hagen, Professor, The wood bury university of California, Clothes department director, E-mail: Kathryn.hagen@woodbur.edu

must be considered as interrelated. Our mature Chinese woman might hesitate to embark on an adventurous road of international travel because of fears stemming from these underlying issues. In addition, she will have to consider the potential pitfalls relating to her maturity, including *ageism*, which renders her potentially invisible or weakened. By adding a fourth "path" to this list of intersecting challenges, we have appropriated this theoretical construct and adapted it to the focus of this paper. Thus we are committed to addressing these constraints as a whole.

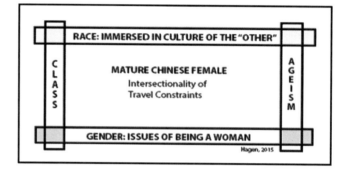

Chart 1 *This chart illustrates the previously defined areas of intersectionality, with the addition of ageism as a fourth dimension of constraint.*

Historical Context

"Women hold up half the sky." Chairman Mao–Zedong, c. 1950

Certainly things have changed enormouslyfor women during the last forty years. Today's over forty Chinese women grew up in a world of war, isolation and frugality. They also experienced an extreme ideological shift, transitioning from celebrated "iron girls"who dressed in military uniforms and could do men's jobs when necessary, to more feminized women who like to shop for pretty clothes and pamper themselves. Despite this change, these women maintain their independence and generally control the family pocketbook. [2]

China has also taken a fascinating ideological journey, shifting from a political emphasis to an astoundingly successful economic reconstruction. In 1978 China opened its doors to the outside world and each decade since has seen an increased emphasis on tourism as a productive service industry. Until the 1980's, however, people traveled into the mainland, but very few Chinese went outward. But increasingly open policies and the growth of personal wealth encouraged internal and overseas investors to offer enticing optionsand Chinese travelers took to the road.

Today huge numbers of Chinese citizens travel abroad, and not just to Hong Kong and Macao. [3] According to the South China Morning Post, 97 million Chinese went on international trips in 2013, and that number is only growing. [4] The booming Chinese economy provides resources for even middle class travelers to expand their horizons, and almost 35% of their budget will go into shopping "Chinese tourists spend so much abroad that some foreigners are calling us 'walking wallets'," says Song Rui, a researcher at the Chinese Academy of Social Sciences.

On-line information is also changing the face of travel. Specialized, convenient services encouragetravelers to branch out. "This new generation of Chinese outbound travellers is making their own decisions...by doing their own research online, going beyond the old stereotype of big buses of group tourists," said Lily Cheng, managing director of TripAdvisor China. Popular destinations range from Thailandto South Africa to Australia to South America. [5] Group visas are easily obtainable for almost anywhere in the world, and many countries have also removed or lowered barriers for individual tourism.Travel is part of the *consumerism movement* in China today, and destinations are *branded products*. [6] Well-heeled young professionals are the primary consumer segment and many travel internationally three or more times a year.

As of 2010, women make up more than 55% of that equation. [7] Not surprisingly, they have become target subjects for travel providers; they are affluent, have more time to travel, they spend more money at their destination, and their opinions often dominate when families make vacation decisions. [8]

Motivations and Constraints of the Mature Female Traveler

In China, much like other countries, the popularity of travel stems most directly from status. International travel has an additional cachet. Travel also speaks to knowledge, adventure, fitness, great shopping, and spending time with family. Youthful Chinese seem fearless, ready to embark on exotic journeys whenever they can get away. But things are not always so clear or easy for our subject focus, the mature Chinese woman. As much as she may want to travel, certain practical or psychological obstacles may stand in her way. More research is being done on these challenges and the "push-pulls" that formulate her attitudes.

In a study of mature Chinese female travelers conducted atAmherst, eight types of constraints were identified that most impact our subject cohort. These include:"limited knowledge of tourism," "health and safety concerns," "culture shock," "lack of travel partners," "low quality of service facilities," "limited availability of information," "negative reputation of tour guide," and "few employer paid vacations." [9]

It is easy to understand any mature woman's concerns about her personal health issues or the safety of an unfamiliar country. Fear of terrorists or disease can be negative considerations. She may be uncomfortable with other ethnicities and hesitant to expose herself to cultures so different from her own. Language issues can also be a consideration. Mature Chinese women may also remember a time when vacations were a job-associated benefit. Trips were organized and financially supported by their employer. As one study subject suggested, "... for ［my］ generation, the idea that ［we have to］ pay for our own trips has not yet rooted in ［my］ mind."

Not having a travel partner is also a common complaint. Chinese women are generally required to retire five to ten years earlier than their husbands. They may have the time and desire to go abroad, but their husbands are still too busy to accompany them. Yet travel may be one of the very things that can help alleviate depression over the loss of youth, good health or professional identity. The mature woman's search for new means of *authentic self-expression* might well include travel, which can provide meaningful experiences to counteract perceived identity gaps. Making personal travel choices can help to shapea new post-career sense of self-determination that can increase confidence and a sense of self. Travel destinations can provide varied and sometimes profound experiences, adding to memory and knowledge in significant ways. The inevitable sense of adventure and freedom can also build self-esteem that otherwise tends to diminish with isolation and age.

Chinese tourism is also entering a new phase that may further motivate our educated mature traveler.Though seeing generictourist sites and shopping in tourist-oriented markets was oncesufficient, Chinese tourists today are in search of more sophisticated travel experiences. They look to services providers to offer something different, transitioning from traditional tour groups to enhanced individual experience.The mature female traveler may be especially motivated to seek deeper meaning in her travel experiences. A search for authenticity can be a key element of the pull to explore the "other". Author Wolfgang Arit describes a "spirit-centered" form of authenticity that is also prevalent in China.[10] Beautiful nature experiences or spiritual sites can add a new level of meaning to travel experience and result in a higher degree of psychological well-being.[11]

Such obvious benefits have led many mature women to develop strategic means to alleviate commonconstraints. For example, they often consult friends and colleagues before making plans or choosing a specific destination. The resulting security in knowing what to expect and an anticipation of positive experiences helps to allay their doubts. Mature Chinese women are also especially interested in organic foods and the chance to experience other cuisines can be highly appealing. Eco-friendly destinations are especially alluring, and traveling with other female retirees to such destinations creates a productive commonality and relieves anxiety. Anotherpopular strategy is to connect with "donkey" friends, who are people willing to organize frequent trips, motivated simply by the search for pleasant travel partners.

Research also indicates that women have different values than men when it comes to travel.[12] Though traveling alone may seem risky at first, there are many potential advantages.Travelauthor Leyla Alyanak asserts that every woman, no matter what her age, should try solo travel even if she has a partner at home. She believes a solitary voyage can make women more independent, less fearful, and build confidence in coping skills. In fact, it can be a life-changing experience.

Specialist vendors also offer some highly appealing options for mature travelers, such as private shopping services, helicopter/ private jet services, educational travel, cooking and culinary experiences, and cultural immersion. There are also volunteering experiences that combine travel with opportunities for worthy causes. How could any woman not come back from a trip like that feeling better about herself? These are all attractive options that speak to self-individuation, and it is likely future travel will become even more customized.

After the Planning

For the mature woman with extra time on her hands, travel provides *four areas of consumption* that can structure otherwise empty hours. These include trip planning, packing, actual travel, and post-travel sharing and communication. Each one of these four can offer satisfying rewards. Our final focus is on the second element, the results of which inevitably impact the actual journey.

Assembling critical elements that will accompany our traveler on her journey is a key part of the identity-building process, but the act of packing for an overseas journey can seem fraught with potential pitfalls. Though this collection of garments and other necessities may seem trivial to some, a poorly conceived set of travel elements can undermine everything.Just as many of us have overflowing closets of unrelated garments, over-packing is almost inevitable ifa list is not carefully considered and executed. A trip that involves multiple destinations, efficient and organized packing becomes even more crucial.

Such issues include the need to travel light, to anticipate weather changes, to look polished in varied circumstances including posted photos, and to be comfortable in uncomfortable circumstances like long plane trips. If it is hard for her to find affordable clothing in China designed specifically for the mature customer, how much harder will it be to find *travel specific* clothing that supports her quest for a more "glamorous" international identity? The *intersectionality* of problematic issues connected to aging, dress, as well as travel can create an overwhelming barrier to participation.

Case Study: Joie Yen

Joie Yen, [not her real name], is a Chinese born, Dartmouth educated professional in her late forties. As one who travels constantly for business and pleasure, she has evolved a highly strategic and disciplinedsystem for packing. Her packing time averages about thirty minutes, which is astounding. Though few women would even try to duplicate her efficiency, it is still worth studying her methods.

1. She does not buy any clothes, anywhere, no *matter how appealing*, that are not functional and suited for travel. No heavy garments, including jeans that take up room. Clothing must be light and suited to layering. No dry-cleaning, no ironing.

2. Accessories are key. A business outfit becomes an evening look with the addition of nice jewelry or a colorful scarf. She takes three to four pairs of shoes that dress an outfit up or down.

3. *She never checks a bag,* even on a trip as long as two months! Five skirts, five lightweight jackets, five or so tops. Everything mixes and matches. Her classic go-to travel clothes were collected over years, so she will not chance losing them through misplaced luggage. This rule also saves time in airports and she always has what she needs with her.

4. Comfort first. Long plane rides are hard enough without wearing uncomfortable clothes or shoes. Wear fabrics that breath on the plane, so any sweat wicks away.

5. Roll everything for fewer wrinkles and space efficiency. Separate elements like underwear or tops into different clear bags that keep them organized.

That's it. Simple, but not for everyone. Having traveled with Joie through multiple cities and situations, I can attest to the functionality and discipline of her system. Since she is not that inclined to buy clothes while traveling, extra luggage room is also not critical. But most women, who spend about 35% of their travel budget shopping, will need some kind of expandable luggage. Cases that start out as minimal garment bags can expand to full-size luggage.

Not everyone has Joie's amazing merchandising skills either. Therein lies opportunity. Creating coordinated, comfortable capsule travel collections, appealing but functional soft luggage, and versatile, light accessories, all with a distinct but subtle Chinese aesthetic, could be highly attractive and practical for mature women who want to travel. Strategic interior pockets for passports and money, as well as cell phones, would spare distracted travelers from carrying handbags that can easily be stolen or lost. Research has also shown that ecologically friendly products are desirable if the buyer believes the benefits are genuine, and they are willing to pay extra for such products. A sustainability emphasis could be an additional positive feature of the collections. Allowing some customization of these elements, such as assorted prints or add-ons, would add even more appeal.

Worldwide in 2015, the Baby Boomers of every country are retiring and traveling. There is no dearth of mature folks cruising, trekking, flying, and exploring the "other", which bodes well for John Lennon's dream of one world, one people. We are all interconnected, with similar intersecting problems, and everyone navigates these elements that threaten to hold us back. Support systems for every means of travel will keep multiplying, as economic opportunities encourage continued growth. Research into ways we creative academics can support the transitioning identity of our mature peers continues, and it is a worthwhile pursuit indeed.

References

[1] An American legal scholar, Kimberle Crenshaw, coined this theoretical concept, in 1989. It has been applied primarily in Feminist Theory.

[2] Jing Lin, Chinese Women Under the Economic Reform Gains and Losses, Harvard Asia Pacific Review, pp. 1–2.

[3] Within a dozen years, the annual number of border crossings has multiplied from less than 10 million in 1999 to almost 70 million in 2011.

[4] South China Morning Post, Thursday, 97 million Chinese tourists went abroad in 2013, posted 1/29.

[5] South China MP: Chinese tourist boom ripples out to more destinations, 9/24/2014.

[6] Wolfgang Arit, China's Outbound Tourism, Routlage, 2006.

[7] Essential China Travel Trends, 2012, China Travel Trends, Dr. Xu Chen ［researcher］.

[8] Mimi Li, Tong Wen & Ariel Leung : An Exploratory Study of the Travel Motivation of Chinese Female Outbound Tourists, Journal of China Tourism Research, 7:4, 411–424, 2011.

[9] JieGao and Deborah Kerstetter, "Older Chinese Women's Perceived Constraints and Negotiation Strategies With Pleasure Travel ［using the theoretical framework of Intersectionality］" (June 5, 2015). Tourism Travel and Research Association: Advancing Tourism Research Globally. Paper 17.

[10] China's Outbound Tourism, Wolfgang Arit, Routledge, 2006, p. 218.

[11] Roger C. Mannell and Ryan Snelgrove, 2011.

[12] Baraban, 1986; DeLuca, 1986; Hawes, 1988.

新时代下银发一族时尚服装产品发展理念

◆ 刘 侃[①]　（上海视觉艺术学院，上海，201620）

【摘　要】　面对快速发展的人口老龄化，潜力巨大的银发产业将迎来黄金时期。我们要转变观念，银发一族需要的时尚产品不是功能简单的傻瓜产品，而是要具有高质量、高科技含量和高智能的产品。本文探讨了银发产业的发展现状、消费特征并对目前功能性服装产品的开发方向做了简单陈述。

【关键词】　银发一族；时尚服装；产品理念

Study on Fashion Clothing Products of Silver Gens under the New Era

◆ Liu kan　*(Shanghai Institute of Visual Art, Shanghai 201620)*

Abstract: *According to the rapid development of the aging population, the potential of the elderly industry will face to a golden era. We need to change our design ideas, the older people need the fashion product is not a simple product, but to have a high quality, high technology content and high intelligence products. This paper discusses the current situation of the development of the elderly industry, consumer characteristics and to present the development direction of functional clothing.*

Keywords: *Silver gens; Fashion; Product idea*

2013年2月发布的《中国老龄事业发展报告(2013)》指出，2013年中国老年人口数量将超过2亿，老龄化水平接近15%。8月16日，国务院召开的常务会议也提出，加快发展养老服务业，既能弘扬中华民族敬老优良传统，满足老年人多层次需求，提高老年人生活质量，又能补上服务业发展"短板"，释放有效需求，催生上千万就业岗位。老龄化加剧的同时，老年人市场作为一个具有巨大消费潜力的市场，被越来越多的创业者所关注。预计到2015年和2050年，

[①] 刘侃，女，上海视觉艺术学院时尚设计学院，博士，副教授。E-mail：iamkanLiu@126.com

老年人潜在的市场购买力可望达到 14 000 亿元和 50 000 亿元。面对这样一个巨大的银发消费市场,首先需要理念更新,不能以传统的"老人"来定义今天或未来的银发一族。新时代的老年产品,不能仅从老年人年龄出发只考虑经济、健康,而应与共俱进,涵盖所有适用于正常生活人群且有利于提高老年人身心健康、生活自主和社会联络方便的普通生活用品和高科技产品。这类产品不但可以提高制造业的可持续发展,还有助于提高老年人的健康水平和生活独立性,缓解独生子女一代的反哺压力,减轻国家社会保障体系的负担。本文拟就当今老龄化的群体发展态势及日常产品设计发展趋势,对人口老龄化社会态势下老年产品的时代特性及研发理念进行创新探究。

1 银发一族的产业发展现状及消费需求特征

1.1 "银发产业"的发展现状

中国老龄科学研究中心的研究报告显示,2010 年我国老年人的退休金总额为 8 000 亿元左右,预计 2020 年和 2030 年我国老年人的退休金总额将分别达到 2.8 万亿元和 7.3 万亿元。国家民政部社会福利和慈善事业促进司的数据表明,2010 年我国老年人口消费规模超过 1.4 万亿,到 2030 年将达到 13 万亿元。与老龄化人口社会需求不断加剧形成鲜明对比的是,国内的"银发产业"尚处于起步阶段,其发展严重滞后于老龄人口的快速增长和老龄消费群体的巨大需求。从全国老龄委提供的数据中看出,当前国内市场每年为老年人提供的产品与服务总价值不足 1 000 亿元,老年人消费市场中的需求与供给之间存在着巨大的逆差。

1.2 银发一族的消费需求特征

如果将 65 岁以上的人群称为银发一族,依据现在良好的生活条件,人均年龄可以达到 80~90 岁,那么这一族群所涵盖的年龄跨度为 20~25 岁。按照年龄结构和身心健康状况,有学者[1]将不同年龄段"银发一族"人群的不同需求划分为以下三个年龄创段层次:1.65~70 岁身心状况健康良好的群体,他们大多数依然与社会联系紧密,注重自身形象。除生活基本消费外,在文化、娱乐、教育、体育以及社交、旅游等方面的活动欲望依然强烈。2.70~79 岁这个年龄段的老人,大多数身心依然健康良好,他们依然有一定的社交活动,但物质消费倾向于对保健、医疗护理、功能产品及相应服务性消费的需求。3.80 岁以上的高龄老年群体,他们中的大部分人由于身体退行性改变或疾患,生活自理能力较差甚或不能自理,对功能性产品、护理性服务及功能性设施、特殊商品需求增加。

每个年龄段的老年人虽然生理特征和生活规律大致相近,但不同文化背景和经济收入者,生活的内容和方式则存在较大差异,呈现为不尽相同的生活态势。受教育程度高和经济收入较好的人群其思想相对开放,易于接受新鲜事物。他们很多人仍然有晚年事业圈和交际圈,乐于使用电脑、智能手机等高科技产品,对新生事物及智能化、功能性产品需求更为迫切。受教育程度和经济收入不高的老年群体,思想相对保守、经济节俭,基本生活范围以社区、公园、家庭为主。他们对周而复始的生活比较适应,但也不乏期待生活中有些新鲜的事物产生。

2 银发一族时尚生活产品设计理念

之前一提到老年产品,人们眼前就会浮现出宽大的衣服、老花镜、拐杖以及行动不便或者行为不能自理的老人的轮椅等带有浓重老年人标签的物品。其实,我们已经不能一成不变的眼光看待现代老人这个群体。现在65岁左右的老人都是新中国成立后出生的,他们经历过国家翻天覆地的经济变革时期,也经历了以全球为背景的人类高科技时代,他们不排斥现代时尚的产品,只是不喜欢华而不实的产品,他们需要的是具有高质量、高科技含量和高智能、多功能性的产品。这里以服装服饰类产品为例,探讨下未来银发一族青睐的产品的设计方向。

2.1 功能型服装产品开发

近二三十年,国内外的纺织科技工作者都在致力于开发具有功能型且舒适安全的服装材料,应用于服装的设计生产中以提高人们的生活质量,并取得了很多科技上的突破。[2]如美国TrapTekt公司研发的火山灰纤维新品Minerale,原料取自火山喷发产生的固体喷发物,经过纳米高温的处理后,聚合在纤维中的多功能性纱线。火山灰因具有微多孔构造,故具有强大的吸附力,能在短时间内消除异味,抑菌防臭,解除瘙痒,穿上这件神奇的抗菌止痒衣后,可大大缓解皮肤瘙痒。而且,火山灰粉末聚合于纤维的高分子中,达到永久性的效能,不会因为穿着或洗涤而逐渐消失功效。

荷兰Advansa公司新近研制出一种紧身减肥服,它由牛奶蛋白纤维和多孔碳燃脂丝纤维构成的暖性纤维织成。当肥胖人穿上这种衣服,暖性纤维能使人体发热出汗,而多孔纤维能吸收热量将汗蒸发,这一过程中就消耗了人体大量脂肪,从而达到减肥目的。该公司新开发的另一种新型调温吸湿排汗减肥服装,其纤维采用杜邦公司PPT纤维与Lenying公司的Lyocell纤维混纺而成。PTT纤维可低温染色、色牢度好、悬垂好,而Lyocell纤维柔软、美观,其快速吸收水分并传递至外部的特性是调温吸湿排汗纤维具有的吸湿排汗的功能。该服装的面料吸湿速度比其他功能面料高49%,而排湿速度高55%。

此外,功能性服装面料的创新开发也为功能服装的发展起到了强大的推动作用。服装面料的"功能性"是指具有特殊作用和超强性能的面料,是一般服装面料所不具备或达不到的性能,如防水性、超保暖性、高吸湿性等。对于老年人而言,他们新陈代谢变慢,对外界环境的自我调节能力变弱,需要依赖面料的功能性来抵御外界恶劣气候,如利用这些功能性面料做成的老人户外服装侧重于耐气候功能,炎热难耐的夏季,吸汗透气的凉爽型面料能使皮肤保持干爽凉快,而严寒、大风、大雨的冬季,由防风保暖透气透湿的轻便纺织材料制成的户外服给老人们提供了一个不冷不湿,又便于活动、伸展自如的时尚造型。

2.2 智能化服饰产品的开发

智能化服饰产品的设计主要分为两个方向：一是对智能服装材料的研究并将之应用到服装设计与制作中；二是结合电子科技技术研发是可穿戴的智能化产品。

2.2.1 智能服装材料在智能服装中的应用

老年人随着年龄的增长体型变化显著，身体机能减弱，身体的灵敏性、柔软性及肌肉力度降低，站立行走不安定性加大，这种机能性退化使得老年人服装更应注重卫生安全和舒适性。对应身体机能特性，老年服装的材料在满足轻便、柔软、吸湿、透气的基础上，应同时能起到良好的保护和协调作用。现在，高机能型的智慧纺织品能更好地适应机能和环境变化的功能，在有附加功能的老年服装产品开发应用上具有广阔的发展前景。

智能服装材料是以智能材料为基础的新型纺织品，先进的材料科学、生物科学、电子技术、纳米技术等高科技推动了智能服装材料的生产技术发展。近年来，随着智能材料的不断创新，智能纤维、智能纺织品、智能服装正成为迅速发展的时尚产品，受到越来越多的关注。目前开发的智能型服装材料主要有形状记忆材料、调温材料、光敏变色材料、抗菌材料、情感变化纺织材料等：

（1）形状记忆材料

形状记忆材料包括形状记忆合金和形状记忆聚合物。英国科研人员研制的一种防烫伤服装，其原理是先将镍钛形状记忆合金纤维加工成螺旋弹簧状，然后再加工成平面状，最后固定复合在服装面料中，当接触高温时，镍钛纤维发生形变，纤维立即由平面状变为宝塔状，使两层织物之间形成较大的空腔，避免高温接触人体皮肤，防止皮肤被烫伤。

（2）调温服装

调温纤维是将相变材料包覆在纤维中，当外界环境变化时，纤维中相变材料发生液、固可逆转变，或从环境中吸收热量储存于纤维内部，或放出纤维中存储的热量，在纤维周围形成温度相对恒定的微观气候，从而实现温度调节功能。美国 Boulder 和 Colorado-based 公司将相变材料微胶囊化后植入纤维中织成面料，制作成服装，穿着者运动时，身体会产生热量，热量由相变材料吸收，当停止活动时，相变材料释放热量，在纤维周围形成稳定的微观气候，从而在一定时间内实现温度自我调节。

（3）光敏变色织物

光敏变色是指某些物质在一定波长的光照射下会发生变色，而在另上波长的光作用下又会回到原来颜色的现象。日本 Kanebo 公司将光敏物质包覆在微胶囊中，用印花工艺制成光敏织物，这种织物在吸收 350~400nm 波长紫外线后可由无色变为浅蓝色或深蓝色。微胶囊化可以提高光敏剂的抗氧化能力，从而延长使用寿命。

（4）智能抗菌纺织品

老年人免疫抵抗力变弱，病菌是其致病的根源。为了保护人体不受细菌、霉菌、微生物等的侵袭，服装用纺织具务清洁、抗菌、抑菌功效是发展途径之一。智能抗菌纺织品多对纤维进行加工以达到抗菌、杀菌、防霉等功效。美国 Nylstar 公司新近制造出的智能聚肽胺纤维，将抗菌剂包藏在纤维内部，保障了纤维的耐久性和安全性，无论人体轻微活动还是剧烈运动，其既阻碍细菌任意繁衍，也不杀死全部细菌，而是控制维持在一个正常水平。

（5）情感纺织品

情感纺织品具有测试人的情绪的功能，可根据人的情绪，变化颜色或释放香味。应用于老年服装能够营造环境气氛、声音、气味等，调节老年人经常因身体机制衰退而产生的失落感，使其心情舒畅、情感活跃。

2.2.2 智能可穿戴服装

智能可穿戴设备一般是指穿戴在人身上、具备智能特性的设备的总称，也可认为是集成微型计算机、通信、感知等装置，为用户提供个性化服务，可穿戴于身上的微型电子设备。通过智能可穿戴设备，可以更好地感知外部与自身信息，能够在计算机、网络甚至其他人的辅助下更为高效地处理信息，以及更好地实现无缝交流。如智能电子服装：

（1）智能电子服装

附有传感和微缩的电子、通讯装置的智能纺织品，可以检查、储存和控制信息，并将测试的数据或功能传递到控制中心进行数据交换。其将织造技术与集成无线遥控技术、GPS 全球定位技术、柔性触摸界面技术等结合一起，应用于服装上，能够远距离监控环境和掌握穿着者的身体状况及所在位置。如美国 Georgia 理工学院的研究人员将光导纤维植入衣料内，可检测到温湿度及穿着者的心率、体温、呼吸、血糖、血压等重要生理指标，可协助医务人员日常监护，维持老年人健康的生活状态。

（2）检测老人平衡状态的防跌智能鞋

麻省理工学院研究生利伯曼利用太空技术开发出一种防跌智能鞋，可以防止老人跌倒，从而避免引发灾难。他们发现平衡感有问题的人的一些特征，他们于是设计了装有六个压力传感器的防跌智能鞋，帮助采集数据以供分析。使用者可将鞋垫交给医生或专家进行分析，依此事前诊断出是否有失衡症状，以便提前治疗，预防摔跤等恶性事故。

（3）检测关节运动的防跌电子裤

美国弗吉尼亚理工大学的工程团队设计出一种电子裤，可帮助确定哪些老人具有高跌倒风险，从而有可能减少由于跌倒事故而受伤的老人数量。该校电子工程系教授汤姆·马丁和马克·琼斯设计的这种电子纺织裤，内嵌有数个电子标签，这些小小的印刷电路板内含有微控制器、传感器及通讯器件。该装置由一个贴在腰间的九伏电池供电。在步行实验中，电子标签收集的数据被传送给一个单独的蓝牙电子标签，再由其通过无线方式将数据传输给一台主机电脑。研究人员通过比较两组测试人员的结果来测试这种电子裤的可行性。由于这种电子裤可以不受察觉地连接至远程医疗设施，将使老年人能够独立生活的时间更长，电子裤就是他的"保护神"。当电子裤检测到个体步态的不稳定性后，系统就能报知医疗设施及本人，提醒他们尽量避开不安全的路面。

除此以外，可穿戴服饰及配件的种类还有很多，如智能手表、智能戒指、智能手环、智能耳机等，弗吉尼亚理工大学运动研究实验室主任瑟蒙·洛克副教授说，发展可穿戴式系统的主要问题之一是人们是否愿意穿用。舒适性和技术创新将是未来穿戴式系统设计的关键。德国海恩斯坦研究院就预测，智能服装是一个数十亿人需要数百万种应用产品的市场。美国著名市场研究公司 VDC 的报告指出，到 2018 年，仅在美国市场，智能型纺织品的产值将达到 30 亿美元，未来具有很大的发展空间。

三、结语

银发产业"钱"景可期，在一定程度上缓解了人口老龄化给社会和资源带来的巨大压力。蓬勃发展的银发产业将不仅推动我国经济的增长，也将实现社会"老有所乐""老有所养""老有所医"的最美夕阳构想。老年人的产品设计不应该是肤浅的功能简单的傻瓜产品，而是需要具有高质量、高科技含量和高智能的产品，老人不排斥时尚现代的产品本身，他们只是不喜欢过度的浮华与装饰，我们应想其所想，与时俱进，更新设计理念，将更多贴近时代生活的设计产品奉献给他们。

参考文献：

[1] 吴国强."银发市场"：对应人口老龄化社会态势的老年产品理念[J].西北人口，2011（5）.
[2] 史密斯·罗曼.功能时尚的智能艺术服装[J]. Science & Technology of Stationery & Sporting Goods, 2015（5）.
[3] 王海毅，冯伟.智能纺织品在老年服装中的应用分析，江苏纺织，2008（12）.
[4] 陈志林.智能服装开发与应用现状[J].纺织科技进展，2010（2）.

当代城市中老年女性化妆品消费观察

◆ 赵宏伟[①] （上海视觉艺术学院，上海，201620）

【摘 要】 时间定在2015年，将今天城市中老年女性的生活背景向前推三十年，她们正是风华正茂的二三十岁的年纪，也是中国经济改革开放、恢复高考、出国留学打工潮风起云涌的时代。三十年的社会变迁，财富积淀，这样一批人现在已经步入中老年人的行列。今天的化妆品消费群体分层分类非常细腻，国门的敞开，旅游业的开放，互联网的兴盛，流行信息传播的迅速，消费模式的改变，让整个地球变成了一个"村"，世界信息几乎同步。今天的城市中老年女性人群是国家积累财富时的创造者，也是个人家庭财富的拥有者。社会的文明程度越高，意味着人们追求美的要求会渗透到生活的各个层面。在化妆品消费需求纷繁复杂的种类中，这样一群城市中老年女性具有一定的生活品质，具备一定的消费能力，她们对化妆品的选择让市场不容忽视。

【关键词】 城市中老年女性；社会变迁；文明程度；化妆品选择；不容忽视

Study on Cosmetic Consumption of Urban Mature People in Contemporary China

◆ Zhao Hongwei *(Shanghai Institute of Visual Art, Shanghai, 201620)*

Abstract: Time is set in 2015. We retrospect our background to 30 years ago when today's urban middle-aged and old women were at life's full flowering age, 20s and 30s. This was the times that China implemented its economic reforming and opening up. Universities restored the college entrance examination and going abroad for further education and a job was so popular among people.

After thirty years of social changes and wealth accumulation, those people have entered middle-aged and old years.

At present, the layering and classification of the cosmetic customer group are fine. The opening up of our country and our travel industry, the flourishing of the internet, the rapid speed

---

[①] 赵宏伟，女，上海视觉艺术学院时尚设计学院讲师，硕士。E-mail: doudou1988zhw@126.com

of popular information transmission route and the change of consumption model make the earth a whole village and the information around the world almost synchronized.

In today's cities, middle-aged and old women are the creators of national accumulated wealth and also the owner of personal wealth. As our social civilization becomes stronger, it means that people's pursuing of beauty has infiltrated every aspect of our lives. Among various and complicated needs of cosmetic consumption, those urban middle-aged and old women have a certain quality of life and certain consumption ability. Their choices for cosmetics cannot be ignored by the market.

Keywords: *urban middle-aged and old women; social changes; social civilization; choices for cosmetics; cannot be ignored*

引言

爱美之心，人皆有之。当今城市中老年女性化妆品消费者对化妆品的需求很大，她们是不容忽视的很大的一个消费市场群。

由于中国社会经济发展和家庭年收入的提高，人们的生活水平也发上了巨大的变化，城市中老年知识女性对化妆品的需求也发生了消费观念上的巨大变化。数据显示整个化妆品市场在社会经济大力发展的 30 年中，化妆品市场销售额平均以每年 23.8% 的速度增长，最高的年份达 41%，增长速度远远高于国民经济的平均增长速度。[1]

1 城市中老年女性的化妆品总体需求

化妆品在城市中老年女性中的使用司空见惯。

通常情况下在选择各类化妆品时的消费习惯有三点是很重要的。

第一点，选择功能。化妆品种类繁多，你要消费这个产品是为什么，产品功能是消费时必须做的选择，是护发还是护肤，是美甲还是彩妆，有目的的选择和分类才能让你在浩瀚的商品海洋中选择你要的产品，不论是在传统模式的商场、超市体验式消费，还是在电子商务的虚拟空间购买消费。

第二点，选择品牌质量和价格。做广告，使用明星的代言是现代化妆品宣传的不二选择，通过大量的广告宣传，商家达到了对商品的推广，选择权却掌握在消费者手里了，中老年城市女性在选择化妆品时对产品质量很关注，要关注生产厂家、生产日期和产品保质期。这是现代消费者自我保护意识提升的表现。城市中老年女性更是关注产品的价格。不用说在实体店中，有时还要货比三家地选一选，学会电视购物、网购的城市中老年女性更是想着怎样最划算。

面对进口的化妆品、中外合资的化妆品和本国制造的化妆品，对于具有不同消费能力的中老年城市女性而言，对产品质量和价格的性价比的选择，往往让产品品质好、价格合理的化妆品更胜一筹。

第三点，对化妆品芳香气味的个性选择。随着化妆品种类的不断更新换代，城市中老年女性选择化妆品的功能，也选择化妆品的品牌与价格，对化妆品的芳香气味也做出相应的固定的喜好选择。这是一个社会文明程度提高，生活品质提升的一种表现。中老年城市女性选择化妆品芳香气味通常都会理智地选择幽幽的果香和淡雅的天然植物香。香艳浓郁的花香不太被中老年城市女性所选择。

化妆品种类纷繁复杂和多样化在当今商业市场中的重要自然不必多说。按照功能，中老年城市女性在选择各类化妆品时也不外乎基本的三大类：

第一类，美发护发产品。城市中老年女性对美发护发用品的依赖是很强的。生理上，由于年龄的原因，到了这个年龄人的毛发开始变白，从两鬓开始，向头顶慢慢延伸。有需求就有市场，各类品牌染发剂充满了市场。

第二类，美容护肤产品。随着年龄的增长，城市中老年女性的脸上或早或晚都会有岁月的痕迹留在脸上，功能性强的去皱抗衰老护肤品自然成了这个年龄层所推崇的产品。

第三类，彩妆产品。社会文明素养的提升，社交活动的不断增加，都影响着中老年城市女性的生活方式和消费习惯与审美变化。淡妆出行，是现代中老年城市女性文明程度提高的一种表现。

今日的零售终端发生着前所未有的变革，改变和更新着人们的消费理念和消费模式。社会发展，消费模式的多样选择让城市中老年女性也面临着在众多产品中选择适合自己的化妆品理智消费，有时会存在在传统体验消费区与虚拟网络消费进行对比后再消费的过程，选择化妆用品的销售购买渠道多样化，也是当代中老年城市女性所面临的新的多重消费模式的新挑战。

2　城市中老年女性的化妆品之染发护发篇

在中老年城市女性中由于工作压力大，生活节奏快，家庭事务多等等原因，很多人的头发早早地就开始变白。白发从两鬓开始，向头顶慢慢延伸，尤其是中年人，有时从心理上真是不愿意接受老年特征的出现。染发，就成了大多数城市中老年女性为了掩饰年龄常常要做的一件事情。去美发店用专业的染发剂染，在家自己买染发剂弄，各种颜色有各种效果。最终的目的都是掩盖两鬓的、头顶的或多或少的白发。由于化学成分对健康有影响的原因，大多数的中老年城市女性对使用染发产品都非常小心翼翼，品牌的选择更挑剔。白发不多的会选择在美发店里染发，白发多的，常常要染发的，很多城市中老年女性会选择在家自己染发。

现在的染发产品色彩琳琅满目，极其丰富，同一品牌的染发剂会有非常多的色彩供消费者选择。

中老年城市女性选择染发剂颜色时要与肤色相称，亚洲人皮肤白皙，染发剂的选择比较多，浅咖啡色、深咖啡色、金棕色、红棕色等都是不错的选择。皮肤偏黄的就要选择比自己肤色深的染发剂，染发剂中偏红色成分多的颜色衬托呼应肤色比较好。皮肤颜色黝黑的就要选择更深色的染发剂来调整发色了。

　　城市中老年女性选择染发剂颜色时要与你的性格身份相符合,大多数的中老年城市女性都是中规中矩地选择不犯错误的大众认可的发色示人(图1)。但在城市中有许多的文艺范儿中老年女性,她们的职业与艺术有关,她们个性鲜明,她们的造型就是这个时代百花齐放的一个符号(图2)。

　　城市中老年女性选择染发剂颜色时也要考虑服装颜色相协调。每个城市中老年女性在对自己身材了解的情况下,服装风格会比较明显,再加上个人偏好,服装色彩都会有一个主基调。选择染发剂要把个人着装颜色因素考虑进去,要选择服装颜色里更深的色彩作为染发剂。例如,偏咖啡色系服装多的要选择深咖啡、深红棕色等;偏冷色系服装多的要选择染发剂中有酒红成分的。

　　新陈代谢缓慢的城市中老年女性发质偏干,发鳞片偏毛,再加上染发烫发双重压力所带来的发质的变差,让中老年城市女性又非常依赖护发产品的大量使用。护发素、焗油护发膏也有很大的市场需求。城市中老年女性通常都会选择同一个染发剂品牌的焗油护理产品,或者与洗发水配套的护发素一起使用,以达到让发质柔顺、光泽有弹性的效果(图3)。

图1　不同颜色的发色

图2　艺术策展人陆蓉之的红发

图3　各类洗发护发染发剂

3　城市中老年女性的化妆品之护肤美容篇

　　中老年城市女性的护肤品使用源于从年轻时就养成的一种习惯,她们有延续性使用护肤品的习惯,多了对护肤类化妆品在功能上的要求。

　　年轻时,会根据个人的肤质选择护肤品。干性皮肤的通常会选择霜类护肤品,混合型皮肤的会选择半流动的乳液类护肤品,油性皮肤的会选择水剂类的护肤品。

　　年轻时,会根据季节的变化选择相应的护肤品。夏天新陈代谢快,皮肤容易出油,选择水剂和乳液护肤品比较常见。春秋两季会根据肤质选择护肤品。冬季来临,通常都会选择保湿的偏

图4 化妆水　　　　图5 乳液　　　　图6 护肤霜

油脂多的护肤品使用(图4~图6)。

随着年龄的增长,对护肤品的功能选择也越来越多样化了。不但要考虑肤质、季节,更要考虑抗皱、去皱、抗衰老等功效。

生物科技的迅猛发展也带动了化妆品行业产品的更新换代,世界各大知名化妆品品牌和本土化妆品品牌都看到了广大的市场前景,抗衰老护肤品、去皱护肤品、换肤护肤品等这些功能性的化妆品给中老年城市女性带来了福音。同时,价格战、质量战也需要中老年城市女性有分辨,在琳琅满目的化妆品中选择适合的产品,并在使用后达到预期的效果。

现在市场上的护肤类化妆品种类很多,选择自然更多,天然无刺激、价格合理、功效显著都是中老年城市女性选择购买的主要条件。在消费观念上,功能性强,价格虽高,只要品牌够大,信誉度够好产品,效果够明显的护肤品在中老年城市女性的消费购买力所能承受的范围都是可以接受的。

4 城市中老年女性的化妆品之修饰化妆篇

中老年城市女性在城市中是不容忽视的一个群体。她们有充满自信的理想、对大方得体的追求,还有赏心悦目的需求。基于这些,彩妆化妆需求也成了中老年城市女性生活中不可或缺的一项内容。

化妆画得好看是要有审美,要训练的。对自身的认知是必要的,化妆这个动作做了,画出的妆面代表着你的化妆技术,也代表着你的个人品位。工作中的中老年城市女性充满着智慧和才干;社交中的中老年城市女性又有着灵活与欢乐;家庭里的中老年城市女性还要有温柔与体贴。在社会中,角色的转换让中老年城市女性对自身的形象更加关注。

化妆可以改变一个人的气质,修饰与不修边幅有着本质的区别,作为城市中的一分子,中老年城市女性的文明程度、素质修养、审美追求都有标准。自然,除了服饰搭配以外的化妆就显得格外的重要(图4)。

化妆,首先要对肤色进行调整,中年以后的城市女性需要调整面部的肤色,选择调整肤色的粉底色,不宜过深更不宜过白。要选择与自己的肤色相近的颜色,只要让肤色看上去均匀就达到目的了。粉底不宜过厚,过油,带有酒精成分的快干型粉底不需要定妆粉,是中老年城市女性很不错的选择。腮红的修饰是中老年城市女性化妆成功的根本,粉底打好了,腮红涂对了,妆面就成功了百分之五十,涂腮红可以使人的气色看上去更健康。

对眉眼的修饰在化妆中也至关重要,"眉清目秀""眼睛是心灵的窗"等词句,就是对眉眼最形象的描述,我们也常形容一人如果有了年纪,尤其是女性,就会有"人老珠黄"这种说法。眉眼的修饰是化妆中的重中之重,在我们脸上,眉毛的形状是有语言的,眉粗眉细,眉高眉低,眉浓眉淡都能代表人的性格,而且好的眉型还可以调整人的脸型。

唇色的改变与调整在化妆的最后会起到画龙点睛的作用。中老年城市女性的服饰丰富多彩,根据服装颜色、季节变化、环境的不同选择适合的唇膏颜色也能看出中老年城市女性的审美品位(图5)。

化妆虽然是外在的、人为的,但对于文明程度很高的当今的中老年城市女性来说,却是不可少的生活品质一种表现。中老年城市女性拥有自己的需求,在有审美、有修养的工作与生活环境中选择着属于自己的美丽。

图7　化妆前后对比　　　　　图8　彩妆化妆品　　　　　图9　化妆前后对比

5　结论

化妆品在不断发展的现代社会继续在更快速地更新换代。新的、更有实效的美容美发化妆品源源不断地随着生物科技的发展,以越来越不可想象的功能性效果更新着产品,引领着市场,带动着中老年女性消费团体的消费新标准。

参考文献：

［1］http://baike.baidu.com/view/3332.htm

［2］http://cn.bing.com/images/search?q=%E6%B4%97%E5%8F%91%E6%8A%A4%E5%8F%91&FORM=HDRSC2

［3］http://cn.bing.com/images/search?q=%E6%9F%93%E5%8F%91%E5%89%82&qs=n&form= QBILPG&pq= %E6%9F%93%E5%8F%91%E5%89%82&sc= 8-3&sp=-1&sk=

［4］*HERS*.2014.9

［5］http://cn.bing.com/images/search?q=%E5%BD%A9%E5%A6%86%E5%8C%96%E5%A6%86%E5%93%81&qs=n&form=QBIR&pq=%E5%BD%A9%E5%A6%86%E5%8C%96%E5%A6%86%E5%93%81&sc=8-5&sp=-1&sk=

The influence of types of SNS advertisement by fashion brand, social distance and involvement on commitment

◆ Dongeun Choi[1] Sunjin Hwang[2][①] *(1. Dept of. Fashion marketing, Sungkyunkwan University, Seoul, Korea; 2. Dept of. Fashion marketing, Sungkyunkwan University, Seoul, Korea) (E-mail: cde88@naver.com, sjhwang@skku.edu)*

Abstract: *The purpose of this study was to interrelate the effect of types of SNS advertisement, social distance and involvement of fashion brand on commitment. For this study based on a formula, three way factorial design was applied 2(Types of SNS advertisement: Direct advertisement, Indirect advertisement) × 2(Social distance: Close distance, Far distance) × 2(Involvement: High involvement product, Low involvement product). The subject of the study was 150 adults in both men and women. The results of study implied what kind of differences are made in the types of advertisement according to the social distance and involvement of the products, and what kind of effect the interaction among the three variables have on commitment of customers. Farther studies of future was also suggested.*
Key words: *Types of SNS advertisement; Social distance; Involvement; Commitment*

时尚品牌社交媒体广告的类型、社会差异与品牌介入对消费者承诺的关系研究

◆ *Dongeun Choi Sunjin Hwang* （成均馆大学，韩国首尔）

【摘 要】 本研究意在探讨社交媒体广告的类型、社会差异与品牌介入对消费者承诺的关系。该研究采用三因子,每一因子为两个变量,公式：2（SNS广告的类型：直接广告,间接广告）*2（社会距离：近距离,远距离）*2（涉及：高涉入产品,低涉入产品）。研究受试者为150位陈年男女性。研究结果表明不同类型广告产生何种差异取决于社会差异与产品介入度并表明了三个变量的相互作用对消费者承诺的影响。同时建议对该课题作进一步的探讨。

---

① Dongeun Choi, Sungkyunkwan University, Graduate Student, Fashion marke*ting*, E-mail：cde88@naver.com
Sunjin Hwang, Sungkyunkwan University, Professor, Fashion marketing, E-mail：*sjhwang@skku.edu*

【关键词】 SNS广告的类型；社会差异；介入；承诺

（中文翻译：谈金艳）

1 Introduction

Advancement of SNS and popularization of smart phones led to expansion of a new advertisement market along with novel marketing techniques. With advantages in subdivision of customers, huge market, fast diffusion and feedback, low cost, and high efficiency, SNS marketing keeps evolving and a number of companies are using it as a method of advancement into new markets(Lee, 2012). The fashion industry, in particular, shows much more active in SNS marketing as the number of non-store companies increases. The reason why the companies are so enthusiastic about social media is that it easily connects between the brand and customers and works as a path of mutual communication to discuss and share the information, products, service, and event in real time(Kim, 2013).

Therefore, we will use Facebook, the most typical SNS media, and make a virtual fashion brand to carry out a research on its substantive effect in advertisement. We will find what kind of differences are made in the types of advertisement according to the social distance and involvement of the products, and what kind of effect the interaction among the three variables have on commitment of customers.

2 Research Method

2.1 Research Model

The research model of this study was show in Fig.1.

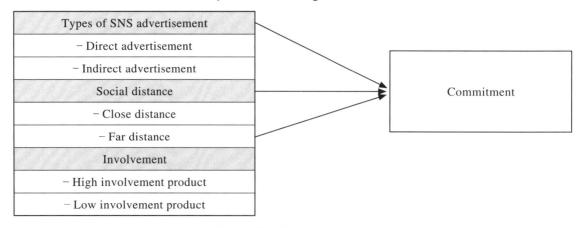

Fig. 1　The Research Model

To find what kind of differences are made in the types of SNS advertisement according to the social distance and involvement of the products, and what kind of effect the interaction among the three variables have on commitment of customers, we set up following hypotheses.

H1. There is significant difference in commitment according to types of SNS advertisement, social distance, and involvement.

H2. There is significant difference in commitment according to types of SNS advertisements and social distance.

H3. There is significant difference in commitment according to types of SNS advertisements and involvement.

H4. There is significant difference in commitment according to social distance and involvement.

H5. Commitment has interaction with types of SNS advertisements, social distance, and involvement.

2.2 Subjects of study and Data analysis

The data for this study were collected from convenient sampling of men and women in 20s and 30s who are living in Seoul and using SNS from January 14 until February 28, 2015. For the total 150 people, we made postings of our virtual brand on the Facebook page with 8 pictures for our survey according to types of advertisement(direct advertisement vs. indirect advertisement), social distance(close distance vs. far distance), and involvement(high involvement product vs. low involvement product). Out of 150, 42(28%) answer to the off-line survey, while 108(72%) to the on-line survey through SNS, and 146 of them were used for the final analysis excluding unfaithful ones.

This study included the demographic analysis, t-test, three way ANOVA, simple interaction analysis, simple main effect analysis for data analysis.

3 Results of the Study

The results of study were shown in Table 1.

Table. 1 The result of 3 way ANOVA analysis(commitment)

Variable	SS	D.F	MS	F-value
Types of SNS advertisement (A)	0.02	1	0.02	0.00
Social distance(B)	54.80	1	54.80	42.31**
Involvement(C)	0.45	1	0.45	1.36
A*B	0.07	1	0.07	0.05

（续表）

Variable	SS	D.F	MS	F-value
A*C	1.17	1	1.17	63.53
B*C	0.31	1	0.31	0.58
A*B*C	2.59	1	2.59	4.83*
Error	1381.801	1167	1.191	
Total		1168		

*: P<.05, **: P<.001

For the major effects of the variables in this study, the types of advertisements ($F_{0.947}=0.00$, N.S.) and involvement ($F_{0.245}=1.36$, N.S.) did not show significant difference, but social distance ($F_{0.000}=42.31$, P<.001) showed significance. Since short distance (M=2.67) presented higher commitment than long distance (M=2.23), the <Hypothesis 1> was partially supported.

In addition, types of SNS advertisement and social distance ($F_{0.817}=0.05$, N.S.), types of SNS advertisement and involvement ($F_{0.62}=3.53$, N.S.), and social distance and involvement ($F_{0.449}=0.58$, N.S.) did not show binary interactions, which led to rejection of <Hypothesis 2>, <Hypothesis 3>, and <Hypothesis 4>.

In the result of <Table 1>, there is ternary interaction ($F_{0.030}=4.83$, P<.05) among the types of SNS advertisement, social distance, and involvement. Thus, <Hypothesis 2> was supported. Therefore, in order to find out the source of the ternary interaction among the three variables, a simple interaction analysis was carried out to have a result as in <Table 2> below.

Table. 2　Social distance*Involvement at Types of SNS advertisement

Variable	SS	D.F	MS	F-value
Social distance*Involvement at Types of SNS advertisement (Direct advertisement)	0.55	1	0.55	2.24
Social distance*Involvement at Types of SNS advertisement (Indirect advertisement)	2.34	1	2.34	5.74*

*: P<.05

As a result of our simple interaction analysis, indirect SNS advertisements of a fashion brand showed significant difference according to social distance and involvement (F0.018=5.74, p<.05). However, direct SNS advertisements of a fashion brand did not show significant difference according to social distance and involvement (F0.137=2.24, N.S.). As we found that indirect advertisement type has interaction with social distance and involvement, we carried out a simple main effect analysis to find the source of the result (<Table 7>). Bothe of the high involvement products (F0.002=9.67, P<.01) and low involvement products (F0.044=4.12, P<.05) showed significant difference.

Table. 3 Social distance*Involvement at Types of SNS advertisement (Indirect advertisement)

Variable	SS	D.F	MS	F-value
Social distance*Involvement (High) at Types of SNS advertisement (Indirect advertisement)	24.16	1	24.16	9.67**
Social distance*Involvement (Low) at Types of SNS advertisement (Indirect advertisement)	7.57	1	7.57	4.12*

*: P<.05

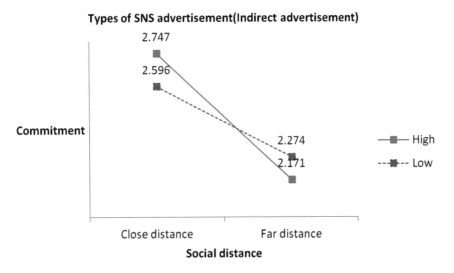

Fig. 2 Social distance*Involvement * Types of SNS advertisement(Indirect advertisement)

In <Table 3> and <Figure 2> above, while short distance showed higher commitment in high involvement products than low involvement products, relatively long distance showed higher commitment in low involvement products than high involvement products. This means that low involvement products are relatively less influenced by social distance, but high involvement products has more commitment in opinions within short distance. In other words, this result conforms to Choi & Kim (2012) that those with short social distance have higher reliability than those with long social distance. Moreover, our result is partially accords with Yang & Na (2013) that customers show higher preference for a product when others are similar to themselves than when they think others are not similar to themselves.

4 Conclusion and Implications

Hence, there is significant ternary interaction effect among the three variables regarding types of advertisement, social distance, and involvement, and it has important influence on commitment when customers face SNS advertisement of a fashion brand. Therefore, our result is very suggestive to SNS marketing strategies of fashion brands because the commitment of customers for the SNS advertisement of fashion brands shows different interactions with social distance and product involvement according to the types of advertisement.

References

[1] Bar Anan, Yoav, Nira Liberman & Yaacov Trope. The association between psychological distance and construal level – Evidence from an implict association test. Journal of experimental psychology. 2006, 135(4): 609–622.

[2] Jayoung Choi & Yongbum Kim. Analyses of persuasion and product evaluation by the social distance of the recommendation in SNS. Korean journal of consumer and advertising psychology. 2012, 13(4): 513–539.

[3] Jungmee Lee. A study of effect of the satisfaction of the SNS on the awareness, satisfaction and loyalty of the festival. ChungAng University, 2012.

[4] Liberman, Nira & Yaacov Trope. The rold of feasibility and desirability considerations in near and distant future decisions – A test of temporal construal theory. Journal of personality & social psychology. 1998, 75(1): 5–18.

[5] Sang a Kim. Case study of using facebook of each type of internal and external fashion brands through customer index. Sookmyung women's University, 2013.

[6] Yoon Yang & Junghar Na. The influence of focus of comparison, social distance, and consumer's need for uniqueness on product preference: Assimilation versus contrast. Korean Journal of consumer and advertising psychology, 2013, 14(1): 69–85.

Emotional Fit Project: Mapping the Ageing Female Form

◆ Anna Maria Sadkowska[①] Katherine Townsend Juliana Sissons (School of Art and Design, Nottingham Trent University, UK)

Abstract: *In this article we present and discuss the preliminary findings from the first phase of the research project entitled "Emotional Fit: Mapping the Aging Female Form" in which we explore mature women's relationship with stylish clothing. The first phase of this qualitative project involved a workshop with participants and a series of in—depth interviews (n=5).*

The research question, which we aim to answer through this project, is that of: how can womenswear be designed more effectively to meet the physical and emotional requirements of an ageing female demographic? Furthermore, through undertaking case studies, interviews and conducting various creative design workshops, it is our aim to respond to the needs of mature fashionable women by creating a series of womenswear fashion solutions. The key methods applied towards the creation of fashion statements and applications are practice—based methods such as creative pattern cutting (by hand and CAD), textile design (print and weave) and manipulation techniques.

Key words: *older women; womenswear; fashion; design for mature women; interviews; creative pattern cutting; emotionally durable design*

情感与着装：中高龄女性服饰搭配

◆ Anna Maria Sadkowska Katherine Townsend Juliana Sissons （诺丁汉特伦特大学，英国）

【摘　要】　该文提出并讨论了"情感与着装"项目第一阶段的初步研究结果。项目讨论了中高龄女性与时尚的关系。该定性项目的第一阶段研究包括小组研讨与深入访谈。该项目意在回答

---

① Anna Maria Sadkowska, Nottingham Trent University, United Kingdom.　E-mail：anna.sadkowska2012@my.ntu.ac.uk

女装设计如何更有效地满足中高龄女性身体与情感的需求？此外，通过个案研究、深入访谈、各种创意设计研讨，意在对中高龄时尚女性的需求作出回应。这些回应恰是打造一系列女装时尚的解决方案。个性化时尚创造的关键方法是基础实践，例如创意样品裁剪（手工和计算机辅助设计）、纺织品设计（印花和编织）与手工技艺。

【关键词】 老年女性；女装；时尚；为成熟女性的设计；采访；创意样品裁剪；情感耐久设计

（中文翻译：谈金艳）

1 Introduction

There are more than 12 million women aged 45–105 in the UK, one fifth of the population. They represent vast economic potential and a wealth of experience and knowledge. However, in most Western societies older individuals, and women especially, often fail to be considered as a prime market by designers and mainstream retailers resulting in a form of socio–cultural invisibility (Church Gibson, 2000). This is often reflected in how businesses and retailers, including the fashion and clothing industry, are overlooking (if not intentionally ignoring) this segment of the British population. This is a missed opportunity for the fashion industry that also results in dissatisfaction and frustration amongst older female customers.

In much the same vein, the Professor of Social Policy and Sociology at the University of Kent Julia Twigg (2013:1), comments that "fashion and age sit uncomfortably together". Further, she defines ageing as "disruption", which highlights the lack of acceptance of this phenomenon within society. In response to this situation, the authors argue that in order to understand and explore the relationship between ageing, fashion, female gender and identity, it is necessary to re–evaluate how fashionable garments are designed for mature women. Consequently, our research objectives are:

• to explore how fashion and clothing is experienced and remembered by a sample of mature British women in their 50's and 60's

• to understand their issues with sizing and fit?

• to discover their aesthetic preferences

• to create a series of womenswear prototypes that reflect their needs and preferences

While fulfilling first three objectives is possible due to a qualitative investigation utilising methods of creative workshops and in–depth interviews, the fulfilment of the last objective will be facilitated through the development of a series of potential design solutions encapsulating aesthetics, innovative fitting and sizing solutions. The project builds on previous research into female ageing and fashion (Sadkowska, 2012), the role of creative pattern cutting in fashion design (Sissons, 2010) and the considered integration of printed textiles and garment shapes

(Townsend, 2004) to contour the female form empathetically. It also considers Chapman's emotionally durable design philosophy (Chapman, 2015) from a fashion perspective: that is clothes as products that meet psychological as well as physical requirements. Significantly, this research triangulates these design approaches with psychological insights into how mature women wear clothes, by considering how fashion products and feelings which once defined the past can potentially become the key to "un–locking" the present (Sadkowska et al., 2014) and facilitate the dialogue between the wearer(s) and designer(s). This involves a conceptual and exploratory fashion practice, where an interdisciplinary methodology is developed through the balancing of theory and practice, which we explain below.

2 Research Context and Rationale

Growing old and the experience of it has become a significant topic in the contemporary social research agenda, due to the increased longevity of human lifespans, which together with the presence of the post–World War II baby boomers, has impacted on the development of an ageing population. The post–industrial economy of improved health care, leisure opportunities and bio–medical technologies have affected both the biological and social spheres of growing old, improving opportunities but also producing new challenges for ageing identities across the gender spectrum (Powell and Gilbert 2009, Fraser and Greco 2005, Featherstone and Hepworth 1991). As Gilleard and Higgs (2005) note, the current ageing generation is the one that created a consumer culture built on youth and sexuality, "*so that their attainment of the Third Age status marks a new stage in the cultural constitution of age*" (Twigg, 2007:300). In this "*contemporary age of aging*" (Powell and Gilbert, 2009, vii) the postmodern approach disrupts the constrained perceptions of growing old, placing the emphasis on the individuals, their bodies and identities, experiences, actions, practices and dynamics.

"[P]*ersons remake themselves over time, and thus their identities change*" (Arxer et al. 2009:46); human biographies have the potential to be translated as the relationships between personal and structural factors; individual and collective experiences, where fashion and clothes, as the communicators and mediators between self and society (Entwistle 2002, Entwistle and Wilson 2001, Crane 2000), can become the key to analyse and particularly understand ageing identities. In the same vein, sociologist Julia Twigg argues that "[clothes] *offer a useful lens through which to explore the possibly changing ways in which older identities are constituted in modern culture*" (2009:93). The phenomenological approach, therefore, with its emphasis on practice and experience, enables "*un–locking an understanding of what it means to be a human person situated within and across the life course*" (Powell and Gilbert, 2009:5). When it comes to fashion and clothing, phenomenology particularly provides the possibility to *"uncover the multiple and culturally constructed meanings that a whole range of events and experiences can have for us"* (Weber and Mitchell, 2004:4), and to establish the interrelation between the stories of individuals, objects and times they inhabit.

Through "Emotional Fit" we exploit these interrelations, in regards to mature women in their 50's and 60's who share common interests and enthusiasm for fashion and clothing. Moreover, as designers and fashion practitioners we aim to utilise our knowledge and skills in order to create a series of fashion prototypes that cater for the stylistic (fashion) and practical/functional (clothing) needs and expectations of mature women as identified by our sample. For the purposes of this project we clearly distinguish between the terms of "fashion" and "clothing". Furthermore, we subscribe to Teunissen's (2013:201) rather conceptual definition of "fashion" as being *"the product of a design that 'attached' to the human body but that also [seeks] to research and explore its own relationship with the body, with identity, self-image, and the environment"*. Consequently, following Eicher (2010:151) we adopt the definition of clothing *"as a noun refer generally to articles that cover the body"*. At the same time, however, we also recognise, following Kawamura (2011:9), the existence of a commonly accepted simplification in which "fashion often functions" as *"clothing fashion, that is, the most trendy, up-to-date clothing that the majority of the people in society adopts and follows"*. This consideration is especially relevant when it comes to analyzing and interpreting of our informants' accounts on their experiences of fashion and clothing.

3 Data and Methodology

Previous investigations into both ageing and fashion have often adopted a qualitative approach through in-depth interviews (Holland, 2004, 2012; Grimstad et al. 2005; Davis 2012) and have focused on specific aspects such as older women's clothing choices (Hurd Clarke et al., 2009; Holmlund et al., 2011), music subcultures through rock fans (Gibson, 2012; Haenfler, 2012), ravers (Gregory, 2012), or professional dancers (Schwaiger, 2012). While these studies have revealed issues of relevance to the current research, they tell little of the meaning of fashion through the individual experience of ageing and identity in the lives of mature women. Few studies have attempted to establish the relationship between memory and clothing (Twigg 2009b, 2010). Some researchers have adapted a phenomenological approach and extended the traditional form of interview with the analysis of artefacts, such as, textiles, garments and photographs (Lerpiniere, 2009; Weber and Mitchell 2004), and workshops for participants (Richards et al. 2012). To date, only a small number of researchers have offered the possibility of combining these methods with equal emphasis placed on both a theoretical approach and a practical response from designer(s) in order to expand existing knowledge and provoke an intergenerational dialogue and associated outcomes.

Accordingly, this project consists of three phases (fig. 1), and includes multiple case studies of members of the UK female population aged between 55–65. The three phases are, in order: Research, Design and Findings Dissemination. Each phase of the study is designed to build on the findings from the previous phase and at each stage we employ different, yet, complementary research methods, as presented below.

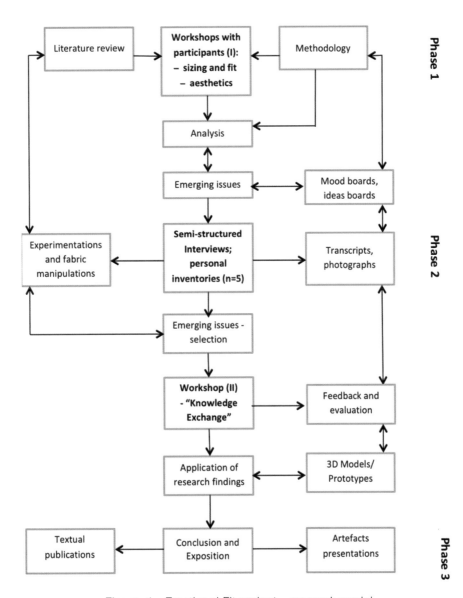

Figure. 1 Emotional Fit project – research model

3.1 Workshop I (21st May 2015, Nottingham Trent University)

In the first stage of the project our aim was clear: we wanted to develop a primary understanding of what problems and issues older women have and share regarding fashion and clothing. In order to fulfil this aim we organised a workshop with 10 participants (tab. 1), with Katherine Townsend, Juliana Sissons and Ania Sadkowska as the facilitators and a research assistant. The workshop lasted 3 hours.

Table 1. Sample characteristics

Name*	Age	Occupation
Anna[1,2]	64	Retired
Barbara[1,2]	65	Semi retired/ part time university researcher
Christine[1,2]	65	Retired
Debbie[1]	62	Retired
Elizabeth[1,2]	67	Retired
Fran[1]	66	Retired
Gwen[1]	65	Retired
Hannah[1,2]	65	Retired
Irene[1]	66	Retired
Joanne	65	Retired

*Pseudonyms were used to protect the participants' anonymity

1 Participants who expressed their interest in being interviewed

2 Participants interviewed

In order to stimulate the process of understanding the complexity of the participants' experiences we first invited them to introduce themselves and freely discuss their personal issues regarding their clothes (fig. 2). Interestingly, without prompting from the facilitators most of the participants discussed their issues to be located within two areas; firstly, that of 'fit', including problems relating to the inconsistent sizing system on the British high street; secondly, that of 'aesthetics'. Here, the issue that caused our participants the most frustration were the recurrent fashion trends nearly all explicitly aimed at young bodies. Our participants felt especially disappointed with the high street, as well as some designer brands, not taking into account the physical changes, naturally occurring to female bodies as they age. The women in our study felt that it was somewhat socially "expected" of them to cover the neuralgic parts of their bodies such as arms and elbows, neck, cleavage and thighs. They also related to the skin, and its changing texture with age density and discolourisation. The participants also felt that the colours widely available in the shops did not compliment their appearances; black and white were their classic/regular choices, however, this was often dictated by the unsuitability of other colours rather than the participants' specific preferences. As indicated by the workshop participants, problems seemed to lie not in the colours per se but in their tonal range. In contrast, all of the women present expressed rather negative attitudes towards colours such as grey or beige through related descriptors of "granny–ish" and "boring". These shades, or 'neutrals' are considered as part of the staple colour palettes for the mature fashion market. Discussion around this issue raised interesting psychological perceptions between ageing and semantics, particularly the notion that once past a certain age women become "invisible" within commercial fashion culture.

Figure. 2   Participants introducing themselves and discussing their clothing preferences

Secondly, we invited our participants to tell us about their favourite and least favourite items of clothing (fig. 3 a~c), which they were asked to bring with them in the flyer sent to every participant prior the workshop. Although this was designed as an individual exercise and we spoke to each woman individually about the items of clothing they brought in (recording their accounts), it quickly developed into a group discussion where the participants had a chance to discuss their preferences amongst themselves as well as compare and contrast their clothing within the sample group (fig. 4 a, b).

Figure. 3 Participants and their favourite garments a. Christine (65); b. Barbara (65) c. Elizabeth (67)

Figure. 4a-b Participants discussing their favourite items

Alongside this activity, the participants had a chance to look at current fashion magazines such as Another Magazine, Vogue and Elle, and relate their needs and tastes to the various images, editorials and adverts presented in these publications. This provided a platform for our participants to directly compare the fashionable clothes on offer with items of clothing that are actually present in their wardrobes. Once again, for many participants this was a chance to express their dissatisfaction with the fashion solutions currently available on the market. On the other hand, these women presented a high level of creativity and widely commented that in fact they would buy some of these products but modify them according to their own needs, for example, by adding sleeves or elongating the shape of a garment. Overall, the women commented that they did not feel there was anything that was entirely suitable for their bodies as presented to them in the magazines, and in fact they reported that they rarely buy fashion magazines themselves. Once again, this highlights the women's disconnection with fashion and clothing as produced and sold by the British fashion industry in the so–called 'grey(ing) market'.

The final element of the workshop was to take the detailed measurements of each participant (fig. 5). The accrued measurements will be utilised in the second phase of the project as data for the production of a series of bespoke pattern blocks, which will inform the development of experimental garment patterns and finally, womenswear prototypes by Juliana Sissons The prototypes will be produced in different size groupings to accommodate different members of the sample. As well as being produced in plain and textured fabrics, Katherine Townsend will develop selected garment prototypes in digitally printed textiles in response to the silhouettes, and information accrued about colour, pattern and surface imagery, by the sample group in Workshop 1 and through the interviews.

Figure. 5　Juliana Sissons taking measurements of the participant

3.1.1 Discussion

As stated in the Introduction, the research question we aim to answer in this project is: how can womenswear be designed more effectively to meet the physical and emotional requirements of an ageing female demographic? Furthermore, alongside developing an in–depth understanding of the various fit and aesthetic preferences of mature women, our objective is to create a series of womenswear prototypes that reflect their emotional and practical needs and preferences. In order to fulfill our research objectives, a key objective is to explore the potential of communicating messages between the project participants (wearers) and the designers.

Similarly, Press and Cooper (2003) identify three areas of design research: understanding of the phenomenon, generating ideas and proposing solutions. The role of Workshop 1 was precisely to allow us to understand the ageing phenomenon as experienced and interpreted by the participants. Furthermore, Lawson (2006:125) describes the design process as "endless" and claims that designing, unlike completing mathematical operations cannot have a predetermined end and, therefore, it should be described as overlapping loops repeated within time intervals allowing for analysis and reflection. Chapman (2015) argues that in order to design products that incorporate longevity, we need to incorporate lasting emotional as well as physical perspectives. Adapting these models is the crucial part of the proposed research model, with a constant evaluation of findings transferability into the design practice, through the testing and sampling of the proposed solutions.

In this vein Workshop 1 had an exploratory as well as generative purpose, and was designed to "*allow the designer [s] to see and understand the relevance of objects in a user's life from the participant's point of view, to inspire design themes and insights*" (Martin and Hannington 2012:130). The gathered visual information i.e. photos and images (tear sheets from magazines – need to clarify this above), are captured and stored in the form of mood and ideas boards, both in hard copy (paper) as well as in digital form, and serve as a direct introduction to the practical work, which will seek to address the some of the participants needs and preferences.

3.2 Interviews

The second element of the first phase of the study was to conduct 5 in–depth interviews with selected participants. The interviewed participants were randomly selected from the workshop 1 participants who declared their interest in being interviewed (table 1). The interviews were designed to be semi–structured and face–to–face informal conversations, "*so that the rapport between researcher and informant will be enhanced, and that the corresponding understanding and confidence between the two will lead to in depth and accurate information*" (Kumar 2005:124). Moreover, these interviews, conducted by Ania Sadkowska, were structured to enable each participant to explicate and articulate as much detail about their individual experience of fashion and ageing as possible.

Each interview was conversational in style and lasted between 60 and 80 minutes, was digitally recorded and transcribed verbatim with consent from each participant. The interview schedule consisted of 16 open–ended questions about the different aspects of women's experiences of fashion and clothing, including questions about the meaning of fashion in the participants' lives, their shopping practices and future expectations towards fashionable clothing. Additionally, the participants were asked to describe critical occasions when they felt really good/ bad about the way they looked. Themes emerging from the interviews have been meticulously analysed in order to build on cultural context via personal histories and experiences of participants, which will be used as inspiration to create fashion prototypes. To date analysis of the material gathered through a series of interviews has revealed various tensions in older women's perceptions of the current fashion and clothing system as well as contradictions regarding their expectations towards it.

Sense of generational belonging

To begin, the majority of the participants asserted that they felt privileged because of the generation were part of. This was present in the narratives of all 5 participants who on numerous occasions highlighted the personal connections with certain fashion practices and behaviours originated in 1960s. Furthermore they felt especially strongly positive about certain British fashion designers who started their creative practices in 1960s and who still are present on the fashion market such as Vivienne Westwood or Paul Smith. From international designers two the most commonly mentioned designers were Yohji Yamamoto and Issey Miayke. Overall, the tone of these cultural connections was that some of these designers, perhaps due to their own ageing, could in a more empathetic manner relate to cater for older women's needs and expectations. In contrast, other, no longer present labels such as Biba, were often discussed with a great level of nostalgia. In the following extract Elizabeth (67) explains the importance of witnessing by her such fashion developments as follows:

So it was the sixties (...) that is when it [fashion] started to really affect, yeah change people really. Yes, I can remember buying my first Biba outfit and my friend got married in Biba. So we were... and of course they had mail order then. Also there was only one Laura Ashley shop, and again at the time Laura Ashley was kind of fantastic. We went to London especially to go to the one Laura Ashley shop. Then when it kind of comes to opening in all of the towns it's not the same. It's like next, once it started to open everywhere it wasn't as interesting. It then became stuff for the masses and it wasn't individual somehow. You had to look very hard to find individual things.

What is compelling in this extract is a certain sense of a shared generational uniqueness as well personal sense of individuality experienced by Elizabeth from being a teenager in a period when developments in fashion were especially exuberant. In this vein, further on in her interview she reflects:

I feel I have been very lucky that there has been the Mary Quant and the Vivienne Westwood and the Paul Smith. (...) And I still think that man [Paul Smith] is a hero (...)

Similar reflections were present in all interviews. Of particular importance, is the shift in cultural and social perceptions relating to how women should present themselves. This was strongly connected to the contemporary fashion solutions at the time, as Christine (65) explains:

I suppose... Somebody like Mary Quant was quite important for my sort of generation. Because she introduced, I mean it came with the development of tights, I think (laughs). You could wear, as I did, we could wear very short skirts and tights. And not feel as we were revealing everything. And she has, I think, introduced more the shift style.

As the above extract shows the interviewees shared a strong generational sense of belonging. Furthermore they shared positive attitudes towards certain fashion designers who they witnessed developing their brands as well as to other designers who are no longer present in the market.

Fashion awareness

As well as discussing their past interests in fashion, all interviewed women expressed their strong current interests in clothing and fashion trends. However, this was often discussed in relation to their own bodily conditions such as height or weight. In the following extract Elizabeth (67) explains how she finds out about changing fashion trends:

I love to kind of look at fashion in magazines and even when it was the fashion show at the time. And I suppose you look and I was interested in what was translated to, from the catwalk into the everyday. And it's interesting, and I like to read in the papers and magazines how people have taken things. And how they have translated it into more everyday things, and it can be colours, it can be shapes, and it could be hemlines. I find all of that very interesting. But I suppose I, being small and round I have never been a fashionable shape.

A similar picture is presented by Christine (65) who explains that her own body type has become a lens through which she filters suitable fashion trends:

So I have always looked at magazines, uhm, I have always been interested in what's been in the shops, but, uhm. In my early days I didn't have a lot of money and I have always been in a way conscious of my body type. So I think that is as much as anything, it's my body type that has determined my interest in fashion

This type of a "targeted" fashion awareness where women exhibit a life long interest in fashion allows them not only to understand their aesthetic preferences but also the impact of their physical/physiological condition on how clothing is presented on their body. Furthermore this is critically important when it comes to developing any potential designs targeting these groups of women.

Bodily changes

Another important theme that emerged from the interviews analysis is that of the bodily changes occurring to women as they grow older. Interestingly, in the interviews one of the most common comments from women was that of certain social limitations linked to exposure of the mature female body. In the following extract Elizabeth (67), who elsewhere in her interview highlighted the importance of being influenced by Mary Quant and wearing mini skirts when she was younger, comments on the unsuitability of such solutions for older women, regardless of their physiques:

(...) when you get to your kind of sixties and seventies it's a bit like" mutton dressed as lamb". I was at the pictures a few weeks ago and I saw a woman who was clearly older than me. And she had on a little mini kilt. And you now, she had little spindly legs and I just thought, no. So there is a bit about making sure that you can be fashionable or stylish but I think it has to suit your age because I think sometimes you see people trying to look young and it doesn't work.

Similar opinions were shared by most of the participants. In the following extract, however, Christine (65) presents a slightly different point of view:

I am dressing for my generation of women. Who are... not wanting to look young. But who just don't want to abandon clothes, which are perhaps more youthful, yeah. So I mean, I probably do dress for my age now, because for example, uhm, I would like my arms covered up. I don't tend to wear... although I could wear lower necks (...) I don't like to show a lot of flesh, let's put it like that. So I wouldn't reveal a lot of flesh. Wherever that's dress for my age, or just dressing for me.

Interestingly, here Christine, yet once again highlights the importance of a certain generational identity amongst women similar to her age. However, despite recognizing certain social limitations as to what older women should and should not wear, Christine attempts to detach herself from being restricted in this way. Instead, she explains her clothing choices in regards to her current lifestyle.

Personal trajectories

The final theme, which emerged from this series of interviews, is that of the importance of our participants' personal trajectories on their current interest and engagement with fashion. The histories our participants shared with us during the interviews differ significantly from each other. For example one of the participants experienced serious health issues affecting what kind of clothing she preferred to wear to conceal some of her body changes. Another participant discussed the great impact that the death of her husband had on the way she currently chooses her clothes. In her interview she explained that she not only lost a great and dedicated clothing advisor in her husband, but also that now, as a widow she does not want to present herself as a woman "searching for a new husband":

It's almost as if they [some of her female friends] think I'll jump on their husbands or something and there is a bit of... I kind of feel I need to be a little bit more conservative about what I wear. I am not looking for anybody else and I don't want people to even think that I am, I was very happy. So it's a silly thing.

Consequently, we argue that these personal trajectories are important when it comes to design for mature women, especially when the aim is to achieve a state of equilibrium: a sense of emotional fit between the design and the wearer.

3.2.1 Discussion

Following from the Workshop 1, the discussed series of semi-structured in depth interviews had an exploratory as well as generative purpose, and were designed to allow the researchers to understand more deeply the psychological aspects of how mature women experience fashion and clothing. In this vein, the conducted interviews revealed many tensions, as well as contradictions in regards to the participants' fashion behaviours and practices. For example, all of the participants expressed a strong common sense of generational belonging. This does influence this group's particular expectations, in terms of their connection with the designer and designing process, which they are keen to be involved in. In this sense, it would seem advisable to make the designer's presence more visible and exposed via different forms of communications with the potential wearers of designed artefacts. This supports the theories associated with emotionally durable design, where products are often 'user tested' prior to production, whereas fashion is generally designed by individuals and teams in response to industry trends and sizing charts, standardized blocks and established lines. Secondly, the participants exhibited strong awareness of current fashion trends, however, always in relation to their own physiques. Once again, this, highlights the need to look for design solutions that allow designers to respond more fully to the wearer's unique bodily features. Furthermore, the proposed solutions should always be considered within the frame of contemporary cultural conditions. In this vein, many women expressed their interest in clothing that can potentially enhance the way their present their mature bodies rather than masking them or creating the false impression of being a younger age. Finally, it is worth re-iterating that all of the interviewed women have had very different life courses resulting in them developing unique systems of values and expectations in regards to fashion and clothing. In order to be successful, the design process needs to acknowledge these personal trajectories.

4 Conclusion

In this paper we have presented and discussed the preliminary findings from the first phase of the research project entitled "Emotional Fit: Mapping the Aging Female Form". The research question, which we aim to answer through this project, is: how can womenswear be designed more effectively to meet the physical and emotional requirements of an ageing female demographic? In this vein our aim is to explore mature women's relationship with fashion and clothing. The first phase of this qualitative project involved a workshop with participants (n=10) and a series of in–depth interviews (n=5).

The initial results of the project have allowed us to develop an in–depth understanding of how the study's participants experience, practice and engage with fashionable clothing on a daily basis. Furthermore the utilized methods allowed us to discover the complexity of this experience both in regards to the participants' aesthetic expectations, often developed throughout their life–long interests in and engagement with fashion and clothing, as well as problems with sizing and fit experienced by these women. Our next step (phase 2) of the project will be to utilized this gathered information and respond to it via creative fashion practice including techniques of geometric pattern cutting, digital printing as well as computerized and traditional fashion and textile crafting techniques.

References

[1] ARXER, S. and MURPHY, J. and BELGRAVE, L. Social Imagery: Aging and, and the Life Course: A Postmodern Assessment. In: POWELL, J. and GILBERT, T. (eds.) Aging Identity: A Dialogue with Postmodernism. New York: Nova Science, pp. 45– 55, 2009.

[2] CHAPMAN, J. Emotionally Durable Design: Objects, Experiences, Empathy. London: Routledge, 2015.

[3] CRANE, D. Fashion and Its Social Agendas. Class, Gender, and Identity in Clothing. London: The University of Chicago Press, 2000.

[4] CHURCH GIBSON, P. No–one expects me anywhere. In: BRUZZI, S. and CHURCH GIBSON, P. (eds.) Fashion cultures: theories, explorations and analysis. London: Routledge, pp. 79–89, 2000.

[5] DAVIS, J. Punk, Ageing and the Expectations of Adult Life. In: BENNETT, A. and HODKINSON, P. (eds.) Ageing and Youth Cultures. Music, Style and Identity. London: Berg, pp. 105– 118, 2012.

[6] EICHER, J. B. Clothing, Costume, and Dress. In: STEELE, V. (ed.) The Berg Companion

to Fashion. Oxford: Berg, pp. 151–152, 2010.

[7] ENTWISTLE. J. The Dressed Body. In: EVANS, M. and LEE, E. (eds.) Real Bodies. A Sociological Introduction, Basingstoke: Palgrave, pp. 133–150, 2002.

[8] ENTWISTLE, J. and WILSON, E. Body Dressing. Dress, Body, Culture. Oxford: Berg, 2001.

[9] FEATHERSTONE, M. and HEPWORTH, M. The Mask of Ageing and the Postmodern Life Course. In: FEATHERSTONE, M. and HEPWORTH, M. and TURNER, B. (eds.) The Body. Social Process and Cultural Theory. London: Sage, pp. 371–389, 1991.

[10] FRASER, M. and GRECO, M. Introduction. In: FRASER, M. and GRECO, M. (eds.) The Body. A reader. London: Rutledge, pp. 1–42, 2005.

[11] GIBSON, L. Rock Fans' Experiences of the Ageing Body: Becoming More 'Civilized'. In: BENNETT, A. and HODKINSON, P. (eds.) Ageing and Youth Cultures. Music, Style and Identity. London: Berg, pp.79–91, 2012.

[12] GILLEARD, C. and HIGGS, P. Contexts of Ageing. Class, Cohort and Community. Malden: Polity Press, 2005.

[13] GREGORY, J. Ageing Rave Women's Post-Scene Narratives. In: BENNETT, A. and HODKINSON, P. (eds.) Ageing and Youth Cultures. Music, Style and Identity. London: Berg, pp. 37–49, 2012.

[14] GRIMSTAD KLEPP, I. and STORM-MATHISEN, A. Reading Fashion as Age: Teenage Girl's and Grown Women's Accounts of Clothing as Body and Social Status. Fashion Theory, 9 (3), pp. 323–342, 2005.

[15] HAENFLER, R. 'More than the Xs on My Hands': Older Straight Edgers and the Meaning of Style. In: BENNETT, A. and HODKINSON, P. (eds.) Ageing and Youth Cultures. Music, Style and Identity. London: Berg, pp. 9–23, 2012.

[16] HOLLAND, S. Alternative Femininities. Body, Age and Identity. Oxford: Berg, 2004.

[17] HOLLAND, S. Alternative Women Adjusting to Ageing, or How to Stay Freaky at 50. In: BENNETT, A. and HODKINSON, P. (eds.) Ageing and Youth Cultures. Music, Style and Identity. London: Berg, pp. 119–130, 2012.

[18] HOLMLUND, M. and HAGMAN, A, and POLSA, P. An exploration of how mature women buy clothing: empirical insights and a model. Journal of Fashion Marketing and Management, 15 (1), pp. 108–122, 2011.

[19] HURD CLARKE, L. and GRIFFIN, M. and MELIHA, K. Bat wings, bunions, and turkey wattles: body transgressions and older women's strategic clothing choices. Ageing and Society, 29 (5), pp. 709–726, 2009.

[20] KAWAMURA, Y. Doing Research in Fashion and Dress. Oxford: Berg, 2011.

[21] KUMAR, R. Research methodology. London: Sage, 2005.

[22] LAWSON, B. How designers think : the design process demystified. 4th ed. Oxford: Architectural, pp. 123–126, 2006.

[23] LERPINIERE, C. The Fabric Snapshot – Phenomenology, fashion and family memory. In: Rouse, E. Fashion & Wellbeing? IFFTI Conference Proceedings held at the London College of Fashion. London, pp.279–290, 2009.

[24] MARTIN, B. and HANINGTON, B. Universal Methods of Design. Beverly: Rockport Publishers, 2012.

[25] POWELL, J. and GILBERT, T. Phenomenologies of Aging – Critical Reflections. In: POWELL, J. and GILBERT, T. (eds.) Aging Identity: A Dialogue with Postmodernism. New York: Nova Science, pp. 5–16, 2009.

[26] PRESS, M. and COOPER, P. The Design Experience: the Role of Design and Designers in the Twenty–first Century. Aldershot: Ashgate, 2003.

[27] RICHARDS, N. and WARREN, L. and GOTT, M. The challenge of creating 'alternative' images of ageing: Lessons from a project with older women. Journal of Aging Studies, 26, pp. 65–78, 2012.

[28] SADKOWSKA, A. M. The Dys–Appearing Body Project: design for sociocultural context of well–being and sustainability. Unpublished MA thesis, De Montfort University, Leicester, UK, 2012.

[29] SADKOWSKA, A., FISHER, T., WILDE, D., and TOWNSEND, K. Interpreting Fashion and Age: Arts–Informed Interpretative Phenomenological Analysis as a fashion research methodology – paper presented at The Fashion Thinking, History, Theory, Practice. 30 October 2014 – 1 November 2014, University of Southern Denmark, Kolding, Denmark, 2014.

[30] SCHWAIGER, E. Ageing, Gender, Embodiment and Dance. Basingstoke: Palgrave Macmillan, 2012.

[31] SISSONS, J. Basics Fashion Design 06: Knitwear, London: AVA Publishing, 2010.

[32] TEUNISSEN, J. Fashion: More Than Cloth and Form. In: BLACK, S., DE LA HAYE, A., ENTWISTLE, J., ROCAMORA, A., ROOT, R., and THOMAS, H. (eds.) The Handbook of Fashion Studies. London: Bloomsbury, pp. 197–213, 2013.

[33] TOWNSEND, K, Transforming Shape: A simultaneous approach to the body, cloth and print for textile and garment design (synthesizing CAD/CAM with manual methods), School of Art and Design, Nottingham Trent University, 2004.

[34] TWIGG, J. Clothing, age and the body: a critical review. Ageing and Society, 27, pp. 285–305.

[35] TWIGG, J. Clothing, Identity and the Embodiment of Age. In: POWELL, J. and GILBERT, T. (eds.) Aging Identity: A Dialogue with Postmodernism. New York: Nova Science, pp. 93–104, 2009a.

[36] TWIGG, J. Dress and the narration of life: Women's reflections on clothing and age. In: SPARKES, A.C. (ed.) Auto/Biography Yearbook 2009, BSA Auto/Biography Study Group. Nottingham: Russell Press, pp. 1–18, 2009b.

[37] TWIGG, J. Clothing and dementia: A neglected dimension? Journal of Aging Studies, 24, 223–230, 2010.

[38] TWIGG, J. Fashion and Age: Dress, the Body and Later Life. London: Bloomsbury, 2013.

[39] WEBER, S. and MITCHELL, C. (eds.) Not Just Any Dress. Narratives of Memory, Body, and Identity. Oxford: Peter Lang, 2004.

浅析我国中老年服装市场发展

◆ 陈 颖[①] （上海视觉艺术学院，上海，201620）

【摘　要】　据社会人口调查，中国目前的老年人已经超过 1 亿人，据统计，到了 2050 年将超过 4 亿人，数据说明中国即将迈进老龄化社会。中老年人的数量占据社会人口的 25%，这是一个庞大的群体。随着社会文明和经济的快速发展，在信息化时代的中老年人对生活的质量要求也逐渐有了新的发展，有了更高的要求。从老年人的心理、生理及社会文明的发展出发，开发中老年服装市场是目前中国服装市场的头等大事，满足中老年人的消费需求，提高生活质量，建设社会主义精神文明，都离不开服装产业。如何更好地为中老年人消费者服务，开发正确的服装产品，培养和引导中老年消费者，是本论文重点分析的方向。

【关键词】　中老年服装；　市场开发；　设计；　购买力；　消费心理

Analysis of the development of China's elderly clothing market

◆ Chen Ying　(Shanghai Institute of Visual Art　Shanghai　201620)

Abstract: According to the social population survey, China's current elderly people have more than 100000000 people, according to statistics, by 2050 will exceed 400000000 people, the data shows that China is about to move into an aging society. The number of elderly people in the community of 25%, which is a huge group. Along with the social civilization and the rapid development of economy, in the information age, the old people, the quality of life requirements also gradually have a new development, with higher requirements. From the development of elderly people's psychological, physiological and social civilization, the development of middle and old aged clothing market is the top priority of China's apparel market to meet consumer demand for the elderly, improve the quality of life, social construction of spiritual civilization,

---

[①] 陈颖，女，讲师，硕士，研究领域：时尚设计与传播。E-mail: cy_siva@126.com

are inseparable from the garment industry. How to better serve for the elderly, to develop the right clothing products, training and guidance of the elderly consumers, is the focus of this paper analysis.

Keywords: *Middle and old aged fashion; Market; Design; Consumption; Psychology*

世界卫生组织,从人口学角度将45~59岁的人划分为中年人,60~74岁划分年轻的老年人,75~89岁为老年人,90岁以上称为长寿老年人。中老年的年龄跨度比较大,因此,中老年服装市场面向的消费者具有多样化、多层次的消费特征。

1 中老年服装市场的现状

服装产品的单调和市场的占有比率过小是目前国内中老年服装市场的现状。一方面,中老年的服装款式多数是以服装基本款为原型的设计产品,款式上单调,让原本步入中老年队列、身材臃肿肥胖的中老年人看起来更加年老颓废,并没有很好地在设计上为中老年人考虑。另外,网络时代的到来,让很多商家放弃了实体店转为网销店铺模式,这让原本在市场上进行消费的中老年服装场所也随之缩小,很多中老年人对网络的认识仍处于一种陌生或排斥的状态,这就造成了中老年消费群只能在有限的空间里选择有限的服装产品,根本无法满足他们的消费欲望。因此,商家、服装设计师和产品营销团队要意识到中老年服装市场的巨大潜力,改变原有的设计和营销方向,从根本上重视中老年这一社会群体,才有可能让中老年服装产业迅速发展起来,进而推动整个服装产业的发展。

2 中老年服装市行的广阔前景

2.1 从人口数量看市场前景

随着我国经济的发展,计划生育政策的推行,我国目前老龄人口占世界的20%,占整个亚洲的50%。据联合国人口基金会公布,自2000年起,我国已经跨入老年国的行列。在未来的50年中,这一比例会继续上升。由此可见,中国的中老年人数量堪称世界之最,因此,中国的中老年服装市场也将是世界上最大的中老年服装市场。做好中国的中老年服装,也就做好了世界的中老年服装。

2.2 从中老年人的购买力看市场前景

在传统的固有思维模式中,我国的中老年人总是喜欢穿子女淘汰的衣服,不舍得去购买新的服装,在经济上有一定的局限性。但是,随着社会历史的进程,未来中老年人的主要群体会集中在"50后""60后""70后"甚至是"80后",都是亲身经历过经济改革发展的,多数是有过接触外来文化影响的经历的,是有文化的一代,对生命的认识和自我价值的认识更加深入。

因此,她们对个性化的追求、年轻化的追求都是很强烈的。此外,1997年,国务院明文规定,每年的7月1日都要上调退休养老金,那么更多的老年人可支配的钱就会提高,实现个人意愿的能力和空间也随之变大。作为商家,应该看到这一点,不要担心中老年人的购买力。

2.3 从中老年人的消费心理看市场前景

从服装社会心理学角度分析中老年人的消费心理会发现,中老年人比青年人更加渴望被社会、被他人认可和肯定,而服装作为一种文化,体现着着装者的文化背景和心理特征,所以,多数中老年人都愿意选择具有个性化的服装,通过服装的款式、材料、色彩等设计元素传递自己的文化品位,而要满足这种消费心理,必然要求更加广阔的服装市场。归纳中老年人的消费心理有以下几个特征:

2.3.1 个性与从众并存

在中老年人的日常生活中,他们希望引起他人的重视,在服装上往往喜欢追求与众不同、能够更好地表现自身特点的服装。同时,根据我国的文化背景,我国的中老年人喜欢群体性的活动,如社区的文化活动、三五好友的外出旅游或是公园早上的集体锻炼等。他们又有着从众心理,希望自己所在的团体受到关注,并有着较强的集体荣誉感,所以,这个时候,在服装的挑选上,又强调统一和谐,这就形成了个性与从众并存的消费特点。

2.3.2 服装的"新"和"旧"

很多中老年人追求服装设计的新颖,希望通过服装潮流的变化让自己也能够耳目一新,给自己注入年轻的血液,而在追求新鲜设计的服装时,又会怀念自己年轻时穿过的一些款式服装,认为那些曾经的美好依然存在,似乎通过一件服装便可以让时光倒流,回到曾经的青春岁月,在心理上得到一种安慰和愉快。

因此,无论是服装设计师,还是服装商家,在开发中老年服装市场的时候,都要尽可能地了解中老年人的心理特征,满足他们的消费需求,才可以做到双方共赢的良性市场发展模式。

3 中老年市场的开发策略

3.1 服装产品的设计创新

在设计中老年服装产品时,除考虑中老年人的体态特征外,应尽量模糊化中老年人的年龄,不要刻意地强调"中老年"。在美国的服装市场中,很难看到"中老年"的服装标语,多数中老年人可以到属于年轻人的商店中购买服装,原因很简单,服装是属于大多数年龄段的,中老年人也可以去选择款式新颖、色彩明亮的服装,而不是仅仅地局限在"中老年"三个字,因此,在设计我国中老年服装的时候,应该放宽思路,更多考虑如何让中老年人更加年轻、有活力、体感舒适,而不是中老年人就应该穿色彩穿单一的黑灰蓝,款式单调的衬衫、西裤,"A"字裙等,去打造所谓的"庄严、稳重和安静等",不然,就无法打破传统固有的中老年服装市场单一的局面。

3.2 中老年服装营销模式的多元化

营销,是对消费者需求的管理,包括售前管理、售中管理和售后管理。营销创新是指企业应尽可能地利用现代高科技手段,最有效、最经济地去谋求和开拓新的消费领域。

在营销模式上,企业除使用传统的营销模式(专卖店、专柜、百货商店等)外,还可以引导中老年消费者采用网络方式进行购物,在宣传自身产品的同时,也教中老年人如何使用网络购买产品,不仅可以让产品走近消费者,也拉近了企业和消费者之间的距离,更加人性化地为消费者服务,让服装品牌深入人心。

3.3 服装产品的准确定价

价格是影响消费者购买欲望的重要因素,我国多数中老年人属于勤俭节约的一代。有些服装的价格很低,但是,都是一些化纤的面料,体感很差,很多老年人不愿意去选择这类服装,而另外一些体感很好的面料,如桑蚕丝类的服装,价格动辄几百,甚至上千,又让很多中老年人望而却步,因此,企业应该尽量地权衡,既能够让中老年人消费者买得起,又能够穿着舒适,实现物美价廉。

综上分析,中老年服装市场有着巨大的潜力和研发价值,无论是服装设计师还是服装企业,如果能够意识到这一点,并进行系统深入的调查研究,分析中老年人的消费心理,以服装设计的三要素为根本,设计出款式造型新颖的中老年服装,一定会开拓出更为广大的消费市场,不仅可以在中国市场上大放异彩,更有可能引领世界的中老年服装文化潮流,打造国际知名的中老年服装品牌。

参考文献:

[1] 梁建芳等.关于中老年服装市行的开发和策略研究[J].苏州大学学报,2002,2.
[2] 田伟.从着装心理看中老年服装市场的发展潜力[J].西北纺织工学院学报,2000,3.

Advanced Fashion

◆ Zeshu Takamura[①]　(Bunka Gakuen university,Tokyo)

中高龄时尚分析

◆ 高村是州　（日本文化学园大学，东京）

1　Background

When senior(over 65 years old) make 7% of the population, they form an elderly people society. Japanese seniors are 22.7% in 2009. Japan is the most "super aging society" in the world. The birth rate of Japan is1.26 person in2005 . Average life expectancy of Japan is 79.19 years old(Male) and 85.79 years old(Female).

By 2030 many countries excluding Africa and the Near and Middle East will plunge into "aging society".

What is an comfortable society for elderly people? The advanced fashion is paid the attention.(Fig.1-Fig.3)

Fig. 1　advanced fashion1

Fig.2　advanced fashion2

Fig.3　advanced fashion3

---

① Zeshu Takamura，Bunka Gakuen university,Tokyo. E-mail:z-takamura@bunka.ac.jp.

What are the unique features of a city in which fashion is prosperous? There are four features Large population, Peace, Wealthy, and four seasons, for example Tokyo and Shanghai.(Fig.4,Fig.5)

Fig.4　Tokyo　　　　　　　　　　　　Fig.5　Shanghai

2　Japanese culture

2.1　Japanese seasons (Fig.6)

Fig. 6　Japanese seasons

2.2 Washoku (Food)(Fig.7)

Fig. 7　Japanese food

2.3 Sadou (Tea ceremony)(Fig.8)

Fig. 8　Japanese Tea ceremony

2.4 Kadou (Flower arrangement)(Fig.9)

Fig. 9　Japanese flower arrangement

2.5 Japanese Textiles(Fig.10)

Fig.10　Japanese textiles

2.6 Twelve-layered ceremonial kimono(Fig.11)

Fig. 11 Twelve-layered ceremonial kimono

3 Advanced style

In Europe seniors have cultures to enjoy fashion.(Fig.12)

Fig. 12 Seniors in Europe

Seniors have difference with young people.(Tab.1)

Table.1 difference between Seniors and young people

	feature	skin	Body type	budget
Young people	Active and lively	Beautiful skin	Sharp body type	No room in the budget
Seniors	Calm	Gradually declining skin	No sharp body type	Have room in the budget

The fashion for seniors is difficult from anti-aging fashion such as young people. The fashion which express internal perfection. The enriched lifestyle becomes the base of the fashion. So what is the fashion for advanced age?(Fig.13-Fig.17)

Mix styles of classic and casual

Similar color gradient sunglasses

Incorporate Men's fashion

Classic and modern

Fig. 13 Advanced Fashion   Fig. 14 Advanced Fashion

Using Headwear

Burning curiosity which lasts

smile which brightens the inner beauty

Wonderful friends

Fig. 15 Advanced Fashion   Fig. 16 Advanced Fashion

Arranging classical items by colorful   Remember humor and elegance

Fig. 17　Advanced Fashion

Advanced Fashion is enjoying their life from now on taking advantage of abundant experience.(Fig.18)

Fig. 18　Advanced Fashion

So how about the Inter-generational exchange?The main point is understanding(Fig.19)

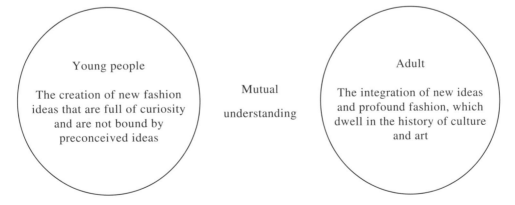

Fig. 19　Inter-generational exchange

To choose your clothes is to choose the way of your own life.　by Coco Chanel

The Possibility of Community Type Clothes Making

◆ Kazuro Koumoto[①]　　(Bunka Gakuen university, Tokyo)

Summary: Japan's apparel market is overflowing with fast fashion, and the quality is good and cheap. It has a wide range of supporters, from children to elderly. However this moderate fashion merchandise seems to be similar all around. For some elderly people and experienced women whom are expecting appropriate fashion for themselves. This thesis is going to introduce a 70-year-old women dyer. She creates unique fashion with vigor which is overwhelming and strong that can't be ignored. Her clothes making is success in cooperating with local small textile factories and also having fun with fellow members.

Key words: fashion creation; textile design; Japanese fiber manufacturer

社区型服装设计工作室的可行性研究

◆ 河本和朗　　（日本文化学园大学，东京）

【摘　要】　日本的服装市场随快时尚的发展而呈过剩状态。快时尚质优价廉，拥有从儿童至老年人的大范围消费群体。然而这些时尚服装看上去是类似的。对于许多中高龄女性来说她们期望为自己找到更为适合的服装。本文将介绍一位70岁女性的染作工作室。她以不可忽视的过人而超强的精力创造了独一无二的时尚。她的创意设计成功与当地小型纺织企业合作，并且得到了资深顾客的喜爱。

【关键词】　服装创意，纺织品设计，日本纤维制造

---

① Kazuro Koumoto, Bunka Gakuen university, Tokyo. E-mail:koumoto@bunka.ac.jp.

1 Introduction

In the early 1990's, Japan's economic bubble had burst and people were suffering in long term of deflation. Recently the economy is showing upturn. To suppress the selling price, some of the Japan's apparel manufacturers have actively immigrated to overseas production. As a result, domestic factories have declined and cheap fast fashion from oversea has increased suddenly. On the other hand, Japan as an aging society, the elderly people over 60 years old constitutes about 33% (41.4 million people) of population (126 million people). However it is difficult to satisfy the elderly people with the polarization apparel market either fast fashion or luxury brands. This thesis is going to introduce about through this 70-year-old dyer's fashion creation, how elderly people can be active and enjoy the fashion creation and business.

2 Fashion of Japan

2.1 Regarding to import penetration ratio of Japan's garments

Most of Japan's garments are imported from aboard. Figure 1, in the last 20 years, import penetration ratio has changed. In 2014, the imported goods increase to 97%. This fact is a very difficult condition for domestic textile manufactures.

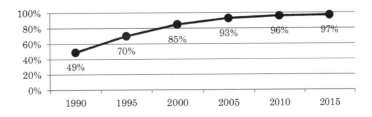

Figure 1   Import of clothing penetration rate

2.2 Japan's Factory

Most of the Japan's fiber manufacturers are small enterprises. As the Table 1 shows that half of the factory's employees are under 20 people. The overseas imported goods have deprived the market thus these small fiber manufacturers are on the verge of crisis. In order to survive, the only thing the domestic factories can do is to be self-reliance. Nevertheless, Japan's factories have taken a job as subcontractor until now; therefor they do not have any experience on product planning and selling methods.

Table 1 The number of the companies according to the number of employees of the fiber manufacture

Number of People	Spinning		Textile		Dyeing rearranging		Sewing	
0～4	43	24.6%	1313	60.3%	906	46.4%	5857	42.5%
5～9	21	12.0%	373	17.1%	394	20.2%	2980	21.6%
10～19	37	21.1%	238	10.9%	292	15.0%	2365	17.2%
20～29	18	10.3%	90	4.1%	131	6.7%	878	6.4%
30～49	19	10.9%	69	3.2%	104	5.3%	767	5.6%
50～99	15	8.6%	48	2.2%	70	3.6%	548	4.0%
100～299	17	9.7%	40	1.8%	46	2.4%	306	2.2%
300～999	2	1.1%	5	0.2%	7	0.4%	56	0.4%
1000～1999	2	1.1%	1	0.0%	2	0.1%	9	0.1%
2000～4999	1	0.6%	1	0.0%	0		4	0.0%
5000～	0		0		0		0	
合計	175		2178		1952		13770	

2.3 Unique Fashion of Elderly People

On the one hand domestic factories has declined, on the other hand the average life span of Japanese people(84-year-old)becomes top of the world. After retire, elderly people remains time and property. How to live an affluent life becomes an important theme. For that reason, elderly people who take time to choose luxurious clothes to keep fashionable as young people are increasing.

Figure 2 Stylish mature people

3 Atelier and Community

3.1 The brand of atelier, "Yabugarashi" (やぶがらし)

The people who want to buy unique garments are increasing; however there is no manufacture to provide the needs. Accordingly, this thesis is going to introduce the garments which create buy the dyer who is popular among the elderly people. A dyer, Chiyo Koyanazu who is creating clothes at the atelier that located at suburb of Tokyo. (Figure 3) She makes the brand and starts to create her works for almost ten years.

Figure 3　Chiyo Koyanazu

3.2 Management of Atelier

The atelier is remodeled from an old farmhouse. (Figure4)The building site is around 200 tsubo. (1 Tsubo=$3.3m^2$)The atelier is based on pattern making, cutting, sewing(Figure 5), area of dyeing (Figure 6) and office etc. Manager and other 4 members who are learning dyeing skill are operating the brand together.

Figure 5　workspace of the atelier Figure

Figure 4　atelier of "Yabugarashi" Figure

Figure 6　area of Dyeing in the atelier

3.3 Clothes Making of "Yabugarashi"

Chiyo Koyanazu is using "Shibori" (tie-die) technique and "Enshuku" (salt shrinkage) technique to create the textiles. The design is simple in order to make use of the textile. Usually the members are doing dress pattern and sewing, however depending on the design, outside order is also possible.

Chiyo Koyanazu is good at using "Shibori" technique. "Shibori" is a technique using thread to tide the texture and dye. This technique has many kind of pattern. (Figure 7, 8)

Figure 9, 10 are two pieces of work by using fine and delicate technique and by dyeing over and over again.

Figure 7 "Shibori" technique

Figure 8 "Shibori" technique

Figure 9 Shibori work

Figure 10 Shibori work

She is also good at using "Enshuku" technique. "Enshuku" technique is using chemicals to shrink the texture and make it into uneven texture.

Figure 11 is a plate which is cutting the Japanese paper into different patterns for placing the chemical glue. The part which is pasted with glue will shrink and will act a gathering (Figure 12)

Figure 13, 14 are producing by "Enshuku" technique. These types of pieces of works are popular for how they are light like air and are comfortable to wear.

Figure 11 "Enshuku" plate

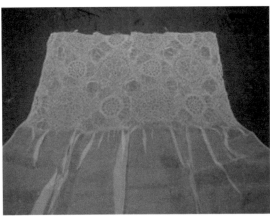
Figure 12 "Enshuku" textile

Figure 13 Enshuku work

Figure 14 Enshuku work

3.4 Flow of the business of "Yabugarashi"

"Yabugarashi" purchases the textures directly from local textile factories, and by processing with "Shibori" and "Enshuku" technique. The atelier staffs will sew and complete the processed textures. The sales partners who are also fan of "Yabugarashi" will sell most of the pieces of works to their friends and local people. (Figure 15) The reputation of the products is spread by word-of-mouth by its originality. Many fans are expecting to the new garments.

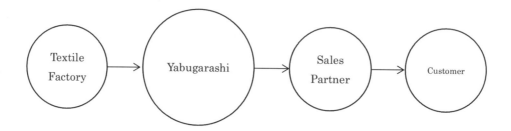

Figure 15 Flow of the business

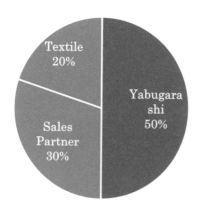

Figure 16 Distribution of sales

The price range of products from "Yabugarashi" is between 30,000 yen to 100,000 yen. The annual amount of sale is about 8 millions. All the garments are handmade thus they are not considering expanding the business. Recently, young members are learning the technique from her. Because of Chiyo Koyanazu's age, she thinks that enjoying in creation is more important rather than expanding the business. As for the distribution of sales, 50% of the selling price belongs to "Yabugarashi", 30% belongs to sales partners, and the rest of 20% belongs to textile factories. (Figure 16)

3.5 Holding event at local community

During spring and autumn, they will hold an event called "Creators' Market" at the atelier. This event is especially for local people. This market is not only selling garments from "Yabugarashi" but also asking their friends who are also creators to sell their accessories. There will be food stalls as well. The local people will make bread, Thai food, Vietnam food, etc. Also their friends will do

Figure 17 event "creators market"

Figure 18 accessories

Figure 19

the jazz performance at the main events. The impromptu jazz performance will hold on a special stage at the veranda of atelier. A lot of local people are looking forward to the event.

4 Expand the possibilities of clothes making by cooperate with the factories

Some of the small Japan's factories take on a job as subcontract thus they have no ability for either planning or selling. Nevertheless, the factories that overcome the difficult situation maintain high skill. As for the individual creators who like to make clothes, it is difficult for them to ask the factories to produce the products, because of the little production. Those small factories are used to depend on big enterprises, but it is time for them to change their thinking. It is necessary for the individual creators to cooperate with the small factories for making unique products and selling the products directly to consumers. At the present, in the university the author is doing a practical experiment, the cooperation between creators and factories. The factories that cooperate with.

"Yabugarashi" are also cooperating with other creators. Those factories were working with big enterprises before, but because of the decreasing of the job, they started to work with creators. Now the president of factory said 「It is better to work with creators to produce different textiles rather than making same products in a large quantity.」About the part of sale, e-commerce is growing rapidly. Japan's small factories will eventually disappear, if they still keep the old system to produce clothes. It is not really attractive that the market is filled with imported products from overseas.

5 Conclusion

As for Japan's aging society in order to live in affluence, unique fashion is needed. The creator who is introduced in this thesis is active and always willing to challenge new things. Her fans are still increasing and hoping her to create more garments. If people accept the unique garments that are made by the creators in different age range, then the small factories in Japan may revive again. Therefore the sales may expand not only in Japan but also in Asia consider in people have similar skin tone, hair color and eye's color.

References

［1］ Ministry of Economy, Trade and Industry 「Industrial statistics of 2014」
［2］ Ministry of Finance 「Foreign trade statistics of 2014」
［3］ MASA&MARI(2014) 『OVER60 Street Snap』 SHUFUNOTOMO Co.,Ltd.

The Universal Fashion Marketing To support customers enjoy fashionable and healthy life

◆ Mori Hideo[1] (The General Merchandise Laboratory, Tokyo, Japan)

Abstract: *The purpose of this thesis is to describe" Marketing of the Universal Fashion" including the thought of the Universal Fashion and the activity of " Non Profit Organization JAPAN UNIVERSAL FASHION ASSOCIATION" (hereinafter: UNIFA)*

Key words: *Customers; Fashion; The Universal Design; Marketing oriented*

人性化时尚营销支持中高龄消费者享受时尚健康的生活

◆ 森秀男 （日本人性化时尚协会，东京）

【摘　要】　本研究意在描述人性化时尚的市场营销。它包括人性化时尚的思考与非赢利组织日本人性化时尚协会(简称：UNIFA)的行动。

【关键词】　消费者，时尚，人性化设计，市场导向

Preface

The Universal Fashion (hereinafter: the UF) is the concept fused of fashion and the Universal Design (hereinafter: the UD). Fashion consists of the clothing which is apparel and the fashion goods (EG: shoes, bags and accessories), however, in this thesis, fashion mainly means apparel.

---

[1] Mori Hideo, the vice chairperson of a board of directors of UNIFA(the Universal Fahion Association), President of GML(The General Merchandise Laboratory), Lecturer of Bunka Fashion College, Tokyo Seitoku University and Kinjo Gakuin University (Nagoya), 401 Polaris Meguro Haramachi , 1-6-7 Haramachi, Meguro-ku, Tokyo, Japan.E-mail: h-mori@ga2.so-net.ne.jp.

Chapter one: What is the Universal Fashion?

1 The concept of the UF

1.1 The definition of the UF as "Fashion for All"

"The UF is the thought and activity to achieve the goal of creating the society, in which as many people as possible can enjoy their fashion regardless of their age, capability (handicaps), body types, size, sex, culture and nationality."

1.2 The four main categories of the UF

(1) The customers of the UF (The customer classification, the customer setting)
(2) To be fashionable (Newness of the fashion and the life style)
(3) The UD (Adaptability to human body and human life behavior-Physical characteristic, Human body functions and more)
(4) Marketing oriented (the UF marketing-mix, products development)

2 The detail of the four categories of the UF

2.1 The customers of the UF

In the process of developing the UF products, persistence to keep the stance of "the customer-centered approach" is needed. In order to do so, it is very important to set a clear picture of customers, listen to their demands and dissatisfactions for the product development and have the attitude to learn from them.

2.2 The Customer classification and the customer setting

To set customers under the certain standards is called "the customer classification". There are three types of it.

(1) "The fashion customer classification" to classify customers under the standards of fashion field.

(2) "The UD customer classification" to classify customers under the standards of human body and human life behavior.

(3) "The Universal Fashion Customer" to be combined of (1) and (2), and to set the kind of customer is called "The customer setting"

A. The fashion customer classification (the detailed standards in fashion field)

This is the standard combined of social and economical characteristics with conditions like personal sensibility in the fashion life of the customers of the UF. It is the basis for "the fashion customer"

(1) Sex

Female, Male

(2) Life stages (Divided life stages based on an assumption of the customer' whole life.)

Young (Students, Young working adults, aged 19-24), Young adult (25-29), Trans adult (30-34), Adult (35-44), Middle age (45-64), Senior (over 65)

(3) Taste (Deference in tastes and senses.)

Avant-garde (innovative, cutting-edge), Contemporary (modern, today), Conservative (traditional, elegant)

(4) Grade (The grade and class based on image, quality, prices and other factors.)

Prestige (the best grade), Bridge (middle), Moderate (normal), Volume (quantitative), Budget (low price) and more

(5) Life occasions (The daily opportunity, purpose, daily scene and sentimental scene)

Official (public / EG: commuting to work or school), Social (social situations / EG: ceremonial events), Private (private situations / EG: travel, week ends, playing sports and more)

B. The UD customer classification (the detailed standards in human body and human life behavior)

This is the standard based on changing of the body functions like aging, capability deference (EG: handicaps) and natural characteristics (EG: height, a bone structure and more) with functions and conditions needed for daily life. This part is directly connecting to the UD support in the UF. It is the basis for "the UD Customer". The typical examples are as follows:

(1) Physical characteristic (Changes in body shape, characteristics of sizes and height, back goes to round, bowed by aging, putting fat on lower abdomen and more.)

(2) Nursing care and support for disabilities and disorders (Support for hemi-paresis sufferers, wheel chair users, rheumatism sufferers and rehabilitation and more.)

(3) Body functions (Physical strength declines, joints stiff, decline of function in activity and working.)

(4) Physiology (One's skin is weak, one's weakness to regulate body temperature and more.)

(5) Safety (Eyesight goes bad, easy to fall down and more. Protection from natural disaster)

C. The Universal Fashion Customer Setting

"The Universal Fashion Customers" is the group which is made of the certain customer characteristics chosen from A and B written above. This is the specific examples:

(1) From "The fashion customer classification"

Choose the sex, the life stage widely from the younger to the seniors with "Trans adult" as a center, the taste as conservative and basic, the grade as "Moderate or Volume" and reasonable., the life occasion as "private" such as "weekends" and "sports scene".

(2) From "The UD customer classification"

Take "support for disabilities and disorders (hemi-paresis sufferers, rheumatism sufferers)" as a sample.

(3) "The UF customer"

Combine A and B, and set C. of the UF customer. When the targeted customer is finally set, the UF product development begins.

3 The UD - Adaptation to human body and human life behavior and the product development

The approach of "human body and human life behavior" was chosen and reconstituted from "the Human Engineering for Quality Life" made by Research Institute of Human Engineering for Quality Life. It has "the human characteristics" and "the life characteristics" on its basis and used for a product developing which has conditions of "ease, safe, comfortable, healthy and useful". Therefore it is used as the compass for a product development for the UD.

4 The examples of the UF product developing

A. "Bee Pant" - Under pants protecting the leaking of the urine for the elder people.

(1) Product by Peace 21 co.

(2) Fashionable 4 types of color design.

(3) Elastic band specially made for tightening the crotch(groin) is used to prevent from leaking of the urine.

(4) Semi permanent deodorization and antibacterial function.

(5) Soft elastic band that has 4 times elastic is used.

(6) Double layer composition of functional material.

(7) The practical new design registered that is collaborated with Tokyo Metropolitan Industrial Technology Research Institute.

B. MIGU Polo Shirts (The UNIFA recommendation product. It would be explained at 2-4 of "Chapter two")

(1) Researched and developed by the study group of Universal Fashion merchandise research and development.

(2) Easy to wear and take off for physical handicapped and for normal.

(3) For hemi-paresis sufferers whose arm is bent, it must be easy to quickly put on and take off the sleeve.

(4) For rheumatism sufferers whose finger or knee is deformed, the sleeve and cuff must be easily released.

C. Multi functional UD Parker jacket to be wearable daily and in town, and in a time of disaster. (a trial product)

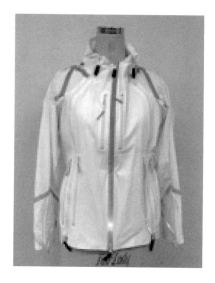

(1) Researched and developed by "UD PROJECT for clothing" that is one of a research subcommittee in IAUD.

(2) Fastener can be opened from the both left and right sides depending on one's purpose.

(3) Increasing the number of pockets with fastener can store more supplies for disaster.

(4) Using more special tapes reflecting light straightly can rescue people in darkness.

(5) "Shock-absorber" put in hood reduces the impact of falling down or collision by hard objects in a time of disaster.

5 Marketing Oriented

The UF is aiming to satisfy customers. To meet this goal, it is required to listen to the voice of customers and use it to plan, design, manufacture and sell the products them through shops, web and TV, Sales-Promotion and publicity are also well-used ways.

5.1 Marketing mix of The Universal Fashion– "MSPSCS"

To define the marketing mix of the UF, there are six types of means.

(1) Merchandise (to be fashionable, the UD, human body & human life behavior, function and quality)

(2) Service (introducing shops. brands, companies to customer through Home Page, seminars, fashion show and the UF exhibition)

(3) Price (reasonability with the grade of the product)

(4) Selling (at real shops, net mall, TV shopping, Joining various trade show and more)

(5) Communication (Home Page, publicity, sales promotion, fashion show, seminars and more.)

(6) Social activity (education at schools, support for the recycle and more)

Chapter two: The activity of JAPAN UNIVERSAL FASHION ASSOCIATION (UNIFA)

1 Profile

The corporate logo of UNIFA

UNIFA started its operation on March 1993 as a non profit group and approved as a Nonprofit Organization (NPO) on December 2001 in Tokyo. The current head office is located in Kyoubashi that is business area near Ginza in Tokyo.

2 The Activities

2.1 Regular and special Seminars

UNIFA normally holds about 10 seminars or lectures in a year. UNIFA invites specialists and business persons from various kinds of fields to give its members a chance to get the newest information.

the meeting of the study group

2.2 Research and development of the UF products

(1) "The study group of the UF merchandise research and development"

UNIFA, Tokyo Metropolitan Industrial Technology Research Institute (Local Independent Administrative Agency), the fashion related companies and the universities are collaborating to produce the UF products. "The study group" for this purpose is very active.

(2) "UD PROJECT for clothing (hereinafter: UDPJ)" that is one of the research subcommittees in International Association for Universal Design (hereinafter: IAUD) of which UNIFA is an associate member of the organization. UDPJ holds work shops and seminars, publishes booklet "Living with clothing" and research and develops UD clothing.

a work shop of "UD PROJECT for clothing" held in 2014

2.3 The UNIFA recommendation system

the UNIFA mark

The UNIFA recommendation system is that if any manufactures apply to UNIFA their products which made with the concept of the Universal Fashion and UNIFA approves the products as "the UNIFA Recommendation products", UNIFA allows the company to put the "UNIFA mark" on the tag of their products. UNIFA is also working as supporting publicity and introducing the products to the retailers to help the manufactures.

A. The UD Polo shirt (The UNIFA recommendation product)

(1) Produced by Descente, Ltd. & Descente Health Management Institute

(2) The quick-release polo shirt for those who have paralysis on their

either arm, rheumatism sufferers and the normal people.

(3) This polo shirt used special pattern-technique and super stretch material "Dow stretch knit material"

B. Men's pajama of universal fashion (The UNIFA recommendation product)

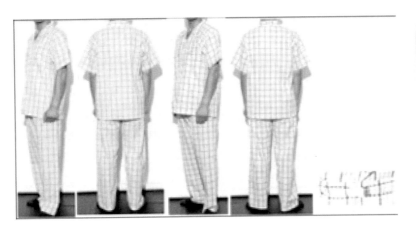
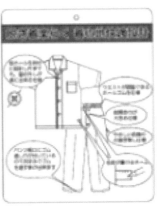
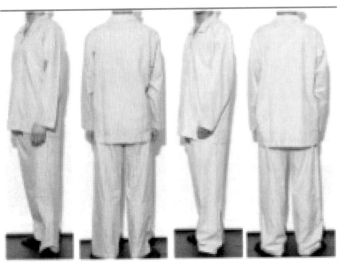
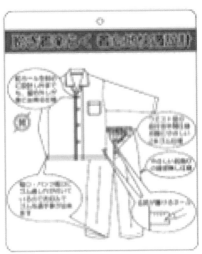

(1) Produced by Hitomi co. that is home wear company.

(2) Button "hole" is tilted to easily button and unbutton the clothing.

(3) West belt is made by "whole elastic band specification" for easy control.

(4) Hem of pant has a hole to insert elastic band to open or close hem.

(5) "Tag to write one's name" is attached at the side seam to judge wearer's name.

(6) Side seam line of pant is "seamless" to be gentle for one's skin.

2.4 The UF Exhibition (trade show)

The newest exhibition joined by about 30 company members of the UNIFA was held in Roppongi Hills in Oct. 2014 and well received. Along with the many apparel products, there

were various other kinds of products like the aging and care products.

2.5 Publishing books

UNIFA published "The declaration of the Universal Fashion part II" in 2009 from the one of the major publishing companies.

2.6 Universal Fashion show

(1) French designers brands were demonstrated at the UF Festa 2014 in Hollywood Hall in Roppongi Hills in Tokyo.

The front page of "The declaration of the Universal Fashion part II

(2) Fashion show held at care facility for the cared elders by the volunteer elders as models.

2.7 To support selling of brands with the UF taste at highly reputable Isetan Department Store Shinjuku by the direction of the UNIFA in 2014.

2.8 Selling the UF brands produced by AIMU co. on QVC TV shopping channel.

"Makana na Kathy" brand for the large size customers.

2.9 UF designer brands for QVC by AIMU co.

"day to day" by HIROMI YISHIDA   "EreuReu" by Eri Matsui

Afterword

I would like to make the final conclusion briefly. After all, the UF has various expansion and potential for the future as the population of elder people grow. It is highly unique concept and thought. I believe that the UF has worth enough role and mission to spread over and be enlightened in not only Japan but whole other world including China as well.

References

[1] UNIFA edited. 2009. "The declaration of the Universal Fashion partⅡ". CHUOKORON-SHINSHA, INC. Tokyo.

[2] Japanese Standards Association. 2009. "The JIS handbook 38: "The accessible design for aged and disabled" Japanese Standards Association, Tokyo.

[3] Japanese Standards Association. 2009. "The JIS handbook 38: "The accessible design for aged and disabled" Japanese Standards Association, Tokyo.

[4] Keiko Imai. 2007. "The power of fashion". CHIKUMA SHOBO Publishing Co., Ltd., Tokyo.

[5] Research Institute of Human Engineering for Quality Life edited. 2006. "The Workshop: Human Life engineering, the second volume Understanding human characteristics and product development." Maruzen Company, Limited, Tokyo.

[6] Naohito Tanaka and Sadako Mitera. 2002. "The Universal Fashion". Chuo houki Publishing Co, Ltd, Tokyo.

[7] Research Institute of Human Engineering for Quality Life edited. 2002. "Human life engineering: The practical guide book for product development". Japan Publication Service. Tokyo.

Understanding Korea Ajumma Way

◆ Haesook Kwon[①]　（*Design College, Sangmyung University, Korea*）

韩国中高龄女性生活方式研究

◆ Haesook Kwon　（祥明大学，韩国）

1. Introduction

Culture defines as "the way of life, especially the general customs and beliefs, of a particular group of people at a particular time" (Culture, Cambridgedictionary online). When used as a count noun, it is the set of customs, traditions and values of a society or community, such as an ethnic group or nation (Culture, From Wikipedia).

Culture awareness most commonly refers to having an understanding of another culture's values and perspective. This does not mean automatic acceptance; it simply means understanding another culture's mind–set and how its history, economy, and society have impacted what people think. Understanding so one can properly interpret someone's words and actions means one can effectively interact with them(Carpenter&Sanjyot, 2012).

Different cultural groups think, feel, and act differently. If one plans to live, work, or do business with other cultures, it's important to understand how they work. In the business world, culture guides decision–making, behavior, thinking patterns and values. Culture influences how people interact with each other.

In Korea a particular type of woman is frequently spotted, alone or in groups. They can be identified by their dress and behavior. The term for these women is 'ajumma'. The Korean word 'ajumma' refers to a stereotypical middle–aged or older woman who showsaunique cultural andbehavioral patterns, which cannot be seenin other countries.Foreigners frequentlyexpressan interestand curiosityabout theculturalcharacteristics of the 'ajumma'(Singleton–Morris, 2014; Warmoth, 2013; Acton, 2013). It is because that looking through the eyes of foreigners, 'ajumma culture' has the extraordinary characteristic. Ajumma represent Korean women. Most Korean

---

[①] Haesook Kwon, Professor, Design College, Sangmyung University, Korea.　E-mail: kwon1004F@hanmail.net

women share one or more life condition with them and it is a universal term that represents Korean womanhood (Rowan, 2014).

The portion of female consumers middle age has emerged as the major driving force in domestic clothing consumption. This phenomenon is expected to be continued along with the current age pyramid for a long time (Seo & Jang, 2009). Most of middle aged womenwould be belonged to an'ajumma' who is one of the most socially influential figures in Korea.As middle-aged women are the nation's most power consumer group in many countries, understanding the major features of 'ajumma'such as fashion style, behaviors, and lifestyle will provide a chance and important data toestablish the marketing strategy aimed at this target market.

2. Data Collection and Methodology

Data were collected through the keywords search of Congress Digital Library and Internet. Main keywords wereajumma, ajumma fashion, ajumma style, ajumma way, ajumma lifestyle..., etc. Total 33 researches, books, article, other materials were used for this study. Data from previous studies, related articles and reports was qualitatively analyzed and partly conductedby the content analysis of research method.

3. Findings

3.1. Definition of Ajumma

3.1.1. Dictionary Definition

Ajumma/Ajoomma is a shortened word of Ajoomeoniwhich is defined in English as: an aunt, an auntie, a middle age lady, a housewife, a wife, a madam or ma'am (Dedal, 2013). But it may not be proper to trust this simple definition when trying to figure out what culturally complex words mean. It would be better off translating the Korean definition which presented in DongA Mate Korean Dictionary (1996):

1. A married woman who is a collateral relation of one's parents
2. The wife of an ajeossi
3. A familiar way of referring to the wife of an elder brother
4. A familiar way of referring to the wife of a person who is the same age as yourself
5. A familiar way of referring to an adult or elder woman

Like this, ajoomeonican be used in both kinshipand non-kinship. The commonmeaning of 'ajoomeoni' in the kinshiphas been mentioned as a woman who is in the samegeneration of parentsin most dictionaries.In other words, the position of ajoomeoni in the kinship is the same level of parents.In the non–kinship, 'ajoomeoni' refers toamarried woman(Kim, 2011).In terms of kinship, it's been used asvery friendly or respectful meaning.

Sometimes ajumma was translated 'aunt' and the recent article in New York Times describedajumma as "aunties". When ajumma is used in the sense of 'aunties', however, it does not refer to a family relationship because aunts in the family are not called 'ajumma' in Korean language(Ajumma,Wikipedia).

The meaning of 'ajumma' in each dictionaryis commonly presented, which is a word that called ajoomeoni lowered, or friendly or intimately (Kim, 2011). In non–kinship, the word 'ajoomeoni' has been substituted by the title 'ajumma'. Sometimes, the term ajumma cause discomfort or rejection depending on the situation or the relationship or social contexts between two. Designation is an index indicating the relationship between Speaker and Hearer. Through the designation, peoplecould be courteously treated or ignored.

As referring to an elder by name without a title in Korea is not socially accepted, the last part of definition–'a familiar way of referring to an adult or elder woman'–is the most common use of the term in Korea and focusing on this study will be investigated.

3.1.2. Social Definition

Going by the dictionary definition alone, there's nothing wrong with the term, but it has a lot of cultural baggage that the dictionary doesn't even hint at(Shure, 2005).Socially accepted idea about ajummahas different and a deeper meaning than thedictionarymeaning.

Under general social circumstance, when a woman referred to as an ajumma, her demographic characteristics such as age, marital status, children are used as an important standard. In the absence of such references, the information about facial features, body type, behavior and the like are used as an alternative reference(Choi, 2000).

–Age range

The age range of ajumma is different depending on researchers from 20s to 60s (Choi, Kim & Kim, 1999; Rowen, 1999; Kim, 2011; Glionna, 2010).In many cases, it has been simply referred to as middle–aged or older/elder women (Kantorfeb, 2014; Park, 2014; Kang, 2012; Wikipedia, 2015; Urbandictionary, 2015).

Sometimes, this term upset young Korean woman in one way or another because ajumma it's like telling–right–on–her–face that she's an old lady (Dedale, 2013).If a woman is called ajumma by a stranger in Korea, she probably won't be happy, thinking her physical appearance, behavior or style was being criticized (Kang, 2012).

Ajumma may be said that a woman of 'certain age', who is neither a young unmarried woman(Agassi) nor grandmother(Halmoni). She is moving from the stage of being young woman to the life stage of grandmother. Typically she would be a married woman with children, although not necessarily so(Rowan, 2014; Ajumma,Wikipedia). Being an ajumma seems to be more about character than just age alone.

–Social status

Ajumma is different from Korean word 'samonim', was originally interpreted as "wife of mentor" or "wife of superior", describes a married woman of high social status. Sometimes ajumma have been described as having low status in the Korean job market, and as often being the last hired and first fired. Ajumma is often a restaurant worker, street vendor, or housewife(Ajumma ,Wikipedia).

–Social awareness

Of course it is used as a term of respect, ajumma in Korea is often mocked or considered a threat because of her tendency to be aggressive or self–centered, mostly in public places(Kang, 2012). The pejorative social awareness of 'shameless&impudent ajumma' had been formed after 1980s. The increasing number of social activities, economic power of ajumma had increased since the 1980s and her social behavior pattern was seen differently from existing ajumma way(Choi & Kim & Kim, 2001). Korean feminists believe the low status of ajumma reflects widespread sexism and classism in Korean culture, particularly the idea that a woman's worth can be assessed mainly on the basis of her age, looks and docility (Ajumma ,Wikipedia).

To summarize the above, themeaning ofajumma is differentdepending on the relati onshipandcircumstancesof two people.In other words, this term can bereferredto express theclosenessandemotion, or also be interpreted as unfavorable behavior to disparage oneself.

3.2. Ajumma style

When used in its pejorative sense ajumma can also have connotations of unfashionableness, with ajumma stereotypically described as wearing baggy pants, mismatched clothing and little makeup,being 'pudgy', having unfashionably short hair in an 'ajumma perm', and 'wearing rubber shoes'(Ajumma ,Wikipedia). Let's look at each of the characteristics in more detail.

3.2.1. Hair style

The hair is a first sign. There are many different styles one can sport though because there are varying degrees and forms of ajumma. Ajumma perm says a lot about her, however, the trademark of the ajumma perm may be mass of tight round curls ppogeuri. Of course, sometimes they have the pyramid, the wavy bob or soft curls cradling the face(Chung, 2011).

How ppogeuri to be the trademark of ajumma? The process overview is as followed.Korea's first hair salon was opened in1920s and the first perm appeared in late 1930s. As the perm was regarded asdecadenttrend, it was prohibited during the Japanese occupation period.Although the perm was in vogue again after the Liberation in 1945, the full scale vogue of perm was accomplished in 1970s. It was because that women's socialactivity was increased and they were required a convenient hairstyle to save their time. Like nowadays, perm expense was expensive at that time, too. Therefore, they were asked to get thesolid perm as much as possible and try to maintain the strong curly waves for as long as possible. The reason behind the unified hairstyle was based on practical reasons. The stronger the curls were, and the longer the hairstyle stayed. It wasthe beginning of theso–called'ajuuma perm'. Since then,through the'80sof itsheyday, it's becomethe trademarkof today'sajumma perm(Oh, 2012; Kang, 2012; Yoon, 2010).

3.2.2. Fashion Style

Another feature of ajumma is their identical clothes style.

–Printed monpe style pants

The tacky blouse/shirts with loose stomach pants which called ajumma pants istypical fashion style.

Originally this pants came from 'monpe' which was Japanese work pant usually worn by fieldworkers or tradesmen, popularized in the war era by villagers. Mompe is distinguished by a looser, low–slung waist and a unique front to back closure, however, sometimes the modern monpe is a tailored, smarter version with an easy to wear flexible button closure and D–ring waist belt (The modern monpe).Asrevealedbyits Japanesename,monpewasfirstintroducedasacolonialimport,andbecamethefirsttypeofpantswornbyKoreanwomen. It is a loosewideleggedpantscinchedattheanklesandtiedatthewaist (Lee & Kim, 2015). Generally, various patterns such as flowers, bright stripes or animal–print are appeared in this pants (Asktheexpa, 2009). When dressing down it's the blouse with this pants, maybe a cardigan or sports jacket to fend off the chill, matched with ultra–comfy shoes, whether slippers or moccasins (Chung, 2011).

–Vibrant patterns and color

For semi–formal occasions, the go–to outfit for ajumma is a simple black jacket with glittering gemstones and a vivid, brightly–colored shirt underneath (Kang, 2012). For dressing up,they are beautiful flowers in full bloom that radiate and shine in the sun, blinding everyone within their presence. Never wear drabcolors unless they're going to a funeral. Also, always match colors. The brighter, the better. Go for prints and patterns, boldly mix and match. Patterned twin sets isa typical style for ajumma. They used to finish off the look with standard pumps and a luxury brand (or faux luxury brand) handbag (Chung, 2011).

-Fluorescent outdoor apparel

Outdoor apparel is another item that is ubiquitousfor ajummas. Whether they are going abroad for a vacation or just visiting a neighborhood park, ajumma pull out their favorite mountain gear. They wear fluorescent outdoor apparelsuch as one red or hot-pink wind breaker with matching pants, especially on a weekend hiking get-together (Kang, 2012). 75% of Korea land is the mountains, therefore, hitting the mountain trails for a good hike is one of Korea's major weekend leisure activities(Kaaloa, 2014). That's why the Korean outdoor apparel industry is flourishing nowadays. From vests to jackets, this outdoor apparel comes in many colors and designs.

-Large sized sun visor

When they get serious and go hiking or jogging, they are dressed head-to-toe in the getup. Cover up all the body parts that can be exposed or simply might be going for the sun visor: wear a "Darth Vader" sun visor to protect sensitive eyes from ultraviolet rays, long sleeves(or short sleeves and separate arm wraps), long pants (the non-ajumma exercising in shorts are just plain clueless), gloves and face mask if necessary. When in doubt of coverage, carry a UV coated parasol. Yellow dust is deadly to ajumma, so a mask that completely covers the face except the eyes is a must (Warmoth, 2013; Chung, 2011; Kang, 2012).

3.3. Ajumma Way of Behavior & Lifestyle

3.3.1.Groups are the best.

Ajummas of a feather flock together, especially on trips or taking hikes in the mountains. They always travel in groups, coordinate their fashion, and carry identical travel bags and/or fanny packs to enhance the group spirit. Excursions to museums and galleries are also group activities, to showoff their camaraderie by talking loudly and gaily amongst themselves (Chung, 2011).

3.3.2. Don't care about the public eye

Many of foreigners reminded ajumma who hog seats on public transportation. Ajummas hunt for an empty seat through the windows while waiting on the platform for the train or bus. As soon as the door opens, they shove their way through to get a seat, dashing like sports star. If a seat is out of reach, they will throw their purse. Even if the space is only big enough for half a bum to be perched, any ajumma worth her salt can squeeze in and widen it enough to seat herself (Dedale, 2013; Chung, 2011; Kang, 2012). Used in its pejorativesense, ajumma has connotations of pushiness, with ajumma described as hard-working and aggressive people who 'grab one by the arm and try to get one to eat at their place,' or 'push' friends and relatives to buy insurance (Ajumma , Wikipedia).

3.3.3. Exercising and Hiking

Many of city ajummas want to do some serious exercising to catch up.They don't want to do anything too heavy, though. Simple jogging is not enough, so they power-walk, swinging their arm back and forth at an almost 90 degree angle. Some experienced joggers walk backward while others walk clapping in front and then behind. Another popular workout is hitting backs against trees. They believe the impact helps improve their blood circulation. Take a brisk, arm-pumping walk around the neighborhood park (Chung, 2011; Kang, 2012).

Hiking is the biggest fad among many ajummas.Hiking is not just for exercise.For tired ajumma, it's a way to connect with others.Though not as popular as mountains, public parks and jjimjilbang (a large, gender-segregated public bathhouse in Korea) serve as good hangout places. The health-conscious ajummas do very unique moves to maintain a toned body on the walking trails in the parks along the Han River (Kang, 2012).

3.3.4. Listen to all, talk to all

It's not being nosy, it's being helpful and considerate; they're worrying on the behalf of others, for others. They are not asking personal questions just for the sake of asking personal questions; they are gatheringnecessary information to use when giving that person the needed advice. The acquired information should be shared with as many other people as possible, because she needs to be open to many other opinions. And other people should be aware of their opinions, too, whether they know each other or not (Chung, 2011).

4. Conclusion

This study is to investigate the characteristics of ajummas, based on term definition, cultural approach to the concept, their way of appearance, behaviors and lifestyle. Findings led to the following conclusions.

(1) Although being an ajumma seems to be more about character than just age alone, ajumma is a woman certain age. Subject being called ajumma indaily life is a marriedwoman who does belong to neither young lady nor grandmother. The term of ajumma is mainly used when special designation such as 'samonim' or 'director' or 'chief' is not available. Ajumma is considered as a middle class or lower class. Many of them are housewife or worker like street vendor or restaurant worker.

(2) Ajumma does not come out with a good social awareness. This pejorative social awareness of 'shameless & impudent ajumma' has formed in 1980s when behavior pattern of ajumma was dramatically changedfrom existing ajumma way. It was caused by their involvement of economic activity.

Push and shove through a crowd to get priority seating, selfish and unstylish—bossy, gossiping magpies with bad perms who pinch pennies and hog seats on the subway.

(3) Beyond the term definition itself, ajummaalso refers to women with special attributes. Ajumma is distinguished by ppogeuri perm, fashion style such as monpe pants, vibrant pattern & color, fluorescent outdoor look and large sizes sun visor. Ajumma way of lifestyle and behavior are another uniqueness of ajumma group. They are always in groups, listen to everyone and willing to share the acquired information with as many other people as possible, and do the power walking and hiking.

In 2014, Chung & Kim analyzed the pursuit of current fashion trends and fashion imagetypes of middle-aged women in Korea. The result showed that middle-aged Koreanwomen were highly conscious of how others perceive them and had a desire to not stand outfrom others. The unified commonality of ajumma group may partly have caused by thiskind of socialbelongings.

Although ajumma is known for being aggressive, loud, and even mean (Becker, 2012), they have shown their power in helping to make Korea what it is today.Rowen (2010) said that the horizon of a society is charted by the growth of its efforts to empower those who are its foundation. The ajumma of Korea fit this speculation.They are part of the foundation of Korea's industrialization and advanced technological and knowledge economy–as mothers, wives, and workers.

Among many Koreans, it's now often used to conjure an image of homemakers who disdain full-time jobs to while away afternoons on park benches, in coffee shops and at social clubs, bragging about their children and, if they've got the money, go on shopping sprees(Glionna, 2010). And today's middle-aged women have turned their backs on the uniform perm and twinset, wearing their hair straight and long and dressing in casual clothes (Chosunilbo, 2005). Perceptions of ajumma are changing, especially in modern metropolitan Korea, but this change seems to come slow.

References

[1] Acton, Dan 101 Korean pop culture words you absolutely MUST know, Retrieved November 6. 2014, from http://www.dramafever.com/news/101-korean-pop-culture-words-you-absolutely-must-know/, 2013.

[2] Ajumma (2014, April 28). Coffee-helps.Retrieved April 6. 2015, from http://coffee-helps.com/2010/04/28/ajumma/

[3] Ajumma, Wikipediathe free dictionary. Retrieved May 6. 2015, from https://en.wikipedia.org/wiki/Ajumma

[4] Becker, Kathleen. South Korea's Unjustly Infamous Ajumma, Pink pangea, Retrieved May 10. 2015, from http://www.pinkpangea.com/2012/12/south-koreas-unjustly-infamous-ajumma/, 2012.

[5] Carpenter, Mason A. & Dunung, Sanjyot P. Challenges and Opportunities in International Business v.10, Retrieved May 15. 2015, from http://2012books.lardbucket.org/pdfs/challenges-and-opportunities-in-international-business.pdf, 2012.

[6] Innocence theory of Korean ajumma. RetrievedMay 16. 2015, from http://blog.naver.com/PostView.nhn?blogId=akekdthkl200&logNo=80112618554), 2010.

[7] Choi, Sangchin Social psychology inside of discourse on Korean ajumma and railroaded the weak person, Cuture& People, The first issue, 2000, 162-184.

[8] Choi, Sangchin, Kim, Jiyoung & Kim , Kibum. Social representation and power of adjumma in Korea. Korean Jounal of psychology:Women, 1999, 4(1), 56-67.

[9] Choi, Sangchin, Kim, Jiyoung & Kim , Kibum. Analysis of Korea ajumma as a psychological construct,Korean Journal of Psychology: General, 2001, 20(2), 327-347.

[10] Chung ,Su-In & Kim ,Young-In. Fashion Image Types and Design Factors for

Middle-aged Korean Women, Journal of the Korean Society of Costume, 2014, 64(5), 91-107.

[11] Chung, Suzy. How to rock Ajumma style, Arts, Lifestyle, Retrieved May 6. 2015, from http://blog.korea.net/?p=5638, 2011.

[12] Culture, CambridgeDictionariesOnline, Retrieved from http://dictionary.cambridge.org/us/dictionary/english/culture

[13] Culture,Wikipedia the free encyclopedia,Retrieved from https://en.wikipedia.org/wiki/Culture

[14] Dedal, Analou. What is an 'Ajumma'? Princess' Attic, RetrieveMay 10. 2015, from http://mysimplyprincess.blogspot.kr/2013/05/what-is-ajumma.html), 2013.

[15] Glionna, John M. South Korea's homemakers don't want to be pegged, RetrieveMarch 11. 2015, from http://articles.latimes.com/2010/apr/21/world/la-fg-korea-ajumma-20100422, 2010.

[16] Jin, Seokjin(2015, April 2). Ajumma-TTMIK Culture Ramblings, RetrieveMay 1. 2015, from http://www.talktomeinkorean.com/shows/culture-ramblings/culture-ramblings-ajumma/

[17] Kang, Michelle(2012, November 12). 'Understanding the ways of the ajumma', Korea Joongang Daily, Retrieved May 7. 2015, from http://koreajoongangdaily.joinsmsn.com/news/article/article.aspx?aid=2964272

[18] Kantorfeb, Jodi(2014, February 7). A Look at Korea's Culture From the Bathhouse, NY Times, Retrieved May 10. 2015, from http://www.nytimes.com/2014/02/09/travel/a-look-at-koreas-culture-from-the-bathhouse.html?_r=0

[19] Kaaloa, Chistine.(2014, September 27). Hiking in Korea:Natural High vs Fashion Heavy, Retrieved April2. Vv2015, from http://grrrltraveler.com/countries/asia/korea/krhikingfashion/

[20] Kim, Jaesun. A study on the semanic and pragmatics of characteristics for Korean kinship terms : focusing on social usage of 'ajoomeoni /ajumma, yimo, unnee, Agasi', Master thesis, The graduate school of Kookmin university, 2011.

[21] Lee,Hyewon and Kim, Alice S. Monpe Modern:Militarized Imperial Uniformity, Laboring Female Bodies, and Class, International Joint seminar-Bridging Britain and the Far East, was held on the 13th and 14th of February, 2015 at Bunka

Gakuen University, Tokyo and welcomed guests from Japan, Britain, China and Korea. Retrieved June 9. 2015, from https://transboundaryfashion.files.wordpress.com/2015/03/synopses.pdf, 2015.

[22] Middle-Aged? Not 40-Something Korean Women.(2005, February 14).The Chosuuilbo. RetrievedApril 7. 2015, , from http://english.chosun.com/site/data/html_dir/2005/02/14/2005021461028.html

[23] Oh My Ajumma!.(2014, July 30). Retrieved April 7. 2015, from https://sojuwanttogotokorea.wordpress.com/2014/07/30/oh-my-ajumma/

[24] Park, Jumin, South Korea's homemakers don't want to be pegged, Los Angeles Times, 2010.04.21. Retrieved from http://articles.latimes.com/2010/apr/21/world/la-fg-korea-ajumma-20100422).

[25] Rowan, Bernard, Care for ajumma! Koreatimes, 2014.02.18., Retrieved from http://www.koreatimes.co.kr/www/news/opinon/2014/02/197_151839.html

[26] Rowan, Bernard. Ajumma — Engine for Korea's Social Progress, posted 2010.02.19. Retrieved from http://www.koreatimes.co.kr/www/news/biz/2015/03/291_61092.html

[27] Seo, Eunkyoung& Jang, Eunyoung. Middle-aged of the British women's apparel purchase situation analysis, Fashion business, 2009, 13(3), 99-108.

[28] Singleton-Morris, Joanne, Ajumma-A Female Korean Role Model With Attitude, Avenue Post, A view from Europe, 2014.01.03., Retreived from https://avenuepost.wordpress.com/2014/03/02/ajumma-a-female-korean-role-model-with-attitude/

[29] Shure, C. La. The concept of ajumma', 2005.09.25., Retrieved from http://www.liminality.org/archives/94/

[30] The Modern Monpe, http://matterprints.com/collections/the-modern-monpe

[31] Warmoth, Brian. 2013. 05. 30, Blog of digital content director, editor &journalist, Retrieved from http://warmoth.org/korea-obervations

[32] Who Wants to be an Ajumma? Saturday, October 24, 2009. Retrived from http://asktheexpat.blogspot.kr/2009/10/who-wants-to-be-ajumma.html

编后记

岁岁重阳,今又重阳。

重阳,是中国传统的登高、远眺、赏菊和插茱萸的日子,也是现今敬老、尊老、爱老、助老的节日。恰在"今又重阳"的佳节前夕,我们又迎来国内外中高龄时尚服饰的专家宾客相聚沪上,就去年的论题继续探讨交流。专家学者的论文亦犹如"登高,远眺"般的高远思路,为本届的论坛成功奠定了基础。

去年,我们《寻找银色光彩——2014中高龄时尚服饰研究》的论坛圆满召开,其论文集亦顺利发行。令我们快慰的是,该项研究的开启立即获得了社会广泛的关注。上海市的相关部门、社会团体、媒体和高校都开始关注该项研究,人们除了对人口老龄化的思考,开始步入对老龄化人群的产品需求、市场预期、服饰产品的具体研究,包括纤维材料和功能设计等等。尤其在论文集付梓前夕,上海市文创基金会又给了该项目充分的肯定、支持和资助。这不能不让我们感到银发事业的高远和温暖。

去年的论坛上我们就中高龄时尚的未来提出了诸多的思考,在中高龄人群的服饰设计、面料、体型、结构、妆容及心理、消费等作了初步的涉猎。今年论文深化了去年的思考,论文开始关注到中高龄服饰的智能化、功能化,这无疑是该项研究的可喜进步,国内外专家以现代科技作为新的思考点,将未来老龄人群时尚服饰产品注入高科技的内涵,让现代科技的成果服务于老龄人群,这不能不是深入研究的结果。

同时,研究也显示人口老龄化也可以转化为中国市场新的"人口红利",巨大的老龄人口,尤其是新生代的中高龄人群,将会催生出新的市场增长点。也为中国纺织服装产业转型、创业的过程中寻觅到新的思路,更为中国中高龄未来市场探究到新的契机。论文集里汇集了中外学者的睿智和有价值的研究,相信本论文集再次为中国的中高龄研究书写了有价值的一笔。

在我们的研究中,我们学院组织了很多大学本科学生参与其中,这些90后的年轻孩子们在研究中表示能为几十年后的父母设计服装,感到无比欣喜,亦更有一番意味在心头。人类步入老龄是必然,让全社会和年轻人共同参与这项研究,也是教育工作者的荣耀与职责。

就此，我们由衷地感谢上海文化创意产业领导小组办公室、中国老年基金会、上海市老年基金会、上海教育电视台、中国纺织服装行业协会、上海服装行业协会等给予该研究的大力支持。

感谢国内外纺织服装院校、科研院所的专家学者对该课题的积极参与和支持。感谢东华大学、东华大学功能防护服装研究中心、上海工程技术大学、苏州大学、江南大学、江南大学生态纺织教育部重点实验室、北京服装学院、汉族服饰文化与数字创新实验室、上海工艺美术职业学院、慧工作室、上海派吉姆数码科技有限公司、上海服装研究所等，感谢海外的院校及科研机构：日本株式会社 MODELLISTA、日本人性化时尚协会、瑞典布罗斯大学纺织学院、美国明尼苏达大学、韩国祥明大学、韩国成均馆大学、英国诺丁汉特伦特大学、美国伍德伯里大学、美国爱荷华州立大学、欧洲设计学院、日本文化服装学园、法国里昂第二工业大学等。

特别感谢上海中外文化艺术交流中心、东华大学兴顺福中老年服装研发设计中心、上海月星集团给予的大力协助。

论文集能得以顺利集结出版，再次感谢东华大学出版社对该研究的一贯支持。

袁 仄
2015 年 9 月于上海视觉艺术学院